Invasive Plant Ecology in Natu
and Agricultural Systems, 2nd Edition

Re
b, .
in ɼ .
by ɼ

Invasive Plant Ecology in Natural and Agricultural Systems, 2nd Edition

Barbara D. Booth

Independent Consultant, Guelph, Ontario, Canada

Stephen D. Murphy

*Department of Environment and Resource Studies,
University of Waterloo, Waterloo, Ontario, Canada*

and

Clarence J. Swanton

*Department of Plant Agriculture,
University of Guelph, Guelph, Ontario, Canada*

www.cabi.org

CABI is a trading name of CAB International

CABI Head Office
Nosworthy Way
Wallingford
Oxfordshire OX10 8DE
UK

Tel: +44 (0)1491 832111
Fax: +44 (0)1491 833508
E-mail: cabi@cabi.org
Website: www.cabi.org

CABI North American Office
875 Massachusetts Avenue
7th Floor
Cambridge, MA 02139
USA

Tel: +1 617 395 4056
Fax: +1 617 354 6875
E-mail: cabi-nao@cabi.org

A catalogue record for this book is available from the British Library, London, UK.

Library of Congress Cataloging-in-Publication Data

Booth, Barbara D. (Barbara Diane), 1961-
   Invasive plant ecology in natural and agricultural systems / Barbara D. Booth,
Stephen D. Murphy, and Clarence J. Swanton. -- 2nd ed.
      p. cm. -- (Modular texts)
   First edition published by CABI in 2003 with title: Weed ecology in natural and
agricultural systems.
   Includes bibliographical references and index.
   ISBN 978-1-84593-605-1 (alk. paper)
   1. Invasive plants--Ecology. 2. Weeds--Ecology. I. Murphy, Stephen D. II.
Swanton, Clarence J. III. C.A.B. International. IV. Title. V. Series: Modular texts.

SB611.B59 2011
632'.5--dc22

2010009986

First edition published by CAB International in 2003. ISBN-13: 978 0 85199 528 1

ISBN-13: 978 1 84593 605 1

Commissioning editor: Rachel Cutts
Production editor: Fiona Chippendale

Typeset by SPi, Pondicherry, India.
Printed and bound in the UK by Cambridge University Press, Cambridge.

Contents

Preface to the 2nd Edition

Our goal in writing this book was to describe how and why plant invasions occur. The book attempts to explain the ecological principles that are important in understanding the potential for a species to become invasive. We think students should understand the ecology of invasive plants and how this knowledge can be used to facilitate management. Ecology is central to our understanding of how and why plant species invade, and yet there are few books that make this connection. This is the niche we hope to fill.

The book was designed as a teaching text for advanced undergraduate students. No extensive ecological background is assumed, although some basic biology is required. At the beginning of each chapter, we have listed the concepts that will be addressed. These can be used as an overview of what is to come, and to assist the reader when reviewing the material. For the instructor, we have designed this book so that the material could be covered in a single-term course by covering approximately one chapter per week. We have tried to write the book and arrange the material so that it is presented in a clear, precise and concise manner, and to include only pertinent information. If we have done our job well, students should be able to read and understand all the information. We have used common names throughout the text, with Latin names given the first time that the species is mentioned in each chapter. We did this because common names are easier to remember when first learning about a species. A species list of common and Latin names is provided at the end of the book.

At the end of each chapter there will be questions that refer to an invasive plant species of your (the student's) choice (i.e. a case study). The plant can be an established invasive species, or perhaps a new species that may have potential to invade your local community You will be asked to summarize information that is known about your plant species in relation to the material discussed in each chapter. There may be a lot or very little information available to you. The idea behind this is to apply the ecological principles you learn in the chapter to an invasive species of interest, and to give you practice researching a topic. Our hope is that, by the end of the book, you will have created a case history of your chosen plant.

We have made a number of changes to this second edition from the first, which was published by CABI in 2003 as *Weed Ecology in Natural and Agricultural Systems*. We have removed the chapters that focused solely on experimental methods and added chapters on the newer fields of landscape and molecular ecology. In addition we have updated the literature.

We thank our publisher CABI for their help and encouragement with this second edition. Of course we accept the responsibility for any errors that occur.

Finally, we thank our spouses, David Beattie, Tara Murphy and Josee Lapierre, who once again throughout the writing of this second edition heard way more about 'the book' than they ever wanted, but kept smiling and nodding their heads anyway. We dedicate this second edition to them.

1 Introduction to Invasion Ecology

Concepts

- The terms invasive, colonizer, weed and others are often used in overlapping and conflicting manners.
- An invasion is the geographical expansion of a species into an area not previously occupied.
- Both native and non-native species can be invasive.
- Most invasions fail.
- The impact of an invasion depends on the area covered by the species, its abundance and the effect per individual.
- Invasion meltdown is the acceleration of impacts on native ecosystems due to synergistic interactions.

1.1 Introduction

It may be tempting for you to start this book with Chapter 2. After all, the *real* information doesn't start until then, and exam questions rarely focus on what you learn in Chapter 1. BUT Chapter 1 is important because it sets the tone for what is to follow. A Shakespearean play or an opera always begins with a prologue. If you walk in after the prologue has finished, you will certainly follow the plot and enjoy the play, but you might not understand the 'why' of the characters' actions. Consider this chapter a prologue. You may already know much of what we are about to say, and you may not be tested on it, but it will put what you are about to learn into context.

In this book, our goal is to provide you with a basic ecological understanding of how plants invade natural, disturbed and agricultural ecosystems. Whether an invasive plant is in a natural community or a highly managed farm, the underlying questions and principles will be the same. A number of books are available on plant invasions; however, they often assume an in-depth understanding of ecological principles (Mooney *et al.*, 2005; Sax *et al.*, 2005; Cadotte *et al.*, 2006), focus heavily on the control and management of

invasive species (Sheley and Petroff, 1999; Myers and Bazely, 2003; Liebman *et al.*, 2007; Clout and Williams, 2009), or provide a detailed description of the biology of individual invasive species without providing a broad background (Czarapata, 2005; Young and Clements, 2009). Here we focus on the underlying ecological principles that explain the process of plant invasions.

There are a number of excellent plant ecology texts (e.g. Barbour *et al.*, 1999; Gurevitch *et al.*, 2006). This book was not designed to replace a good, comprehensive text on basic ecological theory. Rather, its purpose is to examine invasive plant species in systems from agricultural and natural communities. Whether an invasive plant is in a natural community or a highly managed farm, the underlying questions and principles will be the same. All types of ecological systems are controlled by the same processes.

1.2 Levels of Ecology

The word ecology was derived from the German word (*oekologie*), which was derived from the Greek words *oikos*, meaning 'house', and *logos*, meaning 'the study of'. Thus, ecology is the study of organisms and their environment. We can divide the environment into biotic (living) and abiotic (non-living) factors. Examples of biotic factors are competition and herbivory. Abiotic factors can be physical (e.g. temperature, light quality and quantity), or chemical (e.g. soil nutrient status).

The field of ecology is vast. It is concerned with areas as diverse as the dispersal of seeds, competition within and between species, and nutrient cycling through ecosystems. Each of these operates on a different temporal and spatial scale and each has a different focus. Thus, they all ask different types of questions, and require a different protocol to answer them. For convenience, ecological questions can be categorized into different scales:

- Molecular ecology examines how DNA is expressed as traits that may allow plants to invade.
- Individual organisms can be studied to examine how abiotic factors affect their physiology and phenology.
- Groups of individuals of the same species can be studied to look at population-level processes.
- Groups of co-occurring populations of different species can be studied to ask community-level questions.
- Interactions between a community and its abiotic factors can be studied to answer landscape- and ecosystem-level questions.

Each of these categories blends into the next. They are not discrete units of study, rather they are useful, practical and somewhat arbitrary divisions which help to simplify the understanding of invasion ecology. In this book we discuss population ecology (Chapters 2–7), interactions among populations (Chapters 8 and 9), community ecology (Chapters 10 and 11), landscape ecology (Chapter 12), and molecular ecology (Chapter 13). In the final chapter (14) we incorporate all levels of ecology to discuss the process of plant invasions.

1.3 What's in a Name?

Every book on invasive species must first start with an attempt at defining terms. Many attempts have been made to define *invasive* and other terms describing a species' status. Pyšek (1995), for example, reviewed definitions of an *invasive* species and found that it has been described as:

- an alien in a semi-natural habitat (Stirton, 1979; Macdonald *et al.*, 1989)
- a native or non-native entering any new habitat (Mack, 1985; Gouyon, 1990)
- a native or non-native that is increasing in population size (Joenje, 1987; Mooney and Drake, 1989; Le Floch *et al.*, 1990)
- any non-native increasing in population size (Prach and Wade, 1992; Binggeli, 1994; Rejmánek, 1995)
- any non-native species (Kowarik, 1995).

Colautti and MacIsaac (2004) summarized more recent definitions of an *invasive* species as:

- a non-native species (Goodwin *et al.*, 1999; Radford and Cousens, 2000)
- a native or non-native species that has colonized natural habitats (Burke and Grime, 1996)
- a widespread non-native species (van Clef and Stiles, 2001)
- a widespread non-native species that has a negative effect on habitat (Davis and Thompson, 2000; Mack *et al.*, 2000).

Clearly, there is little consensus on what invasive means. Similar problems occur with other terms, such as *weed*. Weeds have typically been defined as 'plants which are a nuisance' (Harper, 1960) or 'a plant where we do not want it' (Salisbury, 1961). Barbour *et al.* (1999) defined a weed as a 'non-native invasive plant' and they distinguished between 'invasive plants' that invade only natural or slightly disturbed habitats, and 'pest plants' that interfere with agricultural or managed natural areas. This definition, however, requires us to further define *non-native* and *invasive*, and to separate natural from disturbed habitats. These definitions are based on our perceptions of the impact of the plant. Similarly, the terms *invader* and *colonizer* have often been used in a conflicting manner. The distinctions between them are quite subtle and result from differing viewpoints. According to Rejmánek (1995), *weeds* interfere with human land use; *colonizers* are successful at establishing following disturbance; and *invaders* are species introduced into their non-native habitat. There is substantial overlap among these terms. A plant may be considered as only one of these, or it may be included in all of these categories (Fig. 1.1). Given the confusion of terminology, we take a broad view of the term invasive, defining it as a species that has a negative ecological or economic effect on a natural or managed ecosystem.

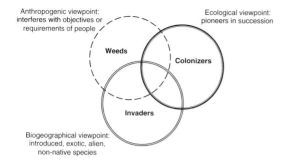

Fig. 1.1. Weeds, colonizers, and invaders are similar concepts but result from differing viewpoints (redrawn from Rejmánek, 1995).

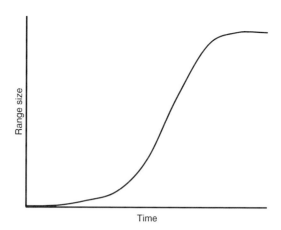

Fig. 1.2. The process of invasion.

It is important to differentiate species that are *invasive* from species that are *introduced*. While most invasive species are introduced, this is not always the case. For example, in the north-eastern USA, the native hay-scented fern (*Dennstaedtia punctilobula*) becomes invasive under conditions of intensive browsing by white-tailed deer and the thinning or removal of overstorey tree canopy (de la Cretaz and Kelty, 1999). In addition, not all introduced species become invasive. Many plant species are intentionally introduced for use as crops, for erosion control, or for aesthetic reasons (e.g. in gardens). While some introduced species have become invasive, many will persist in small self-sustaining populations but will not expand in range or will remain at low density. Furthermore, many introduced species would go extinct in their introduced range without human management (e.g. maize will never become invasive in Canada).

1.4   What is an Invasion?

Elton (1958) was the first person to formalize the study of invasion ecology, recognizing that invasions of introduced species could have global impacts (Richardson and Pyšek, 2008). Since then, interest in invasion ecology has grown. An invasion is 'the geographical expansion of a species into an area not previously occupied by that species' (Vermeij, 1996). The process that a plant species goes through when it invades has a series of stages. Initially, following a species' introduction, its geographical expansion occurs very slowly (Fig. 1.2). This lag period can last as long as centuries. Following the lag period, there may be exponential expansion of the species' range. Eventually, the increase in a species' range

size will level off. In Chapters 2 to 13 of this text we will discuss individual processes that influence one or several of these stages of invasion, and in the final chapter we synthesize these individual chapters and revisit the invasion process.

Most invasions fail, but some will succeed

Most invasions fail. That is, most species when introduced into a community will not survive. A successful invasion is a rare event (Williamson, 1996). How rare? Williamson (1996) proposed the 'tens rule' to describe how approximately 10% of species pass through each transition from being imported (dispersal) to becoming casual (introduced to the wild) to becoming established, and finally becoming an invasive (Williamson 1996) (Fig. 1.3). This is a rough rule, but has been shown to apply around the world, under many situations (Williamson and Fitter, 1996). What is evident from this rule is that successful invasions are rare (Williamson, 1996).

Our knowledge of why an invasion fails can be quite limited (Rejmánek, 1999). Successful invaders are obvious, but failed invaders are not; therefore, it is difficult to study the process of a failed invasion. Some reasons that invasions fail are (Crawley, 1987; Lodge, 1993):

- inappropriate abiotic conditions for the species to survive
- introduced species are outcompeted by other species
- presence of natural enemies such as herbivores and diseases

- lack of species to pollinate, disperse or facilitate the invader
- low-density effects, such as difficulty finding mates
- lack of the genetic diversity to succeed.

If most invasions fail, then what are we worried about? The small proportion of invasions that *do* succeed can have drastic effects on populations, communities or ecosystems. An invasion may cause another species to become extinct, or it could have profound community-wide impacts on many species or on ecological processes. In the following section we consider some of these impacts.

1.5 The Impact of Plant Invasions

Plant invasions have high ecological and economic costs associated with them (Parker *et al.*, 1999; Pimentel *et al.*, 2000, 2001). Some economic costs are easily quantifiable (the cost of control or management, yield loss) whereas other are not (damage to ecosystems, loss of recreational land, aesthetic impacts). Pimentel *et al.* (2000) estimated the costs

of invasive plants to crop and pasture land in the USA as well over 34 billion dollars annually. In India, the cost is even higher, at US$38 billion per year (Pimentel *et al.*, 2001). While the general concepts of evaluating costs and benefits in ecosystems have been criticized (Gatto and de Leo, 2000; Nunes and van den Bergh, 2001), such statistics do show the context of the huge economic impact of invasive species. The ecological costs of invasion are much harder to understand and quantify. What is the cost associated with garlic mustard (*Alliaria petiolata*) invading a forest? If an invasion causes another species to go extinct, what is the lost value?

In natural ecosystems, species such as buckthorn (*Rhamnus* spp.), canegrass (*Phragmites communis*), garlic mustard and kudzu (*Pueraria lobata*) are obvious invaders because they can dominate an ecosystem (Anderson *et al.*, 1996; Reinartz, 1997; Galatowitsch *et al.*, 1999; Pappert *et al.*, 2000). However, even seemingly innocuous species such as Norway maple (*Acer platanoides*), a tree commonly planted along roadways in North America, can devastate a natural ecosystem and severely reduce the diversity of native species (Webb *et al.*, 2000; Webster *et al.*, 2005). The total impact of an invasion is due to three factors (Parker *et al.*, 1999; Hellmann *et al.*, 2008):

- the size of the range
- the abundance within the range
- the effect of each individual invasive plant on the community (per capita effect).

A species' range and abundance are fairly straightforward to quantify, so we tend to use them synonymously with impact (Parker *et al.*, 1999). But they only tell part of the story. For example, a plant could be widespread and abundant but still be fairly benign. Both ox-eye daisy (*Leucanthemum vulgare*) and common mullein (*Verbascum thapsus*) are examples of this. Sometimes the evidence of invasiveness is based solely on these two criteria, with little knowledge of the effect of each individual plant. For example, we know that the range of purple loosestrife (*Lythrum salicaria*) is expanding and that it is highly abundant in some habitats, but it has been difficult to quantify the long-term impacts (Hager and McCoy, 1998). Quantifying the per capita effect of an invasive species is complicated because impacts can affect species' genetics, individuals, population dynamics and communities, or ecosystem and landscape properties and processes (Parker *et al.*, 1999; Hellmann *et al.*, 2008).

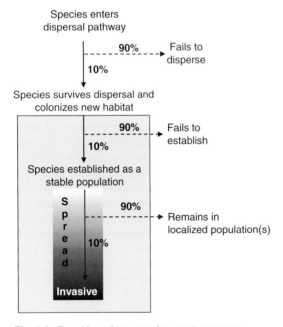

Fig. 1.3. Transitions that a species must overcome to become invasive. Approximately 90% (80–95%) of species are lost at each transition. While a species may fail at some stage, it may re-enter the invasion cycle numerous times, thus increasing its likelihood of success (adapted from Kolar and Lodge, 2001).

Invasion meltdown

The effect of one species' invasion is further complicated by synergistic effects when more than one species invades. That is, the combined effect of two species may be greater than the effect of the two species individually. Synergistic effects may not be predictable. 'Invasion meltdown' is the acceleration of impacts on native ecosystems due to synergistic interactions (Simberloff and Von Holle, 1999; Simberloff, 2006).

Invasion meltdown occurs in two ways (Parker *et al.*, 1999). First, established invaders may facilitate other invaders and therefore each successive invasion becomes easier. For example, the Japanese white-eye bird introduced into Hawaii is the only bird to eat and subsequently disperse seeds of the invasive fire tree (*Myrica faya*) (Woodward *et al.*, 1990). Invasion of this plant was further facilitated by introduced pigs that also disperse seeds and create disturbance that favours the establishment of weeds over native species. Once the fire tree invades, it alters nutrient and water cycles, thus facilitating the invasion of other non-native species (Vitousek *et al.*, 1987). So what is the effect of the Japanese white-eye bird? The second way that invasion meltdown occurs is that each attempted species introduction disrupts the abundances of native species to some extent, and thus the community becomes more and more invasible as the total number of introduction attempts increases. In this way, even unsuccessful invasions can influence long-term community dynamics.

1.6 Summary

In this chapter, we introduced you to the basic concepts of invasion ecology. The term *invasive species* is defined many ways; we prefer to use a loose definition that includes all plants that have a negative ecological or economic effect on an ecosystem. A plant invasion is the geographical expansion of a species. For a variety of reasons, most invasions fail. In this text, our goal is to understand why invasive plants occur where they do. We do not address how to get rid of them. In the next chapter, we begin by looking at plant populations. The first step to investigating populations is to determine their distribution and abundance.

Box 1.1. Invasive species case study: introduction.

At the end of each chapter, there will be a case study box in which you will be asked a series of questions relating to an invasive species of your choice. At this point, you should select the species that you wish to focus on. This may take some thought. Are you more interested in natural or managed systems? Are you interested in wide-ranging common invasive species, locally problematic species or recently introduced species? For some species, there will be a lot of literature available, while for others there may be large gaps in our knowledge. In the first case, you will have more information to read and synthesize. In the second case, you will be asked to suggest what information is needed and how this should be obtained. To get started, you may want to refer to a book on invasive species in your region. It is a good idea to create a bibliography of references and resources you may need.

1.7 Questions

1. How would you define the term 'invasive'?
2. Name a plant that you would consider an invasive species but that someone else would not. Name a plant that you would not consider an invasive species but that someone else would. Explain how this is possible.
3. Why do you think most plant invasions fail? Why are some successful?

4. Describe a habitat near you that has experienced a plant invasion. How has that habitat changed?

Further Reading

Davis, M.A. (2009) *Invasion Biology*. Oxford University Press, New York.

Perrings, C., Mooney, H. and Williamson, M. (eds) (2010) *Bioinvasions and Globalization: Ecology, Economics,*

Management, and Policy. Oxford University Press, New York.

Zimdahl, R.L. (2007) *Fundamentals of Weed Science*, 3rd edn. Academic Press, San Diego, California.

References

Anderson, R.C., Dhillion, S.S. and Kelley, T.M. (1996) Aspects of the ecology of an invasive plant, garlic mustard (*Alliaria petiolata*), in Central Illinois. *Restoration Ecology* 4, 181–191.

Barbour, M.G., Burks, J.H., Pitts, W.D., Gilliam, F.S. and Schwartz, M.W. (1999) *Terrestrial Plant Ecology*, 3rd edn. Benjamin/Cummings, Menlo Park, California.

Binggeli, P. (1994) The misuse of terminology and anthropometric concepts in the description of introduced species. *Bulletin of the British Ecological Society* 25, 10–13.

Burke, M.J.W. and Grime, J.P. (1996) An experimental study of plant community invasibility. *Ecology* 77, 776–790.

Cadotte, M.W., McMahon, S.M. and Fukami, T. (eds) (2006) *Conceptual Ecology and Invasion Ecology: Reciprocal Approaches to Nature*. Springer, Dordrecht, The Netherlands.

Clout, M.N. and Williams, P.A. (eds) (2009) *Invasive Species Management: a Handbook of Principles and Techniques* (Techniques in Ecology & Conservation Series). Oxford University Press, New York.

Colautti, R.I. and MacIsaac, H.J. (2004) A neutral terminology to define 'invasive' species. *Diversity and Distributions* 10, 135–141.

Crawley, M.J. (1987) What makes a community invasible? In: Gray, A.J., Crawley, M.J. and Edwards, P.J. (eds) *Colonization, Succession and Stability*. Blackwell Scientific, Oxford, pp. 429–453.

Czarapata, E.J. (2005) *Invasive Plants of the Upper Midwest: an illustrated Guide to their Identification and Control*. University of Wisconsin Press, Madison, Wisconsin.

Davis, M.A. and Thompson, K. (2000) Eight ways to be a colonizer; two ways to be an invader: a proposed nomenclature scheme for invasion ecology. *Bulletin of the Ecological Society of America* 81, 226–230.

de la Cretaz, A.L. and Kelty, M.J. (1999) Establishment and control of hay-scented fern: a native invasive species. *Biological Invasions* 1, 223–236.

Elton, C.S. (1958) *The Ecology of Invasions by Animals and Plants*. University of Chicago Press, Chicago, Illinois.

Galatowitsch, S.M., Anderson, N.O. and Ascher, P.D. (1999) Invasiveness in wetland plants in temperate North America. *Wetlands* 19, 733–755.

Gatto, M. and de Leo, G.A. (2000) Pricing biodiversity and ecosystem services: the never-ending story. *BioScience* 50, 347–355.

Goodwin, B.J., McAllister, A.J. and Fahrig, L. (1999) Predicting invasiveness of plant species based on biological information. *Conservation Biology* 13, 422–426.

Gouyon, P.H. (1990) Invaders and disequilibrium. In: di Castri, F., Hansen, A.J. and Debussche, M. (eds) *Biological Invasions in Europe and the Mediterranean Basin*. Kluwer, Dordrecht, The Netherlands, pp. 365–369.

Gurevitch, J., Scheiner, S.M. and Fox, G.A. (2006) *The Ecology of Plants*, 2nd edn. Sinauer Associates Inc., Sunderland, Massachusetts.

Hager, H.A. and McCoy, K.D. (1998) The implications of accepting untested hypotheses: a review of the effects of purple loosestrife (*Lythrum salicaria*) in North America. *Biodiversity and Conservation* 7, 1069–1079.

Harper, J.L. (ed.) (1960) *The Biology of Weeds*. Blackwell Scientific, Oxford, UK.

Hellmann, J.J., Byers, J.E., Bierwagen, B.G. and Dukes, J.S. (2008) Five potential consequences of climate change for invasive species. *Conservation Biology* 22, 534–543.

Joenje, W. (1987) Remarks on biological invasions. In: Joenje, W., Bakker, K. and Vlijm, L. (eds) *The Ecology of Biological Invasions*, Proceedings of the Royal Dutch Academy of Sciences, Series C 90, 15–18.

Kolar, C.S. and Lodge, D.M. (2001) Progress in invasions biology: predicting invaders. *Trends in Ecology and Evolution* 16, 199–204.

Kowarik, I. (1995) Time lags in biological invasions with regard the success and failure of alien species. In: Pyšek, P., Prach, K., Rejmánek, M. and Wade, M. (eds) *Plant Invasions: General Aspects and Special Problems*. SPB Academic Publishing, Amsterdam, pp. 15–38.

Le Floch, E., Le Houerou, H.N. and Mathez, J. (1990) History and patterns of plant invasion in Northern Africa. In: di Castri, F., Hansen, A.J. and Debussche, M. (eds) *Biological Invasions in Europe and the Mediterranean Basin*. Kluwer, Dordrecht, The Netherlands, pp. 105–133.

Liebman, M., Mohler, C.L. and Staver, C.P. (2007) *Ecological Management of Agricultural Weeds*. Cambridge University Press, Cambridge, UK.

Lodge, D.M. (1993) Biological invasions: lessons for ecology. *Trends in Ecology and Evolution*, 8, 133–137.

Macdonald, I.A.W., Loope, L.L., Usher, M.B. and Hamann, O. (1989) Wildlife conservation and the invasion of nature reserves by introduced species: a global perspective. In: Drake, J.A., Mooney, H.A., di Castri, F., Groves, R.H., Kruger, F.J., Rejmánek, M. and Williamson, M. (eds) *Biological Invasions. A Global Perspective*. John Wiley, Chichester, UK, pp. 215–255.

Mack, R.N. (1985) Invading plants: their potential contribution to population biology. In: White, J. (ed.) *Studies on Plant Demography*. Academic Press, London, pp.127–142.

Mack, R.N., Simberloff, D., Lonsdale, W.M., Evans, H., Clout, M. and Bazzaz, F.A. (2000) Biotic invasions: causes, epidemiology, global consequences and control. *Ecological Applications* 10, 689–710.

Mooney, H.A. and Drake, J.A. (1989) Biological invasions: a SCOPE program overview. In: Drake, J.A., Mooney, H.A., di Castri, F., Groves, R.H., Kruger, F.J., Rejmánek, M. and Williamson, M. (eds) *Biological Invasions. A Global Perspective*. John Wiley, Chichester, UK, pp. 491–508.

Mooney, H.A., Mack, R.N., McNeely, J.A., Neville, L.E., Schei, P.J. and Waage, J.K. (eds) (2005) *Invasive Alien Species: a New Synthesis* (Scope Series 63). Island Press, Washington, D.C.

Myers, J.H. and Bazely, D. (2003) *Ecology and Control of Introduced Plants*. Cambridge University Press, New York.

Nunes, P.A.L.D. and van den Bergh, J.C.J.M. (2001) Economic valuation of biodiversity: sense or nonsense? *Ecological Economics* 39, 203–222.

Pappert, R.A., Hamrick, J.L. and Donovan, L.A. (2000) Genetic variation in *Pueraria lobata* (Fabaceae), an introduced, clonal, invasive plant of the southern United States. *American Journal of Botany* 87, 1240–1245.

Parker, I.M., Simberloff, D., Lonsdale, W.M., Goodell, K., Wonham, M., Kareiva, P.M., Williamson, M.H., Von Holle, B., Moyle, P.B., Byers, J.E. and Goldwasserr, L. (1999) Impact: toward a framework for understanding the ecological effects of invaders. *Biological Invasions* 1, 3–19.

Pimentel, D.A., Lach, L., Zuniga, R., Morrison, D. (2000) Environmental and economic costs of nonindigenous species in the United States. *BioScience* 50, 53–65.

Pimentel, D.A., McNair, S., Janecka, J., Wightman, J., Simmonds, C., O'Connell, C., Wong, E., Russel, L., Zern, J., Aquino, T. and Tsomondo, T. (2001) Economic and environmental threats of alien plant, animal, and microbe invasions. *Agriculture, Ecosystems and Environment* 84, 1–20.

Prach, K. and Wade, P.M. (1992) Population characteristics of expansive perennial herbs. *Preslia* 64, 45–51.

Pyšek. P. (1995) On the terminology used in plant invasion studies. In: Pyšek, P., Prach, K., Rejmánek, M. and Wade, M. (eds) *Plant Invasions: General Aspects and Specific Problems*. SPB Academic Publishing, Amsterdam, pp. 71–81.

Radford, I.J. and Cousens, R.D. (2000) Invasiveness and comparative life-history traits of exotic and indigenous *Senecio* species in Australia. *Oecologia* 125, 531–542.

Reinartz, J.A. (1997) Controlling glossy buckthorn (*Rhamnus frangula* L.) with winter herbicide treatments of cut stumps. *Natural Areas Journal* 17, 38–41.

Rejmánek, M. (1995) What makes a species invasive? In: Pyšek, P., Prach, K., Rejmánek, M. and Wade, M. (eds) *Plant Invasions: General Aspects and Special Problems*. SPB Academic Publishing, Amsterdam, pp. 3–13.

Rejmánek, M. (1999) Invasive plant species and invasible ecosystems. In: Sandlund, O.T., Schei, P.J. and Vilken, A. (eds) *Invasive Species and Biodiversity Management: Based on papers presented at the Norway/UN Conference on Alien Species, Trondheim, Norway* (Population and Community Biology Series). Kluwer, Dordrecht, The Netherlands, pp. 79–102.

Richardson, D.M. and Pyšek, P. (2008) Fifty years of invasion ecology – the legacy of Charles Elton. *Diversity and Distribution* 14, 161–168.

Salisbury, E.J. (1961) *Weeds and Aliens*. Collins, London.

Sax, D.F., Stachowicz, J.J. and Gaines, S.D. (2005) *Species Invasions: Insights into Ecology, Evolution, and Biogeography*. Sinauer, Sunderland, Massachusetts.

Sheley, R.L. and Petroff, J.K. (1999) *Biology and Management of Noxious Rangeland Weeds*. Oregon State University Press, Corvallis, Oregon.

Simberloff, D. (2006) Invasion meltdown 6 years later: important phenomenon, unfortunate metaphor, or both? *Ecology Letters* 9, 912–919.

Simberloff, D. and Von Holle, B. (1999) Positive interactions of nonindigenous species: invasional meltdown? *Biological Invasions* 1, 21–32.

Stirton, C.H. (1979) Taxonomic problems associated with invasive alien trees and shrubs in South Africa. In: *Proceedings of the 9th Plenary Meeting AETFAT*, pp. 218–219.

van Clef, M. and Stiles, E.W. (2001) Seed longevity in three pairs of native and non-native congeners: assessing invasive potential. *Northeastern Naturalist* 8, 301–310.

Vermeij, G.J. (1996) An agenda for invasion biology. *Biological Conservation* 78, 3–9.

Vitousek, P.M., Walker, L.R., Whiteaker, L.D., Mueller-Dombois, D. and Matson, P.A. (1987) Biological invasion by *Myrica faya* alters ecosystem development in Hawai'i. *Science* 238, 802–804.

Webb, S.L., Dwyer, M., Kaunzinger, C.K. and Wyckoff, P.H. (2000) The myth of the resilient forest: case study of the invasive Norway maple (*Acer platanoides*). *Rhodora* 102, 332–354.

Webster, C.R., Nelson, K. and Wangen, S.R. (2005) Stand dynamics of an insular population of an invasive tree, *Acer platanoides*. *Forest Ecology and Management* 208, 85–99.

Williamson, M. (1996) *Biological Invasions* (Population and Community Biology Series). Chapman and Hall, New York.

Williamson, M.H. and Fitter, A. (1996) The characters of successful invaders. *Biological Conservation* 78, 163–170.

Woodward, S.A., Vitousek, P.M., Matson, K., Hughes, F., Benvenuto, K. and Matson, P.A. (1990) Use of the exotic tree *Myrica faya* by native and exotic birds in Hawai'i Volcanoes National Park. *Pacific Science* 44, 88–93.

Young, J.A. and Clements, C.D. (2009) *Cheatgrass: Fire and Forage on the Range*. University of Nevada Press, Ren and Las Vegas, Nevada.

2 The Distribution and Abundance of Populations

Concepts

- A population is a group of interbreeding individuals of the same species found in the same place at the same time.
- A population's distribution changes over time.
- The distribution of a species can be mapped using historical data, field observations and remote sensing.
- A population's abundance can be measured as frequency, density, cover or biomass.
- The genetic structure of an invading population is influenced by natural selection, genetic drift and founder effects.

2.1 Introduction

A population is a group of potentially interbreeding individuals of the same species found in the same place at the same time. Like many ecological terms, this definition is flexible because it can be used to describe populations at many scales. For example, a population may be the number of individuals contained within a small area, such as a field, or it may refer to the local or regional distribution of the species.

Populations can be studied in a number of ways. A population's density and distribution quantify how it is dispersed over space. Age and sexual structure quantify the demographic characteristics of the population at any one time (Chapter 3). Population dynamics are quantified by measuring the change in births (natality), deaths (mortality), immigration and emigration over time. Population ecologists ask questions such as:

- What determines a species' distribution and/or density?
- How do a species' traits influence its distribution and abundance?
- How do biotic and abiotic factors affect a population's growth and reproductive rate?
- What is the age structure of the population?

The first step in understanding any species is to document its distribution and abundance. Distribution is a measure of the geographical range of a species, and abundance is a measure of the number or frequency of individuals in an area. In the context of an invasive species, these measures give the researcher an idea of the scope of the potential problem that the species might cause. Note that we say *potential* problem. While distribution and abundance provide useful information, more data should be obtained before a decision is made on how invasive a species is. That is why, in Chapter 1, we discussed the change in species distribution and abundance as well as the impacts of a species when discussing the process of invasion.

2.2 Population Distribution

A population's distribution, or range, describes where it occurs. In practical terms, this is a description of where the species has been recorded (Gaston, 1991). It may occur elsewhere but has not yet been recorded or identified anywhere else.

A species' distribution changes over time

A species' distribution is not static; its boundaries are dynamic. A population's distribution will change over

---

time either naturally or through human influence. Following the retreat of the last North American ice sheet approximately 10,000 years ago, trees migrated northward, each species at a different rate and following a different route (Davis, 1981). At a smaller scale, a species' distribution will change during the process of succession over decades (Chapter 11). Human disturbances, such as changing land use, alter the environment so that different species are favoured and therefore population distributions will change. Human actions also aid in the dispersal of individuals, thus further increasing a species' distribution (Chapter 6).

By following changes in a species' distribution, it is possible to tell whether a population is expanding or contracting. Asking what controls a population's distribution, and whether and why a species' distribution changes over time are fundamental questions to invasion biology. To better understand a species, we might want to ask the following questions about its distribution:

- Is the species at its current limit of distribution?
- Will the species continue to expand into new locations?
- Is the species found on specific soil types or land forms?
- Are there likely dispersal routes for this species?

Answers to these questions may warn us where problems are likely to occur or, alternatively, where control measures have been effective.

We can also gain information on species' characteristics, such as dispersal mechanisms or habitat preferences, by looking at distribution changes. Consider these two examples:

- Forcella and Harvey (1988) analysed how the distribution of 85 agricultural weeds introduced into the north-western USA changed between 1881 and 1980. They found that species' migrations patterns were dependent on the species' point of entry and on the type of agriculture (e.g. grain, cattle) that the weed was associated with. Furthermore, migration patterns tended to follow land transportation routes.
- Thompson *et al.* (1987) mapped the expansion of purple loosestrife (*L. salicaria*) from 1880 to 1985. They noted that the species first spread along canals and waterways, and later along roads (Fig. 2.1). Delisle *et al.* (2003) followed the spread of purple loosestrife from 1883 as it spread down the St Lawrence River in Canada and then along roadways, and identified two major periods of invasiveness (from 1890 to 1905 and from 1923 to 1946).

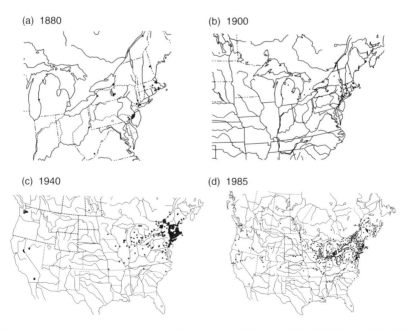

(a) 1880   (b) 1900   (c) 1940   (d) 1985

Fig. 2.1. Distribution maps of purple loosestrife (*L. salicaria*) in (a) 1880, (b) 1900, (c) 1940 and (d) 1985 (from Thompson *et al.*, 1987).

These examples give insight into how future introductions of new plant species might spread depending on their point of origin.

A population's boundaries are rarely sharp unless that population abuts against a geographical feature, such as a mountain, or human-made feature, such as a city. A species' distribution boundaries are limited by both biotic and abiotic factors. The same factor(s) will not necessarily limit all boundaries of the range equally. For example, abiotic factors, such as temperature, are more likely to influence distribution at higher latitudes and altitudes, whereas biotic factors, such as competition, are more likely to be important at lower latitudes and altitudes (Brown *et al.*, 1996). Normand *et al.* (2009), for example, found that, for a third of the European species they tested, the higher altitude and latitude boundaries were determined by abiotic stress, such as cold temperatures and a short growing season, and that other factors, such as biotic interactions, soil or historical factors, were more important in the southern or lower part of the range.

In North America, the northern limit of invasive garlic mustard (*A. petiolata*) tends to be defined by the temperature (garlic mustard needs sufficiently warm springs and summers, and winters that are not too frigid) and the soil types (garlic mustard occurs outside the often acidic soils of the core boreal forest). However, as climate change accelerates, it is likely that garlic mustard will be able to invade the boreal forest when temperatures moderate (Murphy, in press).

Estimating and mapping distribution

Mapping a species' distribution can be done on a number of scales, depending on how the information is to be used. For example, Erickson (1945) mapped the distribution of the flowering shrub Fremont's leather flower (*Clematis fremontii* var. *riehlii*) at several scales. This species was restricted to approximately 1100 km² in the Missouri Ozarks (Fig. 2.2). Individuals, however, are not evenly distributed throughout their range because they live only in sites where the abiotic and biotic conditions are suitable for them. For example, the range of Fremont's leather flower was subdivided into four watershed regions. Within regions, there were groups of glades, which are rocky outcrops on south- and west-facing slopes. Clusters of Fremont's leather flower tended to occur at the bases of these glades. Finally, within clusters, there were loose aggregates

of up to 100 individual plants. A researcher wanting to study how Fremont's leather flower is pollinated would require a fine-scale distribution map showing the locations of aggregates. However, this scale would not be useful to a researcher interested in the broad-scale climatic controls affecting the species; in this case, a large-scale map of the species' entire range would be required.

The traditional method of collecting data on the actual distribution of a species is to consult public records such as government documents, herbaria, field notes or academic journals. These types of data allow for the construction of historical distributions (Thompson *et al.*, 1987; Forcella and Harvey, 1988; Fuentes *et al.*, 2008; Pearman *et al.*, 2008). Maps such as those in Fig. 2.1 give a clear view of a species' regional distribution and any changes that have occurred over time.

Distribution data collected from the sources listed above have the disadvantage that few of them are computerized, so it is time-consuming to collect and organize the information (Peterson *et al.*, 2003). These records are also dependent on the accuracy and precision of the data collected, and this may be difficult to judge. Also, collection bias occurs when all sites and species have not been sampled equally over time and space. This results in areas with less intense sampling being under-represented on maps, and areas along major roads being over-represented (Schwartz, 1997). Sampling bias may take the form of climate bias where samples are collected in only a small region of a species' potential distribution (Loiselle *et al.*, 2007). In addition, there may be time gaps when no samples were collected, making it difficult to follow the invasion process (Delisle *et al.*, 2003). Finally, there will also be a sampling bias towards large or more obvious species. For example, Canada goldenrod (*Solidago canadensis*) has an obvious, yellow inflorescence and is more likely to be observed and recorded than common ragweed (*Ambrosia artemisiifolia*), a co-occurring weed that is much less obvious. This is why many people think they are allergic to goldenrod pollen when ragweed is the actual offender.

Field sampling and herbaria records give us information only about the current or recent past distribution of species because records may go back a few hundred years at most. Thus, the initial invasions of some species cannot be tracked this way. One way to trace early introductions of species and their distribution changes is to use palaeoecological

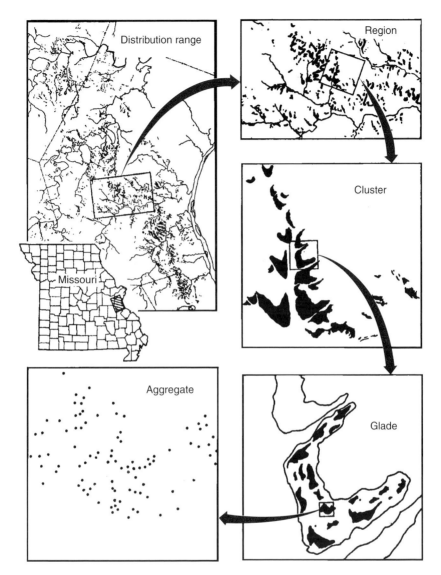

Fig. 2.2. Distribution maps of Fremont's leather flower (*Clematis fremontii* var. *riehlii*) in the Missouri Ozarks showing scales of distribution at the levels of range, region, cluster, glade and aggregate (from Erickson, 1945).

records (Pearman *et al.*, 2008). Data obtained from microfossils, such as pollen grains, and macrofossils, such as leaves and cones, go back hundreds or thousands of years. Pollen grains, for example, are preserved in peat or lake-bed sediments and can be retrieved and then identified (often to species level) to obtain a record of past vegetation. These records can be dated because the sediment is laid down in yearly layers. Furthermore, these soil layers can be radiocarbon dated.

Changes in species composition over time can be traced by constructing diagrams that show changes in pollen composition and abundance over time. Using this method, extended time series can be constructed. For example, the pollen diagram in Fig. 2.3 shows that common purslane (*Portulaca oleracea*), which is found in North America, was not an invasive species from Europe as previously thought; in fact, it had existed in the area from at least *c.*1350 BCE when the Iroquois began cultivating maize (Jackson, 1997).

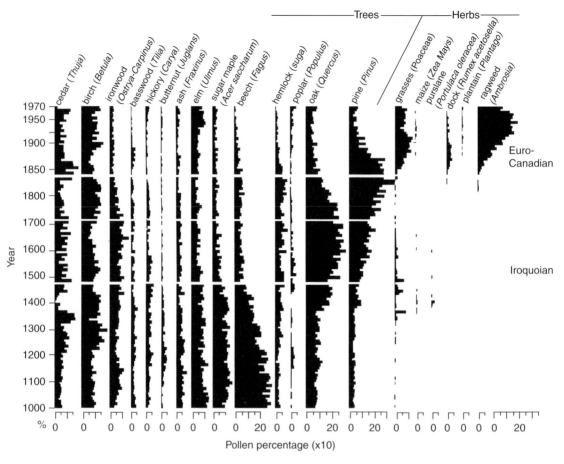

Fig. 2.3. Pollen diagram of Crawford Lake, Canada, showing the percentage of pollen for each species, with open bars representing 10× the percentage of pollen. Note the increase in maize (*Zea mays*), common purslane (*Portulaca oleracea*) and grasses (Poaceae) during the Iroquoian period and increases of weeds such as ragweed (*Ambrosia* spp.), dock (*Rumex acetosella*) and plantain (*Plantago* spp.) following land clearing by European settlers (from McAndrews, 1988).

New methods to map the current distribution of weeds are also being developed (Lass *et al.*, 2005). Such methods use remote sensing with either aircraft or satellite imagery. Photos or videos are taken to record the specific light reflected (spectral reflectance) by plants and ground terrain. To detect and map a species using this method it must be possible to distinguish its reflectance pattern from the background of surrounding vegetation, ground, roads, and other features. Species with particular growth patterns are easier to sense remotely. To date, there has been some success in mapping weeds of rangeland and pasture using remote sensing. Lass *et al.* (1996) were able to map the spatial distribution of common St John's wort (*Hypericum perforatum*) and yellow star-thistle (*Centaurea solstitialis*) in rangeland using multispec-

tral digital images taken from aircraft. Hamilton *et al.* (2006) used air photographs to map the distribution of the invasive Russian olive tree (*Elaeagnus angustifolia*) along riverbanks with 4 × 4 m grids.

Remote sensing has the benefit of covering large patches of land, so it can be used to follow the invasion of a species and monitor whether management practices are working. But, before it can be employed, we must have biological information about the species to be able to detect it properly and interpret the images. A species' spectral reflectance pattern may change over its life cycle. For example, Lass and Callihan (1997) found that yellow hawkweed (*Hieracium pratense*) and ox-eye daisy (*L. vulgare*) could be detected more accurately at the full-bloom stage than at early-bloom or post-bloom stages.

Species distribution models

Potential distribution

A species does not always fill all possible sites that it can survive in. The entire area that lies within the outer boundaries of a range is termed the extent of occurrence (Gaston and Fuller, 2009). This is the entire geographical spread of a species. Within the extent of occurrence is the area of occupancy, which is the area where the species actually occurs. A species' distribution may be limited by its inability to disperse to other sites or by its inability to compete with other species. However, many species thrive after being either intentionally or accidentally introduced into a new habitat. The area in which a species can (in theory) survive is its potential distribution. This can also be thought of as its physiological distribution. The potential distribution is based on the abiotic environment only.

The potential distribution of a species may be far greater than its native distribution. For example, the natural distribution of Monterey pine (*Pinus radiata*) is limited to approximately 6500 ha in the coastal fog belt of California, where it is considered threatened. Monterey pine is also found in Hawaii, Australia, New Zealand, South Africa, Chile, Spain and the British Isles, where it is planted for wood production. It is even considered invasive in some places. How can one species be threatened, ubiquitous and invasive? The answer is that in its native range, Monterey pine has been threatened by development, logging, changing weather patterns and diseases. However, the species is a widely planted plantation tree, such that it covers over 4 million ha (Clapp, 1995; Lavery and Mead, 1998). It is planted extensively in countries with habitats similar to California where it is a fast-growing tree and can be harvested in 25-year rotations. Monterey pine has become invasive in places with a Mediterranean climate and has invaded grasslands and native eucalypt forests (Richardson and Bond, 1991).

Climate envelope models

Predicting where a species is able to survive (its potential distribution) is a useful tool for invasion biologists. It allows managers to focus their efforts on sites most likely to be invaded by a particular species. One way to do this is to look at the abiotic requirements of a species to predict where it is likely to spread. Patterson (1996) and Patterson *et al.* (1997) estimated the potential distribution of a number of agricultural weeds using laboratory-based studies to determine the temperature and light conditions required by each species. From these data, they created a mathematical model to predict where the right combination of conditions exists for the species to survive and reproduce. For example, after growing tropical soda apple (*Solanum viarum*) in growth chambers under a variety of day and night temperatures and photoperiods, Patterson *et al.* (1997) compared their results with climatic conditions in 13 southern states of the USA. They concluded that temperature and photoperiod were not likely to limit the expansion of this species, and suggested that measures to control the expansion of soda apple beyond its current distribution in Florida should be taken immediately. This type of approach uses only abiotic factors that can be experimentally controlled, and does not take into account seasonal temperature extremes or precipitation patterns (Patterson *et al.*, 1997). For this reason, its use is limited to factors that can be looked at in controlled growing conditions.

An alternative way to predict a species' potential distribution is to compare the climatic conditions (e.g. temperature and precipitation) of the species' native habitat with those of a potential habitat (Peterson, 2003; Herborg *et al.*, 2009). Potential distribution is based on simple presence and absence data from the species' native range. These data are then compared with environmental data from the potential habitat. The goal is to identify the area of a potential invasion site that is similar to the species' native range. There are many methods and models that accomplish this.

CLIMEX is one computer model that predicts a species' potential distribution (Sutherst and Maywald, 1986). This model considers measures of growth such as temperature, moisture and day length, and then adjusts this based on stress indicators such as excessive dryness, wetness, cold and heat, to give an ecoclimate index. Scott and Batchelor (2006) used CLIMEX models to predict the distribution of a number of species of asparagus (*Asparagus* spp.) in Australia based on their native distributions in southern Africa. Bridal veil (*Asparagus declinatus*), for example, is a fern-like annual that negatively impacts native vegetation (Leah, 1981). Using parameters from the native distribution (Fig. 2.4a), Scott and Batchelor

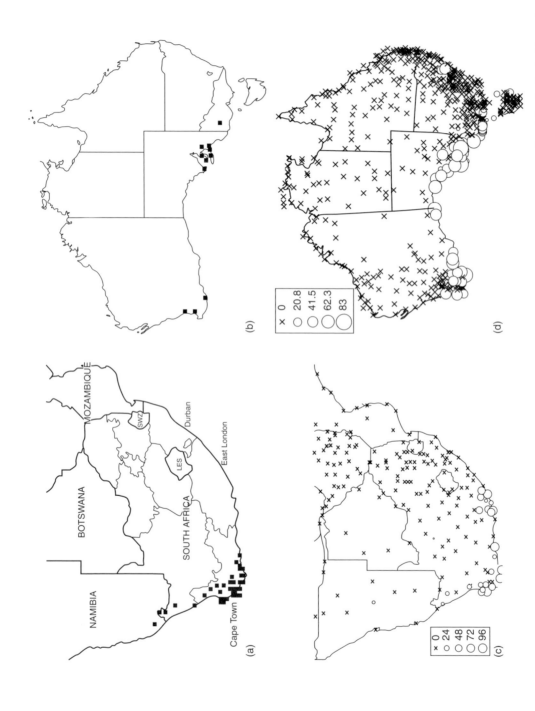

Fig. 2.4. Observed distribution of bridal veil (*Asparagus declinatus*) in (a) southern Africa and (b) Australia, and model-predicted distribution in (c) southern Africa and (d) Australia (from Scott and Batchelor, 2006). In the legend boxes, crosses indicate where the species is not expected to survive, and circles where suitable locations are predicted; the larger circles indicate the most suitable locations.

developed a model for this species using data that best matched its southern African distribution (Fig. 2.4c). The parameters included temperature and moisture variables, such as lower and upper optimal temperatures, and wet and dry stress thresholds, which determine growth and development. The model was then used to predict the distribution of bridal veil in Australia (Fig. 2.4d). Figure 2.4b shows its current distribution in Australia based on herbarium and published data. Similar distribution models were developed for the other species of asparagus, and Scott and Batchelor (2006) ranked these based on the amount of data available, such as the physiological requirements of each species. For bridal veil, they had a medium level of confidence in the distribution model because they lacked some physiological data and biological information from its native range.

Similarly, Holt and Boose (2000) were also able to map the potential distribution of velvetleaf (*Abutilon theophrasti*) in California. They concluded that the distribution of velvetleaf was not likely to increase, because its range was limited by water stress. There are problems associated with using climate envelope models such as CLIMEX. Sax *et al.* (2007) concluded that these models have two major limitations:

- They cannot accurately predict the potential range for species with small natural ranges, such as the Monterey pine, discussed earlier in this chapter.
- Some species, especially those with large native ranges, may occur well beyond climates predicted by their native range. Examples of this climate mismatch are Japanese knotweed (*Fallopia japonica*), garlic mustard, and spotted knapweed (*Centaurea maculosa*).

Climate-based models also assume that the present distribution of a species is at equilibrium; that is, the species' natural range is stable and not changing. We know, however, that a species' distribution constantly changes due to historical events, such as natural or anthropogenic climate change. In addition, the native distribution of a species may not reflect the full range of abiotic conditions it can tolerate. This may occur when the species' distribution is constrained by physical boundaries or historical events that prevent the species from filling its entire potential range (Svenning and Skov, 2004). Furthermore, evolutionary changes may occur in a species after it is introduced. These changes may alter a species' abiotic requirements; therefore, the species distribution model may incorrectly predict where that species will be able to survive. Beaumont *et al.* (2009) stressed that models should include data from the entire range (native and invaded) of the species. They found that three hawkweed species (*Hieracium* spp.) could occur in different climatic conditions in the invaded range from those in the native ranges. Broennimann and Guisan (2008) used this approach, including data from the European and North American ranges, to improve their prediction of the potential distribution of spotted knapweed in North America.

A final problem with climate-based models is that a plant's distribution is controlled by many complex factors beyond climate (Sax *et al.*, 2007). Some species distribution models improved on climate-based models because they added other environmental variables, such as geological and topographical variables, to the framework. For example, Peterson *et al.* (2003) used data describing slope, elevation, aspect and water flow, as well as a variety of climatic data to model the potential distribution of four species, including garlic mustard (Fig. 2.5).

Models that include only abiotic constraints on the distribution of a species and ignore biotic factors, such as species interactions (Chapters 8 and 9), growth rates, and dispersal mechanisms (Chapter 6), may be misleading (Hampe, 2004; Engler and Guisan, 2009). Process-based species distribution models have been developed to include biotic constraints. Using species-specific process-based models, researchers can predict a species' abundance or the probability that a species will occur in a given location (Morin and Thuiller, 2009). However, these models are relatively new and require much more extensive data collection. Thus, their use is limited to species that have been studied in detail. To use these models, information must be known about biotic relationships, land-use history and dispersal abilities that may constrain species migration (Morin and Thuiller, 2009).

Species distribution models are an important tool for predicting where species are likely to occur after they are introduced into a new environment. To make more accurate predictions, however, it is important to select the appropriate model. Pearson and Dawson (2003) stressed that the factors included in the model must correspond to the scale of investigation (Fig. 2.6). Ibáñanez *et al.* (2009) developed a hierarchical model at regional, landscape and local

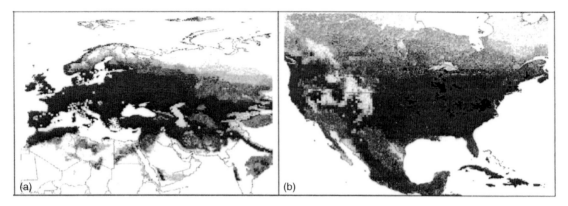

Fig. 2.5. Predictions of (a) the native distribution in Europe and Asia and (b) the potential distribution in the USA of garlic mustard (*A. petiolata*). White symbols in (a) show the occurrence data used to build the model. In (b) black areas indicate known occurrences in the introduced range, with darker shading indicating a greater level of agreement with the model (from Peterson *et al.*, 2003).

Environmental variable	Scale domain						
	Global >10,000 km	Continental >2,000–10,000 km	Regional >200–2,000 km	Landscape >10–200 km	Local >1–10 km	Site 10–1,000 m	Micro <10 m
Climate	████	████	████				
Topography			████	████			
Land use				████	████		
Soil type					████	████	
Biotic interaction						████	████

Fig. 2.6. Environmental variables more likely to be associated with various scales (from Pearson and Dawson, 2003 and based on Willis and Whittaker, 2002).

levels that included climate, landscape structure, habitat type and canopy closure to predict potential distributions of three woody plants in the north-eastern USA. Complex models such as this are desirable but require extensive data collection. Guisan and Zimmermann (2000) and Guisan and Thuiller (2005) have reviewed the process of species distribution modelling.

Climate change

More recently, species distribution models have been used to predict how the distribution of native and invasive species may change as Earth's climate changes in response to the enhanced greenhouse effect. Species distribution models can be used to predict how invasive species' distributions may change in response to changing temperature and precipitation regimes (Pearson and Dawson, 2003; Hijmans and Graham, 2006). Jarnevich and Stohlgren (2009), for example, modelled the potential distribution of kudzu

(*P. lobata*) under current climate and projected climate change by 2035. They concluded that kudzu may expand into the north-eastern USA and decline in other areas.

For climate change models, Engler and Guisan (2009) stressed the importance of dispersal, highlighting the difference between 'potentially suitable' and 'potentially colonizable' habitat. That is, just because a species could survive in a habitat (i.e. the habitat is potentially suitable) does not mean it could disperse there (i.e. the habitat is potentially colonizable).

2.3 Population Abundance

While distribution describes the geographical extent of a population, abundance describes a population's success in terms of numbers. Individuals will not be equally dispersed throughout their entire range; there will be areas of high and low density. It was previously assumed that individuals within a population were most abundant in the centre of their distribution and that they become less frequent towards the limits of their range as abiotic and biotic conditions are less suitable. This was called the abundant-centre hypothesis. More recently, this assumption has been questioned, and studies suggest that it holds for only a small proportion of taxa (Sagarin *et al.*, 2006).

Measures of abundance

The method used to measure abundance will depend on the species in question, the habitat type (e.g. forest, field), the goal of the study and the economic resources. All measures of abundance require field sampling. See Box 2.1 for a brief description of sampling methods.

Density and frequency

Density and frequency are the two simplest methods of measuring abundance. Density measures the number of individuals in a given area (e.g. a square metre or hectare). Frequency is the proportion of sampling units that contains the species. It is easy to measure because only a species' presence or absence is noted for each sampling unit. Frequency is a fast, non-destructive sampling method and is not prone to incorrect estimates by the researcher. Density is also non-destructive and, while it is more complicated to measure than frequency, it provides more information.

While frequency and density are probably the most commonly used measures of abundance, there are some difficulties associated with using them. Density assumes that you are able to separate individuals. This is not a problem in many animals because they are distinct individuals. In plants, however, many species are capable of vegetative reproduction and clonal growth, and it is therefore often difficult to distinguish one genetic individual from another (Chapter 5). Frequency does not have this problem.

A further difficulty in identifying individuals is that individuals of the same species may appear morphologically different depending on their age, stage of growth, or their environment. Many plants differ in appearance from one life stage to another. For example, the seedlings of most species look very different from a mature adult. In addition, a species' appearance may differ depending on its environment. This is known as phenotypic plasticity (Chapter 7). For example, leaves of aquatic plants often appear different depending on whether they are above or below the water, or leaves of terrestrial plants may differ depending on whether the leaf is produced in the sun or the shade. The variable appearance of a species may make it difficult to count. Therefore, measures of frequency and density might exclude individuals that are morphologically different and result in an underestimation of their abundance.

A final problem in using frequency and density as a measure of a population abundance is that they do not distinguish among the sizes of individuals. Therefore, larger individuals are scored the same as smaller ones, even though they will have different influences on the community. Larger plants will be likely to have more effect on the physical environment (e.g. through shading) and they tend to produce more seed than smaller ones, thereby having a greater influence on subsequent generations.

Cover and biomass

Cover and biomass are sometimes used in place of density and frequency when an indication of individual size is important. Cover is the proportion of ground occupied by a given species when viewed from above. It measures the spread of a species but not its height. Therefore, a shrub and a clonal plant may have the same cover value. Cover is another non-destructive sampling method. It may be difficult to get an accurate value of cover because it is subjective and therefore not precise (Kercher *et al.*, 2003) It is typically measured as a visual estimate, so percentage cover estimation is often broadly categorized (e.g. 0%, 1–5%, 5–10%, 10–25%, 25–50%, 50–75%, and 75–100%). Still, this method is widely used and considered valuable because it provides useful information with relatively low effort by the researcher.

Biomass is the weight of vegetation per area. It is useful when an accurate indication of plant size is needed. It is sampled by collecting the shoots and sometimes the roots from a given area. When collecting, the plant can also be divided into parts such as roots, stems, leaves and reproductive structures to observe how plants allocate biomass to different parts. The sample is then usually dried and biomass is given as a dry weight. Collecting actual plant samples to determine biomass is not practical for larger organisms such as trees. Also, this method is destructive as the plants are harvested.

Box 2.1. Sampling a population's abundance.

If we want to describe a population, we cannot count or measure every individual. Therefore, we measure some of the individuals and use that to represent the population as a whole. The samples should be both random and representative of the population being sampling. Fortunately, randomly collected samples are usually representative (Underwood, 1997). One way of measuring abundance is to collect data from randomly placed sampling units called quadrats (not quadra*n*ts).

The optimal number, size and shape of quadrats will depend on the species being studied, the statistical analyses to be carried out on the data, and the financial and physical resources of the researcher (Underwood, 1997; Zar, 1999; Quinn and Keough, 2002). The quadrat size for understorey vegetation is usually $1\,m^2$, quadrats for understorey trees and shrubs are about $10\,m^2$, whereas quadrats for canopy trees are about $100\,m^2$; however, these sizes are just guidelines.

The measures of abundance will be affected by quadrat size in different ways (Krebs, 1999). Frequency is more dependent on quadrat size than other measures of abundance: large quadrats will result in more species having 100% frequency, whereas in small quadrats many frequencies will be zero (Fig. 2.7). The smaller the plot, the more likely you are to miss individuals.

The spatial arrangement of individuals within a population can affect estimates of abundance. Quadrat sampling assumes that individuals are randomly distributed and that the environment is relatively homogeneous.

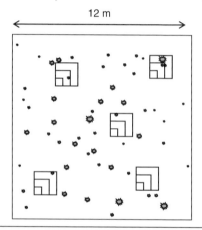

Quadrat size	Frequency = {no. quadrats with species present/no. quadrats} x 100%	Mean density = {(Σ no. present per quadrat)/no. quadrats}/quadrat area	Mean cover = (Σ estimated % cover per quadrat)/no. quadrats
☐ 0.5 m x 0.5 m = $0.25m^2$	0/5 x 100 = 0%	{(0+0+0+0+0)/5}/0.25 = $0/m^2$	(0+0+0+0+0)/5 = 0%
☐ 1.0 m x 1.0 m = $1m^2$	2/5 x 100 = 40%	{(1+1+0+0+0)/5}/1 = $0.4/m^2$	(1.5+.1.5+0+0+0)/5 = 0.6%
☐ 1.5 m x 1.5 m =$2.25m^2$	4/5 x 100 = 80%	{(1+3+1+1+0)/5}/2.25 = $0.5/m^2$	(1.5+13+1.5+1.5+0)/5 = 3.5%
True value: 12 m x 12 m =144 m^2	none	57/144 = $0.4/m^2$	{(4x.1) + (38x0.01) + (15x0.005)/144} x 100% = 0.6%

Fig. 2.7. The effect of quadrat size on estimates of the density, cover and frequency of a population. There is no frequency value for a 'true' population.

Continued

Chapter 2

Box 2.1. Continued.

In fact, this is rarely the case. Figure 2.8 shows estimates of the density, frequency and cover of three populations (calculated using randomly placed quadrats). Estimates are different even though the true values of the three populations are the same. Therefore, under some circumstances, mean density, frequency and cover may be of limited value because of sampling bias. There are ways to sample populations that are highly non-random, but this requires the use of advanced statistics (Cardina *et al.*, 1996; Dieleman and Mortensen, 1999; Gibson, 2002).

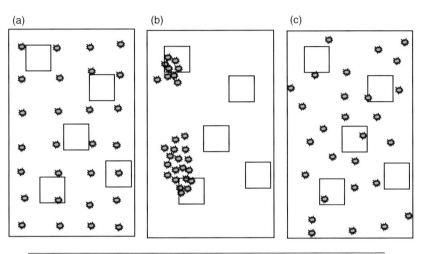

Distribution pattern	Frequency	Density	Cover
(a) Regular	40%	0.4/m²	3.2%
(b) Clumped	40%	2.0/m²	16.0%
(c) Random	60%	0.6/m²	4.6%
True value	–	0.7/m²	5.6%

Fig. 2.8. Influence of plant distribution on the estimation of the density, cover and frequency of a population. Individuals within each population are arranged (a) regularly, (b) in clumps and (c) randomly. The quadrats are the same size and are randomly distributed within the population.

2.4 Natural Selection, Genetic Drift, Founder Effects and Bottlenecks

Populations that are genetically variable (high genetic diversity) can adapt or evolve in response to a changing environment. This is beneficial, especially for newly introduced species that may disperse to an environment different from their own. Thus, genetic structure can influence a species' distribution.

Natural selection

As Gregory (2009) notes, natural selection is much misunderstood. To avoid muddling the issue, we will first quote from his paper:

Natural selection is a **non-random** difference in reproductive output among replicating entities, often due indirectly to differences in survival in a particular environment, leading to an increase in the proportion of beneficial, heritable characteristics within a population from one generation to the next.

So a given environment will end up facilitating differential reproductive success; individuals who reproduce more will be more likely to transmit heritable and beneficial traits, and these will be over-represented in the next generation. This does not mean that there is a massive shift in the genetic composition in a population; the relative increase in representation depends on how much benefit a set of genes and their traits confer and the overall population genetics: is the population small or large, and is there a lot of genetic diversity or a little? In small, low-diversity populations it can be more probable that selection will be stronger. Still, one implication is that if the environment changes, then some traits may not be beneficial anymore and other traits may become beneficial. It is not true that environments must change for natural selection to occur because mutations, while random, may confer beneficial traits even under stable environmental conditions and natural selection will still occur. Natural selection, like evolution in general, is never static for long.

Genetic drift

There is another mechanism that can be more important than natural selection in evolution. This is the stochastic (random) process of genetic drift. Simply put, in each generation, a proportion of alleles from each parent will *not* be transmitted to the next generation simply because of random assorting during meiosis. If a population is relatively small, some alleles may be lost entirely or some may become so dominant that all (or almost all) individuals have them. This occurs despite the lack of any fitness benefit and regardless of environmental conditions.

The 'neutral theory' suggests that the process of genetic drift is more responsible for genomes and evolutionary history than natural selection because drift is more probable. There are cases alluded to above where there is some external influence acting on genetic drift in the sense that some catastrophic event may kill or disable some otherwise successful individuals and their genes and alleles are lost. This process, called a 'founder effect', is still stochastic at first but then is deterministic because the remnant population's genetic composition is the basis for new genetic drift and any natural selection. A sudden change in the representation of genes from generation to generation is not likely unless there has been a catastrophic event (rather than natural selection) that creates a founder effect. When a founder effect occurs, some phenotypes and genotypes dominate because they did not get killed.

For invasive species, there is often an initially small population that is founded stochastically – the genotypes and phenotypes that arrived were not selected but transported by humans. Now they still need to fit to their environment in a manner that allows them to survive, but this too is a founder effect. Founder effects create 'population bottlenecks', i.e. a population's size is severely reduced relative to a source population, genetic variation is reduced, and the potential for natural selection may be reduced immediately, while genetic drift will be exacerbated further.

Examples of natural selection and genetic drift within invasive species

In their review of invading species, Dlugosch and Parker (2008) found that founder effects, loss of genetic diversity, and the potential for drift and bottlenecks increased during invasions. In some cases though, the invasive species had sufficient mutation rates and strong enough selection to experience rapid natural selection that favours its persistence, e.g. Canary Islands St Johns wort (*Hypericum canariense*). In other cases, the genetic variation in a local population was still low long after invasions. However, between the existence of traits such as cold tolerance and multiple introductions, which can mitigate low diversity and/or high genetic drift, some species, such as common mullein (*V. thapsus*), have succeeded.

Natural selection or drift can be exploited to work against long-term persistence with even pernicious species such as garlic mustard, as appears to be the case in southern Ontario (Yates and Murphy, 2008). Here, a lack of selection pressures from herbivores (Chapter 9) and likely genetic drift have reduced the frequency of expression of anti-herbivore chemicals. As a result, the relatively small populations of herbivores that were accidentally introduced along with garlic mustard are now becoming better able to eat the plants. If genetic drift has caused a permanent reduction in the genes that code for anti-herbivore chemicals, and alternative paths or new mutations cannot keep pace, then garlic mustard may well decline. Although this situation appears to be the case for garlic mustard, as we cautioned above, reality is not that simple and the processes are dynamic, so it will take a decade or more to test these hypotheses in full (Murphy, personal communication).

2.5 Spatial Distribution of Individuals Within a Population

Population abundance is an estimate of the average value of the entire population. Within a population, individuals are not evenly distributed throughout their range. They can be arranged at random, in clumps or in a regular pattern. These distribution patterns are the result of the abiotic environment, seed dispersal patterns, the species' biology, interactions among species and management practices.

It is important to consider spatial arrangement within a population when sampling abundance. While this applies to any invasive species, it is more evident in agricultural weeds. Early studies on the effect of weeds on crop yield loss assumed that weeds were randomly distributed; however, it is now clear that this is not the case (Hughes, 1990; Cardina *et al.*, 1997). Crop yield loss due to weeds will be overestimated if weed distribution is not taken into account (Auld and Tisdel, 1988; Jurado-Expósito *et al.*, 2003). If weeds are clumped in a few areas of the field, then crop loss estimates for the entire field will be lower than if they were randomly distributed. Another field with the same overall weed density but a more random distribution will be likely to have a greater yield loss.

2.6 Summary

The first step to understanding an invasive species is to learn about its distribution and abundance.

This information indicates the extent of the possible problem. However, distribution and abundance are not necessarily good indicators of the species' influence on other populations or on the community as a whole. While we gain some information about whether a species is increasing or decreasing from distribution and abundance data, we need to go further to fully understand the species' dynamics and whether it will or will not affect other populations. The genetic structure of a population will change after it invades. Genetic drift and founder effects can reduce or change genetic diversity and may have more influence than natural selection.

Although the concepts in this chapter are simple, they are important. If incorrectly applied, they could lead to the conclusion that a species is a problem when, in fact, it is not. In the next chapter we look at population structure and dynamics. Individuals within populations are not all identical: they differ in age, size, sex and developmental stage.

Box 2.2. Invasive species case study: distribution.

In Chapter 1, you were asked to select an invasive species for study. Research the distribution of your selected species. What resources other than maps are available? Map the distribution of your species using the appropriate scale (e.g. field, regional, continental). You may be able to have several maps, each with a different scale. Consider the following questions:

- At what scale do we know the species' distribution?
- Can we follow changes in its distribution status over time?
- What types of data were used to construct this map?
- What is the potential distribution of your species?

2.7 Questions

1. For each of the following ecosystems, which method of estimating abundance (density, cover, biomass or frequency) would be best and why?

a) forest
b) plantation
c) maize field
d) pasture

2. Why is it important to consider spatial distribution of an invasive species within the following ecosystems?

a) a field of maize
b) a natural forest

3. What biotic and abiotic factors are used to model a species' potential distribution?
4. Explain how genetic drift and founder effects can influence the distribution of an invasive species.

Further Reading

Guisan, A. and Thuiller, W. (2005) Predicting species distribution: offering more than simple habitat models. *Ecology Letters* 8, 993–1009.
Lass, L.W. Carson, H.W. and Callihan, R.H. (1996) Detection of yellow thistle (*Centaurea solstitialis*) and common St. Johnswort (*Hypericum perforatum*) with multispectral digital imagery. *Weed Science* 10, 466–474.

Thuiller, W., Albert, C., Araújo, M.B., Berry, P.M., Cabeza, M., Guisan, A., Hickler, T., Midgley, G.F., Paterson, J., Schurr, F.M., Sykes, M.T. and Zimmermann, N.E. (2008) Predicting global change impacts on plant species' distributions: future challenges. *Perspectives in Plant Ecology, Evolution and Systematics* 9, 137–152.

References

Auld, B.A. and Tisdel, C.A. (1988) Influence of spatial distribution of weeds on crop yield loss. *Plant Protection Quarterly* 3, 81.

Beaumont, L.J., Gallagher, R.V., Thuiller, W., Downey, P.O., Leishman, M.R. and Hughes, L. (2009) Different climatic envelopes among invasive populations may lead to underestimations of current and future biological invasions. *Diversity and Distribution* 15, 409–420.

Broennimann, O. and Guisan, A. (2008) Predicting current and future biological invasions: both native and invaded ranges matter. *Biology Letters* 4, 585–589.

Brown, J.H., Stevens, G.C. and Kaufman, D.W. (1996) The geographic range: size, shape, boundaries, and internal structure. *Annual Review of Ecology and Systematics.* 27, 597–623.

Cardina, J.A., Sparrow, D.H. and McCoy, E.L. (1996) Spatial relationships between seedbank and seedling populations of common lambsquarters (*Chenopodium album*) and annual grasses. *Weed Science* 44, 298–308.

Cardina, J., Johnson, G.A. and Sparrow, D.H. (1997) The nature and consequences of weed spatial distribution. *Weed Science* 45, 364–373.

Clapp, R.A. (1995) The unnatural history of the Monterey pine. *Geographical Review* 85, 1–19.

Davis, M.B. (1981) Quaternary history and stability of forest communities. In: West, D.C., Shugart, H.H. and Botkin, D.B. (eds) *Forest Succession: Concepts and Application.* Springer, New York, pp. 132–153.

Delisle, F., Lavoie, C., Jean, M. and Lachance, D. (2003) Reconstructing the spread of invasive plants: taking into account biases associated with herbarium specimens. *Journal of Biogeography* 30, 1033–1042.

Dieleman, J.A. and Mortensen, D.A. (1999) Characterizing the spatial pattern of *Abutilon theophrasti* seedling patches. *Weed Research* 39, 455–467.

Dlugosch, K.M. and Parker, I.M. (2008) Founding events in species invasions: genetic variation, adaptive evolution, and the role of multiple introductions. *Molecular Ecology* 17, 431–449.

Engler, R. and Guisan, A. (2009) MIGCLIM: predicting plant distribution and dispersal in a changing climate. *Diversity and Distributions* 15, 590–601.

Erickson, R.O. (1945) The *Clematis freemontii* var. *riehlii* population in the Ozarks. *Annals of the Missouri Botanical Garden* 32, 413–459.

Forcella, F. and Harvey, S.J. (1988) Patterns of weed migration in Northwestern U.S.A. *Weed Science* 36, 194–201.

Fuentes, N., Ugarte, E., Kühn, I. and Klotz, S. (2008) Alien plants in Chile: inferring invasion periods from herbarium records. *Biological Invasions* 10, 649–657.

Gaston, K.J. (1991) How large is a species' geographic range? *Oikos* 61, 434–437.

Gaston, K.J. and Fuller, R.A. (2009) The sizes of species' geographic ranges. *Journal of Applied Ecology* 46, 1–9.

Gibson, D.J. (2002) *Methods in Comparative Plant Population Ecology.* Oxford University Press, Oxford.

Gregory, T.R. (2009) Understanding natural selection: essential concepts and common misconceptions. *Evolution Education Outreach* 2, 156–175.

Guisan, A. and Thuiller, W. (2005) Predicting species distribution: offering more than simple habitat models. *Ecology Letters* 8, 993–1009.

Guisan, A. and Zimmermann, N.E. (2000) Predictive habitat distribution models in ecology. *Ecological Modelling* 135, 147–186.

Hamilton, R., Megown, K., Lachowski, H. and Campbell, R. (2006) Mapping Russian olive: using remote sensing to map an invasive tree. RSAC–0087–RPT1. U.S. Department of Agriculture Forest Service, Remote Sensing Application Center (RSAC), Salt Lake City, Utah.

Hampe, A. (2004) Bioclimate envelope models: what they detect and what they hide. *Global Ecology and Biogeography* 13, 469–471.

Herborg, L.-M., Drake, J.M., Rothlisberger, J.D. and Bossenbroek, J.M. (2009) Identifying suitable habitat for invasive species using ecological niche models and policy implications of range forecasts. In: Keller, R.P., Lodge, D.M., Lewis, M.A. and Shogren, J.F. (eds) *Bioeconomics of Invasive Species: Integrating Ecology, Economics, Policy, and Management.* Oxford University Press, New York, pp. 63–82.

Hijmans, R.J. and Graham, C.H. (2006) The ability of climate envelope models to predict the effect of climate change on species distributions. *Global Change Biology* 12, 2272–2281.

Holt, J.S. and Boose, A.B. (2000) Potential for spread of *Abutilon theophrasti* in California. *Weed Science* 48, 43–52.

Hughes, G. (1990) The problem of weed patchiness. *Weed Research* 30, 223–224.

Ibáñanez, I., Silander, J.A. Jr, Wilson, A.M., LaFleur, N., Tanaka, N. and Tsuyama, I. (2009) Multivariate forecasts of potential distributions of invasive plant species. *Ecological Applications* 19, 359–375.

Jackson, S.T. (1997) Documenting natural and human caused plant invasions with paleoecological methods. In: Luken, J.O. and Thieret, J.W. (eds) *Assessment and Management of Plant Invasions.* Springer-Verlag, New York, pp. 37–55.

Jarnevich, C.S. and Stohlgren, J.J. (2009) Near term climate projections for invasive species distributions. *Biological Invasions* 11, 1373–1379.

Jurado-Expósito, M., Lópes-Granados, F., García-Torres, L., García-Ferrer, A., Sánchez de la Orden, M. and Atenciano, S. (2003) Multi-species weed spatial variability and site-specific management maps in cultivated sunflower. *Weed Science* 51, 319–328.

Kercher, S.M., Frieswyk, C.B. and Zedler, J.B. (2003) Effects of sampling teams and estimation methods on the assessment of plant cover. *Journal of Vegetation Science* 14, 899–906.

Krebs, C.J. (1999) *Ecological Methodology*, 2nd edn. Addison Wesley Longman, Menlo Park, California.

Lass, L.W. and Callihan, R.H. (1997) The effect of phenological stage on detectability of yellow hawkweed (*Hieracium partense*) and oxeye daisy (*Chrysanthemum leucanthemum*) with remote multispectral digital imagery. *Weed Technology* 11, 248–256.

Lass, L.W., Carson, H.W. and Callihan, R.H. (1996) Detection of yellow starthistle (*Centaurea solstitalis*) and common St. Johnswort (*Hypericum perforatum*) with multispectral digital imagery. *Weed Science* 10, 466–474.

Lass, L.W., Prather, T.S., Glenn, N.F., Weber, K.T., Mundt, J.T. and Pettingill, J. (2005) A review of remote sensing of invasive weeds and example of the early detection of spotted knapweed (*Centaurea maculosa*) and babysbreath (*Gypsophila paniculata*) with a hyperspectral sensor. *Weed Science* 53, 242–251.

Lavery, P.B. and Mead, D.J. (1998) *Pinus radiata*: a narrow endemic from North America takes on the world. In: Richardson, D.M. (ed.) *Ecology and Biogeography of Pinus*. Cambridge University Press. Cambridge, UK, pp. 432–449.

Leah, J.M. (1981) The impacts of the environmental weed bridal veil *Asparagus declinatus* on native vegetation in South Australia. Honours thesis, Flinders University, Adelaide, Australia.

Loiselle, B.A., Jørgenson, P.M., Consiglio, T., Jiménez, I., Blake, J.G., Lohmann, L.G. and Montiel, O.M. (2007) Predicting species distributions from herbarium collections: does climate bias in collection sampling influence model outcomes? *Journal of Biogeography* 35, 105–116.

McAndrews, J. (1988) Human disturbance of North American forests and grasslands: the fossil pollen record. In: Huntley, B. and Webb, T. III (eds) *Vegetation History*. Kluwer, Dordrecht, The Netherlands, pp. 673–697.

Morin, X. and Thuiller, W. (2009) Comparing niche- and process-based models to reduce prediction uncertainty in species range shifts under climate change. *Ecology* 90,1301–1313.

Murphy, S.D. (in press) The changing natural history of Ontario. In: Nelson, J.G. (ed.) *The Georegions of Ontario*. McGill-Queen's University Press, Montreal, Canada.

Normand, S., Treier, U.A., Randin, C., Vittoz, P., Guisan, A. and Svenning, J.-C. (2009) Importance of abiotic stress as a range-limit determinant for European plants: insights from species responses to climatic gradients. *Global Ecology and Biogeography* 18, 437–449.

Patterson, D.T. (1996) Temperature and photoperiod effects on onion-weed (*Asphodelus fistulosus*) and its potential range in the United States. *Weed Technology* 10, 684–688.

Patterson, D.T., McGowan, M., Mullahey, J.J. and Westbrooks, R.G. (1997) Effects of temperature and photoperiod on tropical soda apple (*Solanum viarum* Dunal) and its potential range in the U.S. *Weed Science* 45, 404–408.

Pearman, P.B., Randin, C.F., Broennimann, O., Vittoz, P., van der Knaap, W.O., Engler, R., Le Lay, G., Zimmermann, N.E. and Guisan, A. (2008) Prediction of plant species distributions across six millennia. *Ecology Letters* 11, 357–369.

Pearson, R.G. and Dawson, T.P. (2003) Predicting impacts of climate change on the distribution of species: are bioclimate envelope models useful? *Global Ecology and Biogeography* 12, 361–371.

Peterson, A.T. (2003) Predicting the geography of species' invasions via ecological niche modeling. *The Quarterly Review of Biology* 78, 419–433.

Peterson, A.T., Papes, M. and Kluza, D.A. (2003) Predicting the potential invasive distribution of four alien plant species in North America. *Weed Science* 51, 863–868.

Quinn, G.P. and Keough, M.J. (2002) *Experimental Design and Data Analysis for Biologists*. Cambridge University Press, Cambridge, UK.

Richardson, D.M. and Bond, W.J. (1991) Determinants of plant distribution: evidence from pine invasions. *American Naturalist* 137, 639–668.

Sagarin, R.D., Gaines, S.D. and Gaylord, B. (2006) Moving beyond assumptions to understand abundance distributions across the ranges of species. *Trends in Ecology and Evolution* 21, 524–530.

Sax, D.F., Stachowicz, J.J., Brown, J.H., Bruno, J.F., Dawson, M.N., Gaines, S.D., Grosberg, R.K., Hastings, A., Holt, R.D., Mayfield, M.M., O'Connor, M.I. and Rice, W.R. (2007) Ecological and evolutionary insights from species invasions. *Trends in Ecology and Evolution* 22, 465–471.

Schwartz, M.W. (1997) Defining indigenous species: an introduction. In: Luken, J.O. and Thieret, J.W. (eds) *Assessment and Management of Plant Invasions*. Springer-Verlag, New York, pp. 7–17.

Scott, J.K. and Batchelor, K.L. (2006) Climate-based prediction of potential distributions of introduced *Asparagus* species in Australia. *Plant Protection Quarterly* 21, 91–98.

Sutherst, R.W. and Maywald, G.F. (1987) A computerised system for matching climates in ecology. *Agriculture, Ecosystems and Environment* 13, 281–299.

Svenning, J.-C. and Skov, F. (2004) Limited filling of the potential range in European tree species. *Ecology Letters* 7, 565–573.

Thompson, D.Q., Stuckey, R.L. and Thompson, E.B. (1987) Spread, impact, and control of purple loosestrife (*Lythrum salicaria*) in North American wetlands. Fish and Wildlife Research Report No. 2. U.S. Department of Interior, Fish and Wildlife Service, Washington, D.C.

Underwood, A.J. (1997) *Experiments in Ecology*. Cambridge University Press, Cambridge, UK.

Willis, K.J. and Whittaker, R.J. (2002) Species diversity – scale matters. *Science* 295, 1245–1248.

Yates, C.N. and Murphy, S.D. (2008) Observations of herbivore attack on garlic mustard (*Allaria petiolata*) in Southwestern Ontario, Canada. *Biological Invasions* 10, 757–760.

Zar, J.H. (1999) *Biostatistical Analysis*, 4th edn. Prentice Hall, New Jersey.

3 The Structure and Dynamics of Populations

3.1 Introduction

In Chapter 2, we discussed ways to describe populations in terms of their distribution and abundance. Populations were treated as whole entities. We then discussed the spatial distribution of individuals within a population and how this would influence estimates of distribution and abundance. For the most part, we treated individuals as identical entities. Populations, however, are made up of individuals that vary in age, size, genetic structure (genotype) and appearance (phenotype). Variation within populations and the repercussions for population dynamics are the subject of this chapter.

Populations are structured by their variation. For example, we can structure a population into age groups or size groups. Similarly, in human populations we could compare the age structure of men and women or we could compare the age structure of populations from two countries. Populations are dynamic; their size and structure change over time. Population structure and dynamics are used to answer questions such as:

- Are populations structured by their age or size?
- Is the population density increasing or decreasing?
- Does the population or species have a particular suite of traits?

3.2 Population Structure

Population structure could be based on any characteristic that is variable within a population. It is not a static feature of a population because individuals age, grow, reproduce and die at different rates depending on their individual characteristics and their environment. In this section, we focus on the age, size and developmental-stage structure of populations, and on the illustration and modelling of population structure.

Age structure

Each species will have a characteristic age structure. For example, annual species, which live for just one year, tend to have many young individuals, with only a few reaching maturity at the end of the growth season. A species' age structure will be influenced by the population's biotic and abiotic environment. Whipple and Dix (1979) proposed five age-class distributions to explain the population trends of trees. These are described in the list below, and shown in Fig. 3.1. They rely on being able to count the annual rings of trees; however, they have been used to examine population age structure in invasive plants such as Amur honeysuckle (*Lonicera maackii*) (Luken, 1988).

- The reverse (or inverse)-J curve shows a population with many more juveniles than adults; this population is likely to be relatively constant or increasing because there are many young individuals to replace those that die. Many annual species are an example of this type of population.
- The bimodal distribution is a result of pulse recruitment (addition of new individuals), in which periods of lower recruitment are followed by periods of higher recruitment. This population is likely to be stable or increase as long as recruitment pulses are frequent enough to replace dying individuals. Some trees that produce large seed crops every few years (masting) are examples of this.
- A decreasing population distribution means that the population is not replacing itself because recruitment is not high enough to replace those individuals that are dying. Failed invasions, where a species invades but does not persist, are examples of this type of distribution.
- If recruitment is zero the distribution will become unimodal as the population ages and no young individuals are added. The population will become extinct unless increased reproduction occurs. This may occur in some managed species where seedling recruitment is controlled but adult individuals remain.
- Finally, a random distribution is typical of a population in a marginal habitat, or one that is responding to disturbance. Populations that have recently invaded a site are also likely to exhibit this type of distribution (Luken, 1990).

Deering and Vankat (1999) found that the age structure of the invasive Amur honeysuckle was a typical inverse-J curve. However, age structures are often difficult to interpret because they do not always fit the theoretical distributions described above, nor

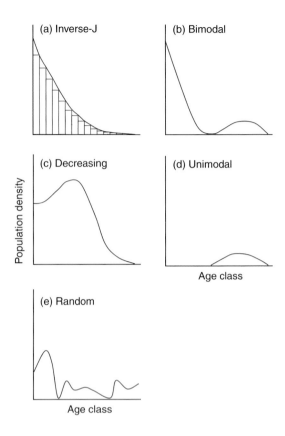

Fig. 3.1. Idealized age structure distribution used to assess population trends (redrawn from Whipple and Dix, 1979).

are they consistent over time and space (Dietz and Ullman, 1998; Paynter *et al.*, 2003; Perkins *et al.*, 2006). Populations of spotted knapweed (*C. maculosa*) in Montana, for example, tended to have inverse-J distributions in 1984, but in 1985 the distribution changed. This occurred following a severe drought in 1984, in which young individuals (2 and 3 years old) experienced higher mortality than old individuals (Boggs and Story, 1987). While overall population density decreased by 40% between 1984 and 1985, the density of younger individuals was reduced by 83%. This resulted in a change of age structure from one year to the next.

Paynter *et al.* (2003) found that the age distribution of the invasive shrub Scotch broom (*Cytisus scoparium*) differed depending on its location and the type and frequency of disturbance (Fig. 3.2). In addition, the maximum age did not differ between native (Europe) and exotic (New Zealand, Australia and elsewhere) habitats. This finding differed from

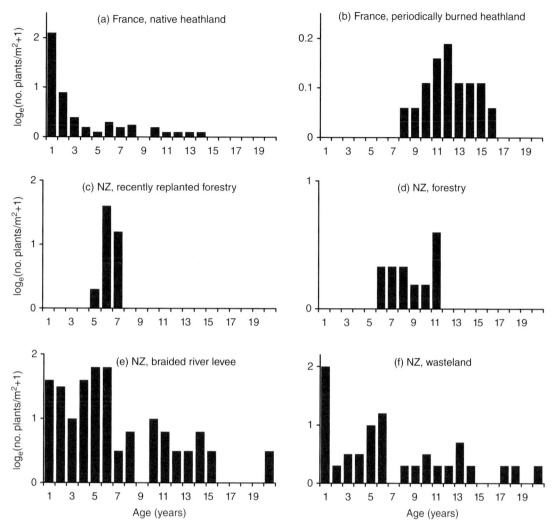

Fig. 3.2. Age structures of Scotch broom (*Cytisus scoparium*) populations in (a, b) native France and (c, d, e, f) invaded New Zealand (NZ) habitats (adapted from Paynter *et al.*, 2003).

previous studies, which suggested that Scotch broom lives longer in its exotic habitat.

The observed structure of a population is the result of abiotic and biotic influences encountered by previous generations of the population. It is important for scientists tracking changes in population density to be aware of age structure, because future changes in abundance depend very much on the current age distribution. As seen in the spotted knapweed example, harsh conditions may differentially affect age groups, causing demographic changes.

Pergl *et al.* (2006, 2007) compared the population age structure of managed (pastures) and unmanaged

populations of giant hogweed (*Heracleum mantegazzianum*) in its native (Russian and Georgian Caucasus) and invaded (Czech Republic) ranges (Fig. 3.3). This species is a huge perennial (up to 5 m tall with 3-m-long leaves) that flowers only once at the end of its life (it is monocarpic, Chapter 4). Populations from the native distribution have fewer young individuals than populations in the invaded habitats. Note that 1-year-old individuals were undercounted because only plants with leaves greater than 20 cm long were counted.

There are complications associated with characterizing populations based solely on age structure.

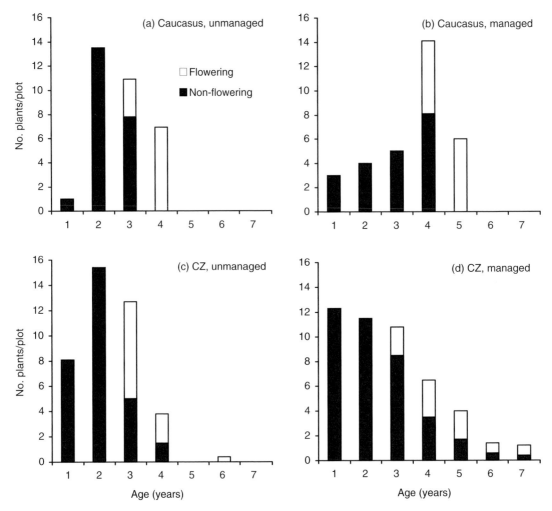

Fig. 3.3. Age structure of giant hogweed (*Heracleum mantegazzianum*) in (b and d) managed and (a and c) unmanaged habitats in its (a and b) native Caucasus and (c and d) invaded Czech Republic (CZ) ranges. Individuals less than 1 year old are under-represented because only plants with leaves greater than 20 cm long were counted (adapted from Pergl *et al.*, 2007).

First, seeds that are persistent in the soil (seed bank) are seldom accounted for when assessing age structure of a population. The seed bank represents potential individuals that replenish the population.

The second complication is that not all plant species can be aged accurately and so age structure data may be suspect. Woody species (most trees and some shrubs) are easier to age than herbaceous species because they often produce annual growth rings which can be counted; however, not all woody plants produce annual rings and some produce more than one ring in a year. Species producing multiple main stems will also be difficult to age.

Some woody species can be aged by counting morphological features such as bud scars. Annual rings in the roots of some herbaceous perennials can also be used (Boggs and Story, 1987; Dietz and Ullman, 1998; Perkins *et al.*, 2006).

The third problem with using age structure data to characterize populations is that age may not be biologically relevant to population processes such as reproduction, growth or death (Werner, 1975). For example, age is not correlated with flower number or plant size in sulfur cinquefoil (*Potentilla recta*) (Perkins *et al.*, 2006). Pergl *et al.* (2006, 2007) noted that giant hogweed in unmanaged sites flowered (and died) earlier in their invaded range than in their native range; however, in managed sites, there was no difference in age to flowering (Fig. 3.3). Therefore management, such as cattle and sheep pasturing, had the effect of delaying reproduction. In the invaded range, the life cycle is accelerated and seeds are produced earlier. These traits may increase the species' invasiveness.

Size structure

One alternative to using age-structured data is to consider the plant's size. Most populations tend to have many small individuals and fewer big ones. However, larger individuals can have a disproportionate effect on the rest of the population because they tend to live longer and produce more offspring than smaller individuals of the same age (Leverich and Levin, 1979). Larger individuals can also directly affect smaller individuals through shading.

Plant size within a population is a measure of success because larger individuals have acquired more resources than smaller individuals. Size may be a better predictor of an event, such as reproduction, than age (Werner, 1975; Werner and Caswell, 1977; Gross, 1981). Werner (1975), for example, found that the rosette size of teasel (*Dipsacus fullonum*) was a better predictor than age of whether a plant remained a vegetative rosette, flowered or died. For example, rosettes attaining 30 cm in diameter had an 80% chance of flowering. Still, size is not a perfect predictor of life cycle events. An example of this is when small, repressed agricultural weeds flower even when they are tiny compared with their neighbours.

The simplest way to measure 'size' is to measure some visible aspect of growth, such as plant height, stem diameter, or the number or size of leaves. Biomass is a more exact measure of size because it is a more direct measure of acquired resources, but biomass measurements require harvesting, drying and weighing the plant, which is a destructive sampling method. Destructive sampling is often undesirable because it means you must remove individuals from the population you are studying and, therefore, change its age structure.

A strong linear correlation between size and age rarely exists in nature. Some species of tree, such as the sugar maple (*Acer saccharum*), remain as slow-growing or suppressed individuals for decades until a canopy gap appears, after which they grow rapidly (Canham, 1985). Alternatively, plants may grow rapidly during the early life stages until they reach a maximum size, and then divert resources to reproduction and maintenance rather than to growth. Size structure develops even in small, annual species. In jewel-weed (*Impatiens capensis*), for example, size structure develops because larger individuals grow faster and have a lower risk of death than smaller ones (Schmitt *et al.*, 1987). This same process applies to the invasive cogeneric, Himalayan balsam (*Impatiens glandiferula*) (Clements *et al.*, 2008). One should never assume that age and size are correlated until the relationship has been tested.

Developmental-stage structure

A plant's stage of development can be used in conjunction with or instead of plant age and size to examine population structure (Sharitz and McCormick, 1973; Werner and Caswell, 1977; Gatsuk *et al.*, 1980; Horvitz and Schemske, 1995; Deen *et al.*, 2001). Stage of development may be more biologically relevant than age or size alone because an individual's developmental stage may be more linked to its likelihood of survival or reproduction.

Horvitz and Schemske (1995) showed that the annual survival and fertility of the prayer plant (*Calathea ovandensis*) varied depending on the individual's phenological stage (Fig. 3.4). Seedlings had less than 10% survival, seeds and juveniles had moderate survival, while other stage classes had over 90% survival. In addition, reproductive individuals produced different numbers of seeds per plant, depending on their size. This type of analysis is used to show how different stage classes of invasive (or any) plants may respond to environmental variation or management (Murphy, 2005; Pardini *et al.*, 2009).

Fidelis *et al.* (2008) found that disturbances such as burning and grazing influenced the stage distribution of eryngium (*Eryngium horridum*), an agricultural weed in Brazil. Using four age-stage classes they found that young individuals occurred more in recently burned areas and that seedlings occurred 1 to 3 years after burning (Fig. 3.5). They suggest that burning may not be a suitable control because it encourages individuals to resprout following fire.

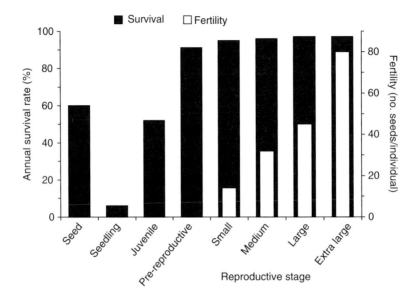

Fig. 3.4. Annual survival percentage and fertility (number of seeds per individual) of the prayer plant (*Calathea ovandensis*). Individuals were classified into five stage classes (seed, seedling, juvenile, pre-reproductive and reproductive) with four size classes of reproductives (small, medium, large and extra large) (redrawn from data in Horvitz and Schemske, 1995).

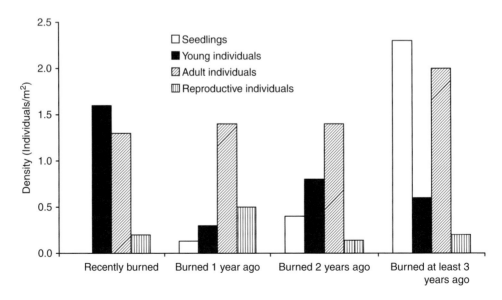

Fig. 3.5. Effect of burning on the age-stage distribution of eryngium (*Eryngium horridum*). Age-stage classes were seedlings, young individuals with rosettes <25 cm across, adult individuals with rosettes >25 cm across and reproductive individuals with an inflorescence axis (Fidelis *et al.*, 2008).

Illustrating population-structured data

Data on age or stage structure can be summarized into a life table (Table 3.1). Box 3.1 summarizes how to calculate survivorship data. Survival data is used to construct survivorship curves that display the proportion of individuals surviving to the beginning of each age class (Fig. 3.6). Survivorship curves are easier to interpret when presented on a log scale. A steeper slope indicates a higher mortality rate.

Pearl and Miner (1935) presented three general survivorship curves (Fig. 3.7). These model curves are often referred to as Deevey curves, after Deevey (1947).

- A Type I curve is typical of species, such as some human populations, with low early mortality and high mortality later in life.
- A Type II curve has a constant mortality rate throughout the lifespan. Some birds have this type of survivorship curve.
- A Type III curve has high early mortality, decreasing later in the lifespan. This is typical of many plant species where seedling mortality is very high.

Deevy curves are less static than Deevey originally envisioned because different curves apply during a growing season or the life of the plant (Mohler and Calloway, 1992; Borger et al., 2009).

Population matrix models

Population modelling allows researchers to hypothesize about future increases or decreases in species (especially invasives) or to back-cast to understand why a population increased or decreased. The models are often calculated in the form of matrix algebra. A matrix is a way of expressing numbers; a simple one would look like this:

$$\begin{bmatrix} 1 & 3 \\ 5 & 4 \end{bmatrix}$$

The numbers in a matrix could represent many things, such as the size of plants or number of offspring produced by a plant. Before computers were common, matrices had the advantage of being a practical way of calculating long series of

Table 3.1. Life table of Drummond phlox (*Phlox drummondii*) (adapted from Leverich and Levin, 1979).

Age at start of interval (days) (x)	Length of interval (days)	No. surviving on day x (n_x)	Survivorship (l_x)	No. dying during interval (d_x)	Mean mortality rate/day (m_x)
0	63	996	1.00	328	0.0052
63	61	668	0.67	373	0.0092
124	60	295	0.30	105	0.0059
184	31	190	0.19	14	0.0024
215	16	176	0.18	2	0.0007
231	16	174	0.17	1	0.0004
247	17	173	0.17	1	0.0003
264	7	172	0.17	2	0.0017
271	7	170	0.17	3	0.0025
278	7	167	0.17	2	0.0017
285	7	165	0.17	6	0.0052
292	7	159	0.16	1	0.0009
299	7	158	0.16	4	0.0036
306	7	154	0.15	3	0.0028
313	7	151	0.15	4	0.0038
320	7	147	0.15	11	0.0107
327	7	136	0.14	31	0.0325
334	7	105	0.11	31	0.0422
341	7	74	0.07	52	0.1004
348	7	22	0.02	22	0.1428
355	7	0	0.00	–	–

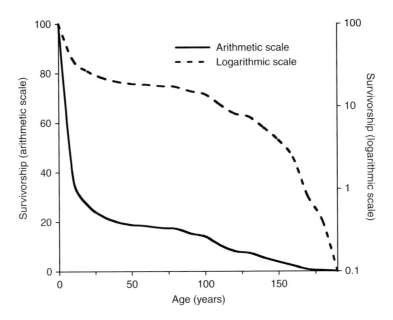

Fig. 3.6. Survivorship curve of Drummond phlox (*Phlox drummondii*) shown on arithmetic and logarithmic scales (data from Leverich and Levin, 1979).

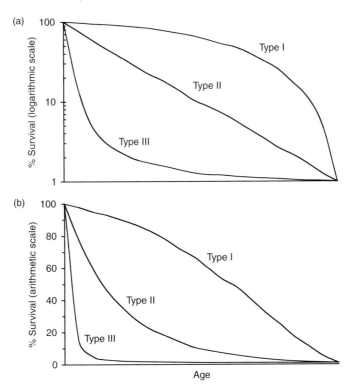

Fig. 3.7. Idealized survivorship curves shown on (a) logarithmic scale, and (b) arithmetic scale. Type I shows low early mortality and high late mortality. Type II shows a constant mortality rate over time. Type III shows high early mortality and low mortality late in life.

Chapter 3

Box 3.1. Collecting survival data and constructing survivorship curves.

Life tables, such as that shown in Table 3.2, are used to summarize survival and mortality rates of a population. There are two types of survivorship tables. The static life table looks at the age structure of a population at one point in time. The cohort life table follows a group of individuals born at the same time. Static life tables are used for organisms, such as humans and trees, that can be aged. Cohort life tables are used more in plants because plants do not run away between sampling dates and are therefore easy to locate each year. Life table calculations are the same for data collected using a cohort or static approach.

Life tables include data on the age class (x) and the number of individuals (n) alive at the start of an age class (n_x) (Table 3.2). To calculate survivorship (l) of each age class (x) (i.e. l_x), the number of individuals alive at the start of an age class (n_x) is divided by the number of individuals in the first age class (n_0). Therefore:

$$l_x = (n_x)/(n_0)$$

Survivorship data can be plotted to visualize them (Figs 3.6 and 3.8). A logarithmic scale is used because it turns a constant mortality rate into a straight line, and it makes the data towards the end of the life cycle easier to interpret. Age-specific mortality rate (m_x) is also included in a life table. This is calculated using the number of individuals dying within the specified age class (d_x), using the equation:

$$d_x = n_x - n_{x+1}$$

and then dividing by the number within that age class:

$$m_x = (d_x)/(n_{x+1})$$

These data are useful when a researcher is concerned with the mortality rate within age classes rather than how many individuals in the population are surviving. From Table 3.2 we can see that 85% of individuals die within the first age class, and that as individuals age, their mortality rate decreases. The mortality rate of the final age class is always one because, alas, everything must die sometime.

Table 3.2. Example of a life table from a hypothetical population.

Age class x	Number alive at start of age class x (n_x)	Proportion alive at start of age class x (survivorship) $(l_x = n_x/n_0)$	Number dying within age class x to $x + 1$ $(d_x = n_x - n_{x+1})$	Probability of death between age class x and $x + 1$ $(m_x = d_x/n_x)$
0	1000	(1000/1000) = 1	(1000 − 150) = 850	(850/1000) = 0.85
1	150	(150/1000) = 0.15	(150 − 50) = 100	(100/150) = 0.67
2	50	(50/1000) = 0.05	(50 − 20) = 30	(30/50) = 0.60
3	20	(20/1000) = 0.02	(20 − 10) = 10	(10/20) = 0.50
4	10	(10/1000) = 0.01	(10 − 5) = 5	(5/10) = 0.5
5	5	(5/1000) = 0.005	(5 − 0) = 5	(5/5) = 1.0
6	0	(0/1000) = 0	–	–

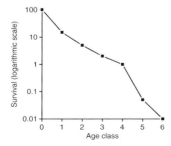

Fig. 3.8. Survivorship curve of the hypothetical data in Table 3.2.

numbers from empirical data. That tradition continues, and matrices are still effective ways of expressing ecological data, although they are less necessary now.

A simple form of population matrix model uses a Leslie matrix. This type of matrix can be expressed as a general equation in a non-matrix form which represents a population that changes over time:

$$n_0(t+1) = \Sigma n_i(t)B_i$$

where $n_i(t)$ = population size in age class i and time t, $n_0(t+1)$ = population size at a future time $(t+1)$, and B_i = no. of offspring/individual in age class i.

In population modelling, you take an initial population with a set of observed values and multiply it by the Leslie matrix to show how the population will change over time – this is repeated for as many stages or observation times (in days, months, years) as there are organisms still alive. The Leslie matrix is the simplest form of matrix because it makes several often unrealistic assumptions. These are:

- Each age class has its own defined survival and fecundity rates but each member in an age class is identical to its cohort.
- Populations never fluctuate – they grow or decline according to linear geometric equations.
- The age structure of a population determines its growth, and earlier reproduction in any age class will contribute more to total population growth.

More advanced matrix models (starting with what became known as the Lefkovitch matrix) have been developed to account for the above limitations, especially in plants, where there can be stage-based rather than age-based changes, including long periods of existence as a non-germinated seed (Chapter 6).

One can test the accuracy of population models by examining sensitivity. Sensitivity is the change in matrix values that represent variables (such as birth rate or probability of flowering), and determine which one affects populations the most in predicted versus realized scenarios. In addition, certain types of variables have restrictions on numbers; for example, survivorship is expressed as a range between 0 and 1, or 0% to 100%. Other variables are free from such limits; for example, fecundity is expressed

as actual discrete values that can have values from zero to hundreds or even hundreds of thousands. If these numbers are treated equally, it distorts their relative importance. Fecundity would be likely to appear to be a more important variable because of the potential size of the values (hundreds of thousands). The importance of survivorship would be likely to be underestimated because of its small potential value (a maximum of 1). It is often necessary to test for such proportional effects of variables; this is called elasticity. When we examine elasticity in a model, we are asking if a detected change and importance of a variable is different from expected from the raw value of the number we measured, or at least how we expressed it mathematically.

There are many ways other than matrix form in which population models can be expressed. However, regardless of how they are expressed, models are now better able to handle large numbers of variables and interactions, as both knowledge and technology (computers) have advanced. None the less, for the ability to compare current models with older ones, and the utility of expressing complex ideas to readers, matrix forms of models are still used today.

Example of the use of population modelling

Shea and Kelly (1998) used population models to examine the impacts of biological control agents on invasive nodding thistle (*Carduus nutans*). They started with a basic model of how the stages of nodding thistle may change over time, and expressed this model in the form of a matrix and a life cycle graph (Fig. 3.9). They had four stage classes: seed bank (SB), small plants (S), medium plants (M) and large plants (L). All three size classes could produce seeds, and some of the seeds would become dormant and enter the seed bank, while others would germinate immediately. Larger plants were most likely to flower and produced the most seeds.

The authors had a lot of data – data on growth, flowering, seed production, seed germination and seed bank changes. From these data, they determined that there were two key transition stages for populations (Fig. 3.10):

- plants that remained small and produced seeds that germinated immediately (S–S) (Fig. 3.10a)
- plants that remained small and either became part of the seed bank or one day germinated to produce small plants (Fig. 3.10b).

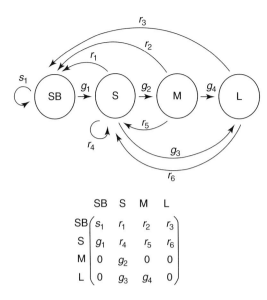

$$\begin{array}{c|cccc} & \text{SB} & \text{S} & \text{M} & \text{L} \\ \hline \text{SB} & s_1 & r_1 & r_2 & r_3 \\ \text{S} & g_1 & r_4 & r_5 & r_6 \\ \text{M} & 0 & g_2 & 0 & 0 \\ \text{L} & 0 & g_3 & g_4 & 0 \end{array}$$

Fig. 3.9. Life cycle graph and structure of the size-based population matrix for nodding thistle (*Carduus nutans*). The four classes shown are seed bank (SB), small (S), medium (M) and large (L) plants. The yearly transitions between classes are shown in the transition matrix (from Shea and Kelly, 1998).

Populations were most sensitive to these two key transition stages, which explained almost all of the variation in the population growth rate. This information was important because the researchers wanted to test and isolate a biological control agent that was most effective at disrupting these sensitive transitions. They determined that the nodding thistle receptacle weevil (*Rhinocyllus conicus*) reduced population growth rates by about 25%. However, this is insufficient to have much impact on nodding thistle populations.

Why does population structure matter?

Interpreting population structure can be difficult and time-consuming. Why, then, do we do it? Why not simply calculate population means (e.g. mean age or height) and use these simple numbers to describe a population? The answer is that there is a lot of valuable information in the variability of a population, and by reducing this to a mean value we lose information (Hutchings, 1997). Populations may have the same mean value of a particular parameter, but differ in structure; for example, the four populations shown in Fig. 3.11 have the same

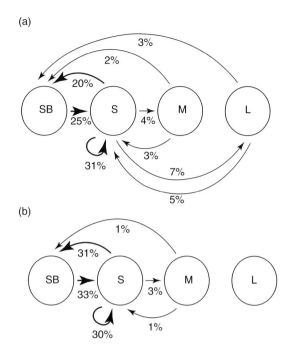

Fig. 3.10. Life cycle graphs for nodding thistle (*Carduus nutans*) at two sites showing elasticity values >1%. The four classes shown are seed bank (SB), small (S), medium (M) and large (L) plants. The most important transitions between classes are shown in heavier arrows (from Shea and Kelly, 1998).

mean stem diameter of 20 cm, but the proportion of individuals in each size class differs. By considering structure we may identify specific individuals of interest. For example, we may be only interested in plants of a certain size (age or stage). If we know, for example, that only individuals above a specific size will produce seed, then we can focus our research on the larger size classes. Recognizing population structure helps us to focus on specific individuals within a population, and gives us a glimpse of possible future population dynamics.

3.3 Life History Strategies in Plants: Population Structure and Life Cycles

The term 'life history' can generally refer to all of the traits expressed during the existence of an individual. It is common to refer to a population or species life history. Examples of traits include the stages of growth (seed, seedling, juvenile, adult), phenology of stages (when the stages appear and for how long) and the familiar traits such as height,

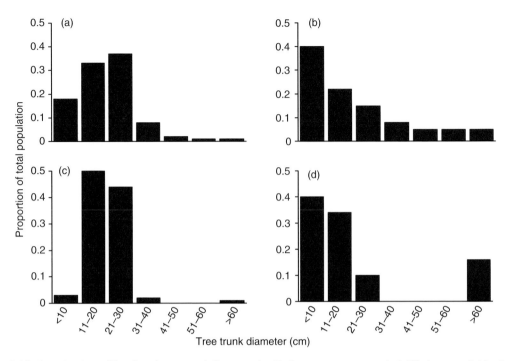

Fig. 3.11. Age structure of four imaginary populations, each with the same mean age, but differing age distributions.

leaf area, number of branches, number of flowers, colour of flowers, mechanism of pollen transfer, how seeds are dispersed and seed dormancy (Chapter 12).

'Strategy' refers to the general trends in expression of traits. For example, when does a plant usually germinate and emerge, does it usually grow tall or does it expand its canopy breadthwise, does it contain toxins or save its energy and produce more seeds? For example, purple loosestrife (*L. salicaria*) may undergo natural selection and experience a change in life history: the plants get smaller but longer-lived in stressful habitats and this may presage a new wave of invasion (Moloney *et al.*, 2009). We note that life history strategies have also been called sets of traits or syndromes. These alternative terms avoid implying that there is a conscious purpose to expression. They also indicate that traits are not static – not all individuals in a species express the same suite of traits in the same way (Grime *et al.*, 1990).

r- and *K*-selection

One way to classify plants by life history strategy is to refer to them as being '*r*' or '*K*' selected (e.g.

Beeby, 1994; Closset-Kopp *et al.*, 2007). Following disturbances, the species that will recolonize most rapidly are generally small annuals that have a rapid growth rate, reproduce early and produce many small seeds. This set of traits allows the species to arrive, germinate, establish and reproduce quickly. Therefore, if further disturbance occurs, there will be genetically variable seed available to re-establish. This specific set of traits is called an '*r*-strategy'; the '*r*' refers to the high intrinsic rate of population growth displayed by a species ('*r*' is discussed later in this chapter).

In situations where disturbance is infrequent and environmental conditions are relatively stable, traits such as large size, longevity and delayed reproduction are favoured. Plants with this set of traits are '*K*-strategists' because the populations are theoretically maintained at or near the carrying capacity (*K*) of the environment ('*K*' is also discussed later in this chapter). Table 3.3 summarizes the characteristics found in *r*- and *K*-strategists.

Agricultural weeds are often characterized as being *r*-selected because they are adapted to frequent disturbance through tillage, herbicides or other agronomic practices. Their lifespan is short, reproduction is early, fecundity is high and seeds are

Table 3.3. Features of r- and K-selected species (adapted from Pianka, 1970).

Feature	r-selected species	K-selected species
Climate	Unpredictable and/or variable; uncertain	Predictable or constant; more certain
Mortality	Occasional catastrophic mortality, density dependent	Mortality rate lower and more constant, density independent
Survivorship	Usually Deevy Type III	Usually Deevy Type I or II
Population size	Variable over time, often below carrying capacity	Constant over time, often at or near carrying capacity
Lifespan	Short, usually <1 yr	Long, usually >1 yr
Body size	Small	Large
Competition	Often low	Often intense
Rate of development	Rapid	Slow
Reproduction	Usually early, monocarpic	Usually late, polycarpic
Offspring	Produce many, small offspring	Produce few, large offspring
Leads to:	Productivity	Efficiency

small (Pianka, 1970). Nevertheless, it would be wrong to state that all agricultural weeds are r-selected. There is a degree of stability in the regularity of disturbance, and so some K-selected species also persist. Such species may be perennials that reproduce annually and have few seeds, with abundant nutrient reserves. With the increase of no-till farming, for example, K-selected species, such as common milkweed (*Asclepias syriaca*), may increase in agricultural systems (Swanton *et al*., 1993; Buhler *et al*., 1994). In fact, agricultural weeds may be anywhere on the spectrum between r- and K-selected. For example, Johnson grass (*Sorghum halepense*) and common cocklebur (*Xanthium strumarium*) are two of the world's 'worst' agricultural weeds; however, Johnson grass is K-selected and cocklebur is r-selected (Holm *et al*., 1977; Radosevich and Holt, 1984). Additionally, despite being r-selected (in general), cocklebur is an effective competitor (for water) and undergoes both early and late germination, characteristics that are not traditionally associated with r-selected species (Pianka, 1970; Scott and Geddes, 1979).

The contrast between r- and K-selection is clearly illustrated by two different varieties of barnyard grass (*Echinochloa crus-galli*) in California (Barrett and Wilson, 1983). *E. crus-galli* var. *crus-galli* has numerous small, dormant seeds. This allows it to survive in unpredictable, heterogeneous habitats, and hence it is more cosmopolitan. *E. crus-galli* var. *oryzicola* does not exhibit dormancy; it has large seeds that germinate with the rice crop (*Oryza sativa*), and large, vigorous seedlings. It is, therefore, more K-selected as it is adapted to homogeneous,

predictable environments (rice paddies) and it is the more noxious variety of weed in rice paddies. However, it is restricted to this habitat and is less of a problem worldwide than *E. crus-galli* var. *crus-galli*.

Most plants cannot be divided neatly into r- or K-strategies because they represent ends on a continuum; most plants have some traits of both strategies. American black cherry (*Prunus serotina*) has traits of both r- and K-strategists and this has allowed it to invade European forests (Closset-Kopp *et al*., 2007). As a seedling and sapling it is a K-strategist that persists in the deeply shaded forest. When a gap in the canopy appears, saplings exhibit characteristics of an r-strategist, growing rapidly to monopolize the newly created space and reproducing early compared with other tree species. Although valuable as a tool, it is naive to use r- and K-selection as the sole criterion in predicting the potential colonization ability or the invasiveness of a plant.

C-S-R selection

Because many plants exhibit a mixture of r/K-selected traits, a modified theory of plant strategy and selection was developed (Grime, 1977, 1979). Grime used traits of the adult phase of the life cycle to characterize plants based on their ability to withstand competitors, disturbance and stress. In his conceptual model, the corners represent ruderals (R) which tolerate frequent disturbance, competitors (C) or stress tolerators (S) (Fig. 3.12).

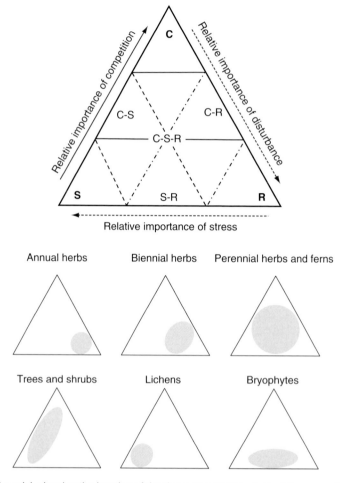

Fig. 3.12. The C-S-R model, showing the location of the three-way strategy types (C = competitors, S = stress tolerators, R = disturbance-tolerant ruderals) and secondary strategies, and the placement of various types of vascular and non-vascular plants along the three axes (redrawn from Grime, 1977).

- C-strategists maximize resource capture in undisturbed but productive habitats by increasing vegetative production and reducing allocation to reproduction.
- R-strategists maximize reproduction and growth and are adapted to disturbed but historically productive environments. These two strategies are somewhat analogous to K- and r-selection, respectively.
- S-strategists are adapted to stressful, harsh environments where disturbance is rare and competition is unimportant. By reducing vegetative growth and reproduction, they maximize their survival.

The characterization of C, S and R species is based on a plant's morphology, physiology, and life history (Table 3.4). Intermediate species are shown in the central region of the triangular model in Fig. 3.12. Invasive species are usually classified as ruderals (R), or competitive ruderals (CR). Both strategies are adapted to productive habitats, but CR-strategists would be found in less frequently disturbed habitats than R-strategists, which have short lifespans that allow species to re-establish after disturbance.

Lambdon *et al.* (2008) used C-S-R strategies to categorize functional groups comprising native or invasive (naturalized alien) species on Mediterranean islands. Functional groups are groups of species

Table 3.4. Characteristics of competitive, stress-tolerant and ruderal plants (adapted from Grime, 1977).

	Competitive (C)	Stress tolerant (S)	Ruderal (R)
Morphology			
Life forms	Herbs, trees, shrubs	Lichens, herbs, trees, shrubs	Herbs
Shoot morphology	Leaves form high, dense canopy, extensive lateral spread of roots and shoots	Variable	Small stature, little lateral spread
Leaf form	Robust	Often small, leathery or needle like	Various
Life history			
Longevity of established phase	Variable	Long	Short
Longevity of leaves and roots	Relatively short	Long	Short
Frequency of flowering	Usually every year	Variable	Produced early in life history
Annual production allocated to seeds	Small	Small	Large
Structures persisting in unfavourable conditions	Dormant buds and seeds	Stress-tolerant leaves and roots	Dormant seeds
Regeneration strategies	Vegetative growth, small seeds, persistent seed bank	Vegetative growth, persistent seedling bank	Small seeds, persistent seed bank
Physiology			
Maximum potential relative growth rate	Rapid	Slow	Rapid
Response to stress	Rapid response to maximize vegetative growth	Slow, limited response	Rapid response to divert from vegetative growth to flowering
Storage of mineral nutrients from photosynthesis	Vegetative structures, some nutrients stored for new growth in following season	Leaves, stems and/or roots	Seeds

with similar traits. The functional groups used by Lambdon *et al.* (2008) are described in Table 3.5. Lambdon *et al.* calculated the positions of the 28 functional groups for both the native and invasive species (Fig. 3.13). The researchers expected that the position of the groups would be the same for both native and alien species. They found, however, that the functional groups of native species were more stress tolerant and that alien species tended more to be ruderals or competitors. In addition, the functional groups of alien species tended to be positioned at the extreme corners of Grime's C-S-R triangle (Fig. 3.13), whereas functional groups with native species were in intermediate positions.

While Grime's strategies have been discussed widely in reference to invasive species, there are some limitations to the scheme that are similar to those bedevilling the use of classifications based on *r*- and *K*-strategies (Tilman, 1987). The problem is

that any fixed classifications are useful only if a plant species is an extreme case, such as when all individuals are expressing only one (*r*, *K*, C, S or R) strategy. Usually, there will be a wide range of expressions of these strategies within a species. A range of expressions do not fit well into a two-dimensional projection of how plants survive. Thus, survival is an outcome of the many processes that exist – there are perhaps millions of traits and interactions that contribute to success. To expect any two-dimensional graphic, whether it is *r*/*K* or C-S-R, to represent anything but a broad means of classification is naive (Lambdon *et al.*, 2008). The ideal for a researcher is that plants would fit neatly in categories. If they did then one would expect the location of functional groups on the C-S-R triangle to remain the same, regardless of which species were included in them. But, as Fig. 3.13 illustrates, this does not happen and there are other species

Table 3.5. Functional groups used by Lambdon *et al.* (2008) for native or invasive (naturalized alien) species on Mediterranean islands. Each species was assigned to one functional group.

Group no.	Functional group
1	Alpine shrubs
2	Alpine herbs
3	Ferns on cliffs and other rocks
4	Aquatic plants – fully submerged
5	Aquatic plants – emergent
6	Dicotyledons with bulbs
7	Trees – wind pollinated
8	Clonal grasses
9	Clonal non-grasses
10	Salt-tolerant shrubs
11	Salt-tolerant herbs
12	Non-clonal grasses
13	Parasitic, saprophytic plants
14	Orchids
15	Other monocotyledons from bulbs and tubers
16	Perennial herbs – wind-dispersed seeds
17	Other shrubs
18	Perennials pollinated by wind or easily accessible to insects
19	Perennials pollinated by vertebrates, bees or long-tongued insects
20	Shrubs and trees pollinated by vertebrates, bees or long-tongued insects
21	Shrubs – succulent
22	Annuals – wind dispersed
23	Annuals – specialized dispersal or pollination
24	Annuals – animal dispersed
25	Other annuals
26	Vines
27	Shrubs with fleshy berries
28	Perennial herbs – wind pollinated

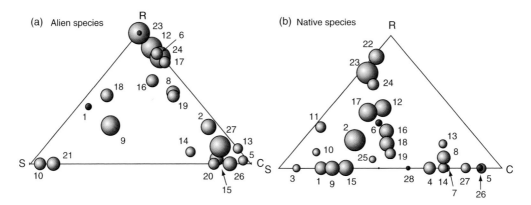

Fig. 3.13. Mean position of functional groups of (a) alien and (b) native species (group numbers given in Table 3.5) on Grime's C-S-R triangle. The areas of circles are proportional to the percentage of species in the functional group.

traits, not considered in the C-S-R model, that account for the success of invasive species and result in their change of position on the triangle.

3.4 Population Dynamics: Density Changes over Time

The density of a population will change over time. Within a short time frame, population density may remain stable, steadily increase or decrease, or fluctuate in a regular or unpredictable fashion. The rate of population change is dependent on the ratio of individuals entering the population through births (B) or immigration (I) to individuals leaving through deaths (D) or emigration (E). Demography is the study of how these processes influence how populations change over time (population dynamics).

Births (or natality) is the addition of individuals to the population. For plants, births may refer to the number of seeds produced, the number of seeds germinating (Chapter 6), or the number of individuals produced via vegetative reproduction (Chapter 5). Many plants have the potential to produce a large number of offspring. This is especially true for some invasive species where a single individual may produce more than a million seeds per season. Some examples of seed production by various different-sized invasive species are given in Table 3.6. The million seeds per season number is extreme, and actually hard to confirm, because of sampling limits and projections for individuals but, certainly, pernicious invasive species produce masses of seeds, e.g. over 30,000 seeds/m^2 each season for garlic mustard (*A. petiolata*), and over 50,000 seeds/m^2 each season for dog-strangling vine (*Vincetoxicum*

rossicum) (Rebek and O'Neil, 2006; Smith *et al.*, 2006).

Given that plants produce so many seeds, why then don't their populations increase continuously? Many seeds will not be viable, while others will not germinate because environmental conditions are not appropriate, or because the seed dies as a result of predation or disease (Chapter 9). In spite of this, there can still be many viable seedlings produced per adult plant. Mortality is the loss of individuals from the population through death.

The change in a population's size (N) from one time period (t) to the next ($t + 1$) can be represented by the equation:

$$N_{(t+1)} = N_t + B - D + I - E$$

In the following sections we look at population growth curves, first using the exponential and logistic models of growth and then by looking at real populations. For now we will ignore the processes of immigration and emigration (Chapter 12).

Exponential and logistic growth curves

As long as births outnumber deaths (i.e. $B > D$), population growth will be positive. Over generations, a population with a constant positive growth rate will exhibit exponential growth (Fig. 3.14a). The greater the difference between birth rate and death rate, the more rapid the increase.

The difference between birth rate and death rate is the instantaneous rate of population increase (r). We discussed this variable earlier in reference to r- and K-selection. The exponential population growth can be shown as:

Table 3.6. Plant size and seed production of various invasive weed species (from Holm *et al.*, 1977).

Species	Common name	Plant height (cm)	No. seeds per plant
Amaranthus spinosa	Spiny amaranthus	to 120	235,000
Anagallis arvensis	Scarlet pimpernel	10–40	900–250,000
Chenopodium album	Common lambsquarters	to 300	13,000–500,000
Digitaria sanguinalis	Large crabgrass	to 300	2,000–150,000
Echinochloa crus-galli	Barnyard grass	to 150	2,000–40,000
Eleusine indica	Goosegrass	5–60	50,000–135,000
Euphorbia hirta	Garden spurge	15–30	3,000
Polygonum convolvulus	Wild buckwheat	20–250	30,000
Solanum nigrum	Black nightshade	30–90	178,000
Striga lutea	Witchweed	7–30	50,000–500,000
Xanthium spinosum	Spiny cocklebur	30–120	150

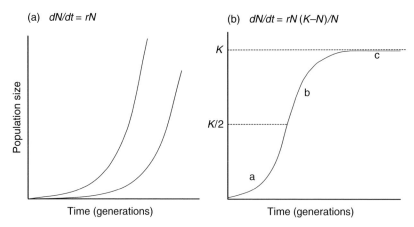

Fig. 3.14. Exponential (a) and logistic (b) growth curves. K = carrying capacity.

$dN/dt = rN$

where dN/dt is the change in N during time (t), or, alternatively, as:

$N_{(t+1)} = N_t \, e^{rt5}$

During the early stages of population growth, density may increase exponentially but, at some point, population growth will slow and density may even begin to decrease. Why is this so? Exponential growth cannot be maintained because at some point a population will become limited by a lack of resources. Eventually there will not be enough resources (e.g. nutrients, light, water, or space) to satisfy the needs of every new individual and so population density will level off. Therefore, the exponential growth curve is idealized only.

The logistic curve is a model of population growth under limiting resources and is therefore more realistic than the exponential model. Once a seed germinates, there are many biotic factors that cause mortality and reduce population growth rate. Every individual requires resources to survive and those that fail to acquire adequate resources will fail to reproduce or may die. The lack of adequate resources will cause the population growth curve to level off. The carrying capacity (K) is the maximum population size that the environment can support. Again, we saw this variable earlier in reference to r- and K-selection. To incorporate K into the population growth equation, the exponential equation can be modified to include the carrying capacity. It looks like this:

$dN/dt = rN \, (K-N)/N$

This logistic growth curve equation incorporates limits to population growth over time. When population density (N) is less than the carrying capacity (K), the term '($K - N$)/N' will be positive and population growth will be positive. As the value of N approaches K, the rate of growth decreases until $N = K$, when the rate of population growth (dN/dt) becomes zero. The population size is stable because the number of births and deaths is equal.

There are three parts to the logistic growth curve (Fig. 3.14b). Initially, population size increases at an exponential rate. The maximum rate of growth occurs at half the value of K ($K/2$). Beyond this, the rate of population increase slows down but is still positive. This occurs because not all individuals will be affected by limiting resources at the same time because of differences in size, age, health and reproductive status. Over time, the proportion of individuals affected by limiting resources will increase and this causes the curve to level off at K.

Real population growth curves

The exponential and logistic growth curves are idealized mathematical descriptions of how population size will change over time. They provide a conceptual framework on which to base more complex approaches to population growth. In real situations, population growth is more variable over time. There are a number of reasons why population size fluctuates. We will address a few here and you will see other examples in the rest of this text. First, the logistic growth model assumes that the environment is stable over time, and therefore the carrying capacity (K) remains stable. This is unrealistic because the abiotic environment is naturally variable: temperature, nutrients, water and light change over time. Even small changes in one factor can affect the number of individuals the environment can support. Second, there is random variation in birth and death rates. This is termed demographic stochasticity. An occasional low birth rate or high death rate can cause the population to decline or even go extinct. Third, the logistic and exponential growth curves assume that populations are independent of other populations. Populations, however, interact (e.g. through competition, herbivory), and this causes population size to fluctuate. Population interactions are addressed in Chapters 8 and 9.

3.5 Summary

Describing population dynamics, population structures, life cycles and life history strategies is difficult because of genetic and environmental variation, and the complex interactions and combinations that can occur. This complexity is the reason why our convenient measures and descriptions of populations are often not adequate even if they do a reasonable job of approximating the real world. This complexity explains why:

- classifying plant population structure by age, growth stage, size and life cycle can be difficult
- life history strategies are good rules of thumb but not all that accurate in predicting the population dynamics and impact of plants, especially invasive species
- simple logistic and exponential equations do not adequately describe populations.

Population dynamics and structure are useful concepts, but they need to be developed and studied in the context of ecological interactions and genetic variation. This means that it is not enough to understand the general patterns of populations. We should also understand how populations change with genetic diversity, variation in reproduction, and with the presence of competitors, herbivores and disease. In short, population dynamics and structure are influenced by many other factors that we will be discussing in future chapters.

Box 3.2. Invasive species case study: structure and dynamics of populations.

- What is known about the population structure and dynamics of your selected species? Suggest ways that your species can be structured, i.e. by age, size or developmental stage.
- Describe the life history strategy of your species.
- Is your species an *r*- or *K*-selected species – or somewhere in between? Place it on Grime's C-S-R model and explain why you placed it there.

3.6 Questions

1. Describe the size distributions of the four populations shown in Fig. 3.11. Assuming that age is correlated with size, what do you think is the likely fate of each of these populations? Explain why. Would your answer change if age were not correlated with size? Explain why.
2. Explain what it means to have a Type I, II, or III survivorship curve.
3. How might the carrying capacity (K) of an invasive species be modified by changes in management practices?
4. Explain why the population size of an invasive plant cannot increase indefinitely.

Further Reading

Cousens, R. and Mortimer, M. (1995) *Dynamics of Weed Populations.* Cambridge University Press, Cambridge, UK.
Neal, D. (2004) *Introduction to Population Biology.* Cambridge University Press, Cambridge, UK.
Vandermeer, J.H. and Goldberg, D.E. (2003) *Population Ecology: First Principles.* Princeton University Press, Princeton, New Jersey.

References

Barrett, S.C.H. and Wilson, B.F. (1983) Colonizing ability in the *Echinochloa crus-galli* complex (barnyardgrass). II Seed biology. *Canadian Journal of Botany* 61, 556–562.

Beeby, A. (1994) *Applying Ecology*. Chapman and Hall, London, New York.

Boggs, K.W. and Story, J.M. (1987) The population age structure of spotted knapweed (*Centaurea maculosa*) in Montana. *Weed Science* 35, 194–198.

Borger, C.P.D., Scott, J.K., Renton, M., Walsh, M. and Powles, S.B. (2009) Assessment of management options for *Salsola australis* in south-west Australia by transition matrix modelling. *Weed Research* 49, 400–408.

Buhler, D.D., Stoltenberg, D.E., Becker, R.L. and Gunsolus, J.L. (1994) Perennial weed populations after 14 years of variable tillage and cropping practices. *Weed Science* 42, 205–209.

Canham, C.D. (1985) Suppression and release during canopy recruitment in *Acer saccharum*. *Bulletin of the Torrey Botanical Club* 112, 134–145.

Clements, D.R., Feenstra, K.R., Jones, K. and Staniforth, R. (2008) The biology of invasive alien plants in Canada. 9. *Impatiens glandulifera* Royle. *Canadian Journal of Plant Science* 88, 403–417.

Closset-Kopp, D., Chabrerie, O., Valentin, B., Delachapelle, H. and Decocq, G. (2007) When Oskar meets Alice: does a lack of trade-off in r/K-strategies make *Prunus serotina* a successful invader in European forest? *Forest Ecology and Management* 247, 120–130.

Deen W., Swanton C.J. and Hunt L.A. (2001) A mechanistic growth and development model of common ragweed. *Weed Science* 49, 723–731.

Deering, R.H. and Vankat, J.L. (1999) Forest colonization and developmental growth of the invasive shrub *Lonicera maackii*. *American Midland Naturalist* 141, 43–50.

Deevey, E.S. (1947) Life tables for natural populations of animals. *Quarterly Review of Biology* 22, 283–314.

Dietz, H. and Ullman, I. (1998) Ecological applications of 'herbchronology': comparative stand age structure analyses of the invasive plant *Bunias orientalis* L. *Annals of Botany* 82, 471–480.

Fidelis, A., Overbeck, G., DePatta Pillar, V. and Pfadenhauer, J. (2008) Effects of disturbance on population biology of the rosette species *Eryngium horridum* Malme in grasslands in southern Brazil. *Plant Ecology* 195, 55–67.

Gatsuk, E., Smirnova, O.V., Vorontzova, L.I., Zaugolnova, L.B. and Zhukova, L.A. (1980) Age-states of plants of various growth forms: a review. *Journal of Ecology* 68, 675–696.

Grime, J.P. (1977) Evidence for the existence of three primary strategies in plants and its relevance to ecological and evolutionary theory. *American Naturalist* 111, 1169–1194.

Grime, J.P. (1979) *Plant Strategies and Vegetation Processes*. John Wiley and Sons, Toronto.

Grime, J.P., Hodgson, J.G. and Hunt, R. (1990) *The Abridged Comparative Plant Ecology*. Unwin Hyman, Boston, Massachusetts.

Gross, K.L. (1981) Predictions of fate from rosette size in four "biennial" plant species: *Verbascum thapsus, Oenothera biennis, Daucus carota*, and *Tragopogon dubius. Oecologia* 48, 209–213.

Holm, L., Plucknett, D.L., Pancho, J.V. and Herberger, J.P. (1977) *The Worlds's Worst Weeds: Distribution and Biology*. University Press of Hawaii, Honolulu, Hawaii.

Horvitz, C.C. and Schemske, D.W. (1995) Spatiotemporal variation in demographic transitions for a neotropical understory herb: projection matrix analysis. *Ecological Monographs* 65, 155–192.

Hutchings, M.J. (1997) The structure of plant populations. In: Crawley, M.J. (ed.) *Plant Ecology*, 2nd edn. Blackwell Science, Oxford, pp. 325–358.

Lambdon, P.W., Lloret, F. and Hulme, P.E. (2008) Do alien plants on Mediterranean islands tend to invade different niches from native species? *Biological Invasions* 10, 703–716.

Leverich, W.J. and Levin, D.A. (1979) Age-specific survivorship and reproduction in *Phlox drummondii*. *American Naturalist* 113, 881–903.

Luken, J.O. (1988) Population-structure and biomass allocation of the naturalized shrub *Lonicera maackii* (Rupr.) *maxim* in forest and open habitats. *American Midland Naturalist* 119, 258–267.

Luken, J.O. (1990) *Directing Ecological Succession*. Chapman and Hall, London.

Mohler, C.L. and Calloway, M.B. (1992) Effects of tillage and mulch on the emergence and survival of weeds in sweet corn. *Journal of Applied Ecology* 29, 21–34.

Moloney, K.A., Knaus, F. and Dietz, H. (2009) Evidence for a shift in life-history strategy during the secondary phase of a plant invasion. *Biological Invasions* 11, 625–634.

Murphy, S.D. (2005) Concurrent management of an exotic species and initial restoration efforts in forests. *Restoration Ecology* 13, 584–593.

Pardini, E.A., Drake, J.M., Chase, J.M. and Knight, T.M. (2009) Complex population dynamics and control of the invasive biennial *Alliaria petiolata* (garlic mustard). *Ecological Applications* 19, 387–397.

Paynter, Q., Downey, P.O. and Sheppard, A.W. (2003) Age structure and growth of the woody legume weed *Cytisus scoparius* in native and exotic habitats: implications for control. *Journal of Applied Ecology* 40, 470–480.

Pearl, R. and Miner, J.R. (1935) Experimental studies on the duration of life. XIV. The comparative mortality of certain lower organisms. *Quarterly Review of Biology* 10, 60–79.

Pergl, J., Perglová, I., Pyšek, P. and Dietz, H. (2006) Population age structure and reproductive behavior of the monocarpic perennial *Heracleum mantegazzianum* (Apiaceae) in its native and invaded distribution ranges. *American Journal of Botany* 93, 1018–1028.

Pergl, J., Hüls, J., Perglová, I., Eckstein, L., Pyšek, P. and Otte, A. (2007) Population dynamics of *Heracleum mantegazzianum*. In: Pyšek, P., Cock, M.J.W., Nentwig, W. and Ravn, H.P. (eds) *Ecology and Management of Giant Hogweed*. CAB International, Wallingford, UK, pp. 92–111.

Perkins, D.L., Parks, C.G., Dwire, K.A., Endress, B.A. and Johnson, K.L. (2006) Age structure and age-related performance of sulfur cinquefoil (*Potentilla recta*). *Weed Science* 54, 87–93.

Pianka, E.R. (1970) On *r*- and K-selection. *American Naturalist* 104, 592–597.

Radosevich, S.R. and Holt, J.S. (1984) *Weed Ecology: Implications for Management*. John Wiley, New York.

Rebek, K.A. and O'Neil, R.J. (2006) The effects of natural and manipulated density regimes on *Alliaria petiolata* survival, growth and reproduction. *Weed Research* 46, 345–352.

Schmitt, J., Eccleston, J. and Ehrhardt, D.W. (1987) Dominance and suppression, size-dependent growth and self-thinning in a natural *Impatiens capensis* population. *Journal of Ecology* 75, 651–665.

Scott, H.D. and Geddes, R.D. (1979) Plant water stress of soybean (*Glycine max*) and common cocklebur (*Xanthium pensylvanicum*): a comparison under field conditions. *Weed Science* 27, 285–289.

Sharitz, R.R. and McCormick, J.F. (1973) Population dynamics of two competing annual species. *Ecology* 54, 723–740.

Shea, K. and Kelly, D. (1998) Estimating biocontrol agent impact with matrix models: *Carduus nutans* in New Zealand. *Ecological Applications* 8, 824–832.

Smith, L.L., DiTommaso, A., Lehmann, J. and Greipsson, S. (2006) Growth and reproductive potential of the invasive exotic vine *Vincetoxicum rossicum* in northern New York State. *Canadian Journal of Botany* 84, 1771–1780.

Swanton, C.J., Clements, D.R. and Derksen, D.A. (1993) Weed succession under conservation tillage: a hierarchical framework for research and management. *Weed Technology* 7, 286–297.

Tilman, D. (1987) On the meaning of competition and the mechanisms of competitive superiority. *Functional Ecology* 1, 304–315.

Werner, P.A. (1975) Predictions of fate from rosette size in teasel (*Dipsacus fullonum* L.). *Oecologia* 20, 197–201.

Werner, P.A. and Caswell, H. (1977) Population growth rates and age versus stage-distribution models for teasel (*Dipsacus sylvestris* Huds.). *Ecology* 58, 1103–1111.

Whipple, S.A. and Dix, R.L. (1979) Age structure and successional dynamics of a Colorado subalpine forest. *American Midland Naturalist* 101, 142–158.

4 Sexual Reproduction

Concepts

- Plants can have complex combinations of gender expression; they can be exclusively male or female, or both male and female at the same time.
- Pollination by animals is more accurate than wind pollination, but more energetically expensive because the floral structures are elaborate.
- Self-compatibility guarantees some degree of mating success because pollination occurs even when only one individual is present. However, it may result in inbreeding depression, which is prevented by self-incompatibility.
- Pollen limitation and Allee effects limit reproductive success.
- Sexual reproduction results in the production of genetically variable and mobile offspring; however, it disrupts well-adapted genotypes and requires investment in reproductive structures.
- Hybridization may create new invasive species, which may displace native species.

4.1 Introduction

Plants use sexual and asexual reproduction. Sexual reproduction requires the fusion of two gametes (a sperm and an ovum) to form a zygote. Each gamete normally contains one set of chromosomes (n) and the zygote will normally have two sets of chromosomes ($2n$): one from each parent. Therefore, sexually produced offspring possess a unique recombination of their parents' genes and are genetically different from their parents.

In contrast, asexual reproduction generally involves the replication of chromosomes without the production of gametes or the need for sex. Asexual reproduction produces offspring that are genetically identical to their parents. We will discuss asexual reproduction in the next chapter; in this chapter, we focus on sexual reproduction.

4.2 Pollination Ecology

In flowering plants, sexual reproduction is facilitated by pollination. Pollination occurs when pollen is transported from the stamen to the stigma on a flower (Fig. 4.1). Once on a compatible stigma, pollen produces a pollen tube that delivers the sperm to the female gametes (ova) and, ultimately, a seed will develop. Through sexual reproduction, there are many ways plants express sexuality. We will examine some of them below.

Gender expression

In most animals, an individual is either male or female. Defining gender in a plant, however, is more complicated. In plants, gender can apply to individual flowers or to the individual as a whole (Table 4.1). The reason for this complexity is related to sexual selection. These are the factors that influence the relative ability of individuals to obtain mates and reproduce offspring (Willson, 1994). Generally, the more options plants have to express gender, the more likely they are to reproduce successfully no matter what environment they encounter. We will also show, however, that there can be risks for an individual in expressing many combinations of gender and,

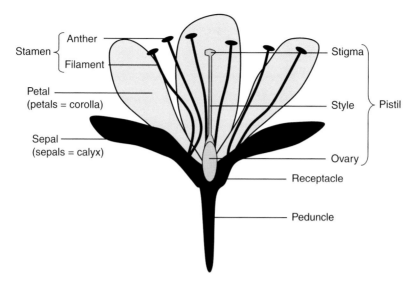

Fig. 4.1. Drawing of archetypal flower.

Table 4.1. Gender expression in plants. Note that some terms apply to the gender expression of the *individual* as a whole, while others refer to gender expression within the *flowers*. Cruden and Lloyd (1995) give alternative terminology for gender expression.

Term	Description
Monocliny	Each *individual* is genetically capable of expressing both genders. Whether both genders are actually expressed can be influenced by genetic and environmental factors
Sequential monocliny	Each *individual* expresses only one gender at a given time. Gender changes over a growing season or from year to year, e.g. saltbushes (*Atriplex* spp.) (Freeman and McArthur, 1984)
Simultaneous monocliny	An *individual* expresses both genders at a given time, but not all *flowers* necessarily express both genders, at the same time or ever
Sexual monomorphism	Male and female gender will be expressed in the same *flower*, though not necessary at the same time
Protandry	Male gender expressed before female gender, e.g. wild carrot *(Daucus carota)* (Dale, 1974)
Protogyny	Female gender expressed before male gender, e.g. common mullein (*V. thapsus*) (Gross and Werner, 1978)
True monomorphism	Both genders expressed at the same time in the same *flower*, e.g. sow thistles (*Sonchus* spp.) (Hutchinson *et al.*, 1984)
Monoecy	In at least some *flowers*, only one gender is ever expressed
Gynomonoecy	Female flowers and perfect flowers exist, e.g. plantains (*Plantago* spp.) (de Haan *et al.*, 1997)
Andromonoecy	Male flowers and perfect flowers exist, e.g. horsenettle (*Solanum carolinense*) (Steven *et al.*, 1999)
True monoecy	All flowers are either male or female; no perfect flowers exist, e.g. nettles (*Urtica* spp.) (Bassett *et al.*, 1977)
Dicliny	Each *individual* is genetically capable of expressing only one gender during its existence. However, species or populations are not always totally diclinous
Dioecy (true dicliny)	All *individuals* are either entirely male or entirely female, e.g. poison ivy (*Rhus radicans*) (Mulligan and Junkins, 1977)
Gynodioecy	Some *individuals* are entirely female; others are monoclinous, e.g. viper's bugloss (*Echium vulgare*) (Klinkhamer *et al.*, 1994)
Androdioecy	Some *individuals* are entirely male; others are monoclinous, e.g. annual mercury (*Mercurialis annua*) (Pannell, 1997)

consequently, there can be benefits to expressing only one or a few combinations of gender.

Gender based on flowers

Most people recognize that flowers can be male and female at the same time because we are taught to recognize the basic structures of a typical flower, i.e. petals, sepals, stamens and pistils. However, some or all flowers on an individual may express only one gender. The sexual expression of a flower also can be separated in time. Male structures (anthers and pollen) may mature first, followed by the female structures (stigma) becoming receptive, or vice versa. There are many complex variations of this with equally complex terminology (Table 4.1).

Gender based on the individual

The gender of an individual plant can be influenced by the environment. Plants often have the genetic ability to be male and female, but the relative expression of male and female traits varies with the short-term environmental conditions and perhaps also with long-term selection pressures (Barrett, 1998; Campbell, 2000). Plants where individuals are genetically one sex are called 'diclinous'; plants that can genetically express more than one sex are called 'monoclinous'. Like its flowers, an entire individual plant can be: male, female, both male and female at the same time, male and then female, female and then male, or continually changing from female to male and vice versa.

Allocation strategies for expressing genders in flowers and individuals

Environmental stress tends to increase the expression of the male gender in plants (Freeman *et al.*, 1980; Escarre and Thompson, 1991). In a resource-poor environment, it is better to be male than female. This is because male structures such as pollen require fewer resources to develop, whereas female structures, such as ovules, are where offspring develop and require the allocation of more resources for nurture and dispersal. As we shall see, males have little control over their reproductive success as pollen can go astray. Because plants cannot predict their future environment, any allocation and gender expression strategy is risky and generally depends on what previous and current selection pressures favour. Some plants use 'bet-hedging' by

allocating equal amounts of resources to both male and female genders; however, even this may reduce fitness if the environment changes to favour the expression of one gender rather than both.

Pollination mechanisms

Because of its microscopic size, pollen is usually produced in large quantities in order to increase the chance of reaching the non-mobile ova. For a plant to mate successfully, pollen must be transferred from the anther to a genetically compatible stigma. Pollen generally is delivered via three mechanisms: by animals (zoophily), wind (anemophily), or water (hydrophily).

Animal pollination (zoophily)

Animal-pollinated species must allocate resources to create floral structures to attract the animals that carry the pollen. Floral morphology varies with the type of animal pollinator, as do the pigments used to colour flowers, the height and breadth of the inflorescence, and the provision of nectar (Wyatt, 1983). These structures can be very resource expensive (Harder and Barrett, 1995). Plants may use many types of pollinators or very few but specialized pollinators (Johnson and Steiner, 2000), but some general trends do exist (Table 4.2). Of most relevance here, animal-pollinated invasive species tend not to need elaborate floral morphologies because they are not usually co-adapted to their pollinators (Baker, 1974). For example, wild carrot (*Daucus carota*) has an open, flat inflorescence that enables a variety of insects to access pollen (Dale, 1974).

There is no guarantee of successful pollination because the inflorescence can be eaten, pathogens or parasites can infest the flowers, or animals can rob nectar without transferring pollen. Plants may increase the likelihood of successful animal pollination by:

- deceiving pollinators, for example by using chemicals that resemble nectar to lure them
- trapping pollinators in a flower to ensure that they are covered in pollen
- forcing pollinators to specialize by hiding rewards such as nectar
- having specialized flowers that require specific structures on pollinators, such as a uniquely shaped proboscis
- flowering occurring only when other species are not flowering.

Table 4.2. Suites of floral traits associated with pollination syndromes (adapted from Howe and Westley, 1997).

Pollinating agent	Anthesis	Colour	Odour	Flower shape
Insect pollination				
Beetles	Day and night	Dull	Fruity or aminoid	Flat or bowl-shaped; radial symmetry
Carrion or dung flies	Day and night	Purple-brown or greenish	Decaying protein	Flat or deep; radial symmetry; often traps
Bees	Day and night or diurnal	Variable, but not pure red	Usually sweet	Flat to broad tube; bilateral or radial symmetry; may be closed
Butterflies	Day and night or diurnal	Variable; pink very common	Sweet	Upright; radial symmetry; deep or with spur
Vertebrate pollination				
Bats	Night	Drab, pale, often green	Musty	Flat 'shaving brush' or deep tube; radial symmetry; much pollen; often upright, hanging outside foliage, or borne on trunk or branch
Birds	Day	Vivid, often red	None	Tubular, sometimes curved; radial or bilateral symmetry, robust corolla; often hanging
Abiotic pollination				
Wind	Day or night	Drab, green	None	Small, sepals and petals absent or reduced; large stigmata; much pollen; often catkins
Water	Variable	Variable	None	Minute; sepals and petals absent or reduced; entire male flower may be released

Wind pollination (anemophily)

Wind-pollinated flowers are often less energetically expensive (Whitehead, 1983). They may be drab, have small or absent petals and no nectar. Wind-pollinated plants must produce vast quantities of pollen to ensure success because there is no vector to directly deliver the pollen. This type of pollination is risky because most of the pollen may not reach the proper stigma, and successful pollination depends on appropriate environmental conditions, such as precipitation, temperature, relative humidity and wind direction (Whitehead, 1983; Murphy, 1999). Examples of wind-pollinated invasive species include ragweeds (*Ambrosia* spp.), quack (or couch) grass (*Elymus repens*) (and other members of the grass family – Poaceae), and Monterey pine (*P. radiata*). It may be advantageous for invasive species to be wind pollinated because they do not rely on specialized animal vectors.

Water pollination (hydrophily)

Water pollination is unique to submergent aquatic plants (Les, 1988). Submergent invasive species that are water pollinated include horned pondweeds (*Zannichellia* spp.) and pondweeds (*Najas* spp.). Generally, water pollination is inefficient because pollen (or sometimes the entire male parts of a flower) must float on the water or be transported in the water to reach stigmas. We emphasize that many familiar aquatic invasive species are not water pollinated. Emergent aquatic plants like cattails (*Typha* spp.) are wind pollinated while other emergents (pickerel-weed, *Pontederia cordata*) and floating plants (water hyacinth, *Eichhornia crassipes*) are animal pollinated.

Self-compatibility and self-incompatibility

Some individuals can successfully mate with themselves if pollen is transferred from anther to stigma either within the same flower or between flowers on the same individuals. This is called self-compatibility. Self-*in*compatibility occurs when genetically identical pollen is rejected (Franklin-Tong and Franklin, 2003). Approximately 60% of flowering plants have self-incompatibility systems (Hiscock and Tabah, 2003).

Some individuals have a mechanism called histochemical incompatibility. This is a bit like a pollen grain causing an allergic reaction on the stamen so that the tissues change and fertilization cannot occur. The basis for histochemical incompatibility is specialized glycoproteins which are expressed in various combinations in pollen, the stigma and the style. The glycoproteins are signals that identify incompatible mates – usually genetically related or genetically the same pollen (Sims, 1993).

There are two types of histochemical self-incompatibility systems: sporophytic and gametophytic. Sporophytic incompatibility occurs when pollen is on the stigma; the glycoproteins signal the stigma not to exude the water needed for pollen to germinate. Gametophytic incompatibility usually occurs as the pollen tube is trying to grow in the style towards the embryo sac within the ovules; this type of incompatibility is the more common.

Plants also have structural mechanisms to avoid incompatible mates. Pollen may not physically adhere to certain stigmas, or it may be too big, too small, the wrong shape or the wrong texture. Additionally, the stigmas of a flower may be located above the pollen-bearing anthers so that pollen cannot fall on top of the stigma. A more elaborate mechanism to ensure cross-pollination is called heterostyly, where different types (or morphs) of flowers have stamens and styles of distinct lengths. Self-compatibility is further reinforced because the male and female functions have genetically based barriers as well. Figure 4.2 illustrates heterostyly in purple loosestrife (*L. salicaria*).

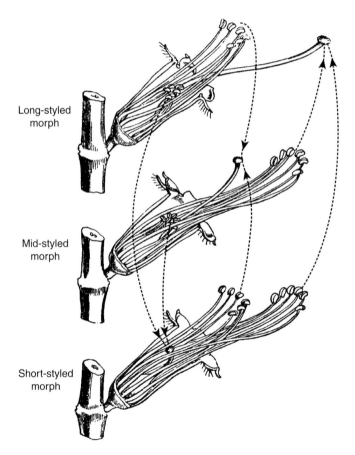

Fig. 4.2. Illustration of heterostyly in purple loosestrife (*L. salicaria*). The three floral forms are: long-styled (with short and mid-anthers), mid-styled (with long and short anthers) and short-styled (with long and mid-anthers). Petals and calyx to the front of the illustration have been removed to reveal flower parts. Dotted arrows show direction that pollen must be transferred from anther to stigma to ensure fertilization (from Briggs and Walters, 1984; originally from Charles Darwin, 1877).

The benefits and costs of self-compatibility

Self-compatibility can be important to colonizing species because it means that a single individual can invade a site and be able to self-fertilize and produce seed. With this advantage, it is not surprising that many invasive species are often self-compatible (Mulligan and Findlay, 1970; Baker, 1974; Barrett, 1992). A second advantage of self-compatibility is that it can be less costly if resource allocation to floral structures is reduced (i.e. pollinators may not be important).

Though self-compatibility might be advantageous, not all invasive species use this, e.g. Jimson weed (*Datura stramonium*) (Motten and Antonovics, 1992). This is because self-compatibility has costs as well as benefits. The main costs of self-compatibility are:

- offspring are more likely to accumulate harmful alleles
- offspring will not have new genetic material incorporated into their genome and so they may not be adapted to changing environments.

Reduced fitness caused by mating with a close relative and accumulating multiple copies of deleterious alleles is formally termed 'inbreeding depression'. Even self-incompatible plants have no guarantees of avoiding costs as their genetically recombined genotypes may not be adapted to the environment. Additionally, a self-incompatible individual may still mate with a close relative because its likely mates (close neighbours) are often close relatives (see Madden, 1995; Lefol *et al.*, 1996; Nunez-Farfan *et al.*, 1996; Guttieri *et al.*, 1998; Sun and Ritland, 1998; see also Stanton, 1994 and Wilson and Payne, 1994 for discussion of mate selection to avoid this problem).

Pollination problems

Usually, fertilization occurs only when pollen lands on a stamen from the same species (i.e. a conspecific species). Improper pollen transfer refers to situations where pollen from an individual of a different species (i.e. a heterospecific species) lands on a stigma. This is a problem for the pollen donor because a lot of pollen is wasted on individuals where fertilization will not occur. The pollen recipient is also affected if heterospecific pollen contains toxins (allelochemicals), pathogens or parasites (Murphy, 1999). For both donors and recipients, the result is lower pollination success and production of fewer viable seeds.

Plants also suffer reduced pollination success and seed set if pollen limitation occurs (Lalonde and Roitberg, 1994; Collevatti *et al.*, 1997; Knight *et al.*, 2005). In a review of the topic, Knight *et al.* (2005) showed that pollen limitation caused reduced fruit set in over 60% of species. Pollen limitation occurs when there are few compatible mates nearby (and therefore not enough pollen is produced) or when there are few appropriate animal pollinators in the community. Environmental conditions influence pollen limitation when poor growing conditions result in insufficient flowers being produced in a given growing season, or when inclement weather prevents animal or even wind pollination from occurring. This is more of a transient (environmentally stochastic) pollen limitation rather than a demographic limitation related to available mates or pollination vectors in the long term.

A number of life history traits are associated with high levels of pollen limitation (Knight *et al.*, 2005). Self-incompatible species, for example, are more likely to be pollen limited. Long-lived species or species that reproduce repeatedly (polycarpic species) are more likely to be pollen limited than short-lived species or species that only reproduce once (monocarpic species). Long-lived polycarpic species, however, are less likely to show demographic effects of pollen limitation because seed production occurs repeatedly, and periods of pollen limitation will not affect the abundance of the population (Ashman *et al.*, 2004; Knight *et al.*, 2005); other vital factors, such as adult survivorship rate, are more likely to influence lifetime seed set in these species.

Invasive species are at particular risk of pollen limitation if they rely on specialized pollinators for pollen dispersal. The spread of the yellow star-thistle (*C. solstitialis*), an outcrossing species, may be limited by the lack of non-native honey bees (*Apis mellifera*) that are required for pollination (Barthell *et al.*, 2001).

Allee effects in smaller populations

Pollen limitation can be linked to the broader phenomenon of the Allee effect. An Allee effect occurs because small populations tend to have lower rates of survival and reproduction than large populations. It results in a positive relationship between a population's density (or abundance) and its growth rate

(Stephens *et al.*, 1999). Allee effects are caused by a number of factors, but the difficulty in finding a mate is most prevalent (Taylor and Hastings, 2005) and, in plants, one cause of this difficulty is pollen limitation. When exotic species first invade a new habitat they are often quite small in population and suffer from pollen limitation, limited genetic variation (see Chapter 13) and possible competition from established species (see Chapter 8). As a result, the Allee effect may be so strong that the colonizers become extinct. Ironically, as far as invasive exotic species are concerned, this is what humans would like. When Allee effects are weak or non-existent, exotic species become pernicious invasive species (e.g. Stephens *et al.*, 1999; Taylor and Hastings, 2005). We will consider Allee effects in the context of biological invasions in Chapter 14.

4.3 The Benefits and Costs of Sex

Sex costs resources. It requires fats, proteins, carbohydrates and water to produce reproductive structures such as petals, pollen and seeds. Resources used in sexual reproduction will not be available for anything else, such as making leaves that will increase photosynthesis. Therefore, in order for sex to exist, its benefits must outweigh its costs. In general, this 'trade-off' between the benefits and costs of sex relates to the concept of fitness.

Fitness

Fitness is a measure of how well an individual succeeds at continuing its lineage. Individuals that are 'fit' are the ones that survive and reproduce successfully because they pass their genes on to the next generation. In any population, the genotype with the highest relative fitness is the one that produces the most offspring that will survive and reproduce themselves. Relative fitness is often measured by testing for significant effects of any phenomenon (e.g. low nitrogen concentrations in the soil) on specific fitness components. These components are usually tangible traits of plants that can be measured empirically. For example, a scientist might measure:

- the number and mass of seeds produced
- the amount and rate of seed germination
- the mass, height and growth rate of seedlings
- how resources are allocated to roots, shoots and flowers.

The effective measure of relative fitness is how much of the original parental genotype survives from generation to generation. This can be measured in terms of the genetic composition of direct descendants or as how much of the population eventually contains some portion of a parental genotype. If the environment generally remains constant, then individuals continue to produce offspring that are very close copies of themselves. The most fit offspring genotypes will be those that are the most similar to the parental genotype. This is what happens with agricultural weeds when farming practices do not change over time: weeds adapt to these specific practices and produce many similar offspring because these offspring encounter an equally favourable and constant environment. When environmental conditions change, however, the fitness of these genotypes may decrease.

Costs of sexual reproduction

Sex disrupts well-adapted genotypes

Plants that reproduce sexually risk breaking up well-adapted genotypes because this type of reproduction results in genetic recombination. In a relatively unchanging environment, offspring that are similar to the maternal genotype are usually better adapted than ones with recombined genotypes. This fitness disadvantage of a recombined genotype is called outbreeding depression (Waser and Price, 1989, 1993; Parker, 1992).

Cost of producing reproductive structures

To reproduce sexually, plants must allocate resources to produce sexual organs and floral structures that increase the chances of pollen dispersal. These structures can be quite resource expensive. Common milkweed (*A. syriaca*), for example, allocates 37% of its photosynthate to nectar production (Southwick, 1984). Plants that are monocarpic, reproducing only once in their lifespan, must maximize reproductive output per unit of resource expended. Even in polycarpic plants, which have repeated reproductive events during their lifespans, the costs of sexual reproduction are important because this may result in resources being directed away from growth and maintenance. Sexually reproducing plants often have to commit resources to reproduction early in the growing season. This increases the risk associated with sexual reproduction because if the weather prevents pollination, or if seeds are destroyed, the plant may not have enough resources left to survive.

Benefits of sexual reproduction

Getting away from the parents: the mobility of offspring

Sexual reproduction generally has an ancillary benefit of producing mobile offspring because seeds or fruits are dispersed away from the maternal parent plant. This benefits the offspring because it means they are less likely to have to compete with their parents, siblings or other relatives. When the environment is not favourable to the parent, and hence the offspring are also likely to suffer, dispersal away from the parent is important. The benefits of dispersal are discussed in more detail in Chapter 6, but it is useful to keep in mind that dispersal is an indirect benefit of sex.

New, better-fitted and adaptable genotypes

The main benefit of sexual reproduction is the potential for genetic combinations that may be better fitted to the current environments, while producing genetically variable offspring that can adapt to changing or new environments. Each seed is a unique genotype containing different alleles of different genes. Hence, whatever type of environment the resulting offspring encounter, there is a high probability that some of them will survive to reproduce. Formally, we sometimes refer to genetic variation as hybrid vigour or heterosis.

Intraspecific (within species) hybrids are very common in sexually reproducing organisms. In plants and some animals (e.g. fish and birds), there is another reasonably common form of hybridization: interspecific (between species) (Rieseberg *et al.*, 2003). Interspecific hybrids are formed when individuals from two different species mate. Hybrids usually form from mating between closely related species because their genomes must be similar enough to successfully produce offspring capable of reproducing themselves. Nevertheless, while hybridization is rare on a per individual basis, it is common on a per species basis (Mallet, 2005). The evolution of at least 25% of plant species involves hybridization at some time (Mallet, 2005). Therefore, hybridization is an important mechanism of evolution. Examples of some invasive species that are hybrids are listed in Table 4.3.

4.4 The Role of Hybridization with Invasive Species

Hybridization between native and introduced species is considered to be one of the biggest threats to native species (Ellstrand and Schierenbeck, 2000; Ellstrand, 2009; Schierenbeck and Ellstrand, 2009). Hybrids can influence native species in a number of ways. An introduced species can hybridize with a native species to create a new highly invasive species (Briggs and Walters, 1984; Schierenbeck and Aïnouche, 2006). Such an event may weaken the gene pool of the native species, especially if it is rare (Rieseberg *et al.*, 2003). This is occurring in Ontario, Canada, where the native red mulberry (*Morus rubra*) is threatened by the exotic white mulberry (*Morus alba*) (Burgess and Husband, 2006). Red mulberry is less fit than both white mulberry and the hybrid mulberry, but only when the white mulberry is the maternal parent of the hybrid (Fig. 4.3).

When two non-invasive species hybridize they may create a novel invasive species that displaces

Table 4.3. Invasive species created through hybridization.

Common name	Latin name	Parent species
Bitter yellow dock	*Rumex × crispo-obtusifolius*	*Rumex crispus × Rumex obtusifolius*
False leafy spurge	*Euphorbia × pseudo-esul*	*Euphorbia cyparissias × Euphorbia esula*
Goat's bladder	*Tragopogon × mirus*	*Tragopogon dubius × Tragopogon porrifolius*
Hybrid goat's beard	*Tragopogon × miscellus*	*T. dubius × Tragopogon pratensis*
Hybrid cordgrass	*Spartina alternifolia × foliosa*	*Spartina alternifolia × Spartina foliosa*
Oxford ragwort	*Senecio squalidus*	*Senecio aethnensis × Senecio chrysanthemifolius*
Tall cattail	*Typha × glauca*	*Typha angustifolia × Typha latifolia*

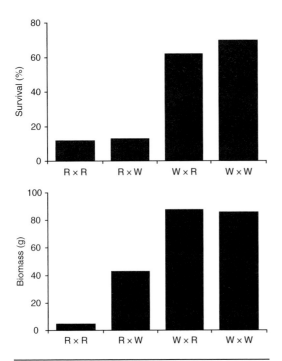

Hybrid[1]	Survival (%)	Biomass (g)	Cumulative fitness
R x R	12	5	914
R x W	13	43	571
W x R	62	88	5486
W x W	70	86	6343

[1]Female x Male parent

Fig. 4.3. Fitness differences among the native red mulberry (*Morus rubra*) (R x R), the introduced white mulberry (*Morus alba*) (W x W) and their hybrids with the red maternal parent (R x W) and white maternal parent (W x R). Fitness differences are shown in survival and biomass (redrawn from Burgess and Husband, 2006).

one or both of the parent species or invades new habitat (Briggs and Walters, 1984; Rieseberg *et al.*, 2003; Hegde *et al.*, 2006; Schierenbeck and Aïnouche, 2006; Ross and Auge, 2008). For example, the range of hybrid goat's beard (*Tragopogon* ×

miscellus) has increased substantially beyond the range of at least one of the parent species in Washington state (Novak *et al.*, 1991). Thus, this hybrid species may have a greater potential to invade new habitat than its parents.

Hybridization between a native species and a related invasive species can be more serious than hybridization between two invasive species. It can cause extinction of the native species if the hybrid species has greater fitness. For example, Freas and Murphy (1988) determined that the widespread Australian saltbush (*Atriplex serenana*) appeared to be hybridizing with the one remaining population of Bakersfield saltbush (*Atriplex tularensis*). Several native sunflowers (*Helianthus* spp.) in the southern USA are vulnerable to extirpation or extinction because of hybridization with the introduced annual sunflower (*Helianthus annuus*) (Rhymer and Simberloff, 1996). The displacement of a native species by a hybrid species can occur rapidly (Huxel, 1999).

4.5 Summary

Sexual reproduction can be energetically expensive because of the floral structures needed, but is usually necessary to produce offspring that are better fit because they are genetically capable of adapting to changing environments. To maximize the benefits of sex, plants have evolved elaborate ranges of gender expression, floral morphologies and pollination mechanisms. However, sometimes, self-mating, as an extreme form of inbreeding, can be more reliable as means of producing offspring and does allow plants to produce new offspring rapidly, even if their choice of mate is limited. The risk of inbreeding depression may be less than the risk of not producing offspring at all. None the less, inbreeding can pose such risks as reduced genetic variation and relative fitness in offspring, so individuals may have self-incompatibility mechanisms to prevent it. Sexual reproduction creates novel

Box 4.1. Invasive species case study: sexual reproduction.

- Research what is known about the sexual reproduction of the invasive species you selected.
- Describe its method(s) of pollination, whether it is self-compatible or self-incompatible, and its form(s) of gender expression.
- What is not known about the sexual reproduction of this species?

genetic combinations that may enhance invasive potential. However, as we shall see in Chapter 5, asexual reproduction has alternative mechanisms that also increase invasibility.

4.6 Questions

1. Plants are sessile (they don't move). What are the implications of this in terms of sexual reproduction?
2. If all pollinators were eliminated from an ecosystem, do you think alien or native plants would suffer more? Why?
3. Design the sexual reproductive system of an imaginary invasive species.

Further Reading

Dafni, A., Kevan, P.G. and Husband, B.C. (2005) *Practical Pollination Biology*. Enviroquest, Cambridge, Ontario, Canada.

Ellstrand, N.C. and Schierenbeck, K.A. (2000) Hybridization as a stimulus for the evolution of invasiveness in plants? *Proceedings of the National Academy of Sciences of the USA* 97, 7043–7050.

Glover, B. (2007) *Understanding Flowers and Flowering: an Integrated Approach*. Oxford University Press, Oxford, UK.

Wheeler, M.J., Franklin-Tong, V.E. and Franklin, F.C.H. (2001) The molecular and genetic basis of pollen–pistil interactions. *New Phytologist* 151, 565–584.

References

Ashman, T.-L., Knight, T.M., Steets, J.A., Amarasekare, P., Burd, M., Campbell, D.R., Dudash, M.R., Johnston, M.O., Mazer, S.J., Mitchell, R.J., Morgan, M.T. and Wilson, W.G. (2004) Pollen limitation of plant reproduction: ecological and evolutionary causes and consequences. *Ecology* 85, 2408–2421.

Baker, H.G. (1974) The evolution of weeds. *Annual Review of Ecology and Systematics* 1, 1–24.

Barrett, S.C.H. (1992) Genetics of weed invasions. In: Jain, S.K. and Botsford, L.W. (eds) *Applied Population Biology*. Kluwer, Dordrecht, The Netherlands, pp. 91–119.

Barrett, S.C.H. (1998) The evolution of mating strategies in flowering plants. *Trends in Plant Science* 3, 335–341.

Barthell, J.F., Randall, J.M., Thorp, R.W. and Wenner, A.M. (2001) Promotion of seed set in yellow star-thistle by honey bees: evidence of an invasive mutualism. *Ecological Applications* 11, 1870–1883.

Bassett, I.J., Crompton, C.W. and Woodland, D.W. (1977) The biology of Canadian weeds. 21. *Urtica dioica* L. *Canadian Journal of Plant Science* 57, 491–498.

Briggs, D. and Walters, S.M. (1984) *Plant Variation and Evolution*, 2nd edn. Cambridge University Press, New York.

Burgess, K.S. and Husband, B.C. (2006) Habitat differentiation and the ecological costs of hybridization: the effects of introduced mulberry (*Morus alba*) on a native congener (*M. rubra*). *Journal of Ecology* 94, 1061–1069.

Campbell, D.R. (2000) Experimental tests of sex-allocation theory in plants. *Trends in Ecology and Evolution* 15, 227–232.

Collevatti, R.G., Amaral, M.E.C. and Lopes, F.S. (1997) Role of pollinators in seed set and a test of pollen limitation in the tropical weed *Triumfetta semitriloba* (Tiliaceae). *Revista de Biologia Tropical* 45, 1401–1407.

Cruden, R.W. and Lloyd, R.M. (1995) Embryophytes have equivalent sexual phenotypes and breeding systems: why not a common terminology to describe them? *American Journal of Botany* 82, 816–825.

Dale, H.M. (1974) The biology of Canadian weeds. 5. *Daucus carota*. *Canadian Journal of Plant Science* 54, 673–685.

Darwin, C. (1877) *The Different Forms of Flowers in Plants of the Same Species*. John Murray, London.

de Haan, A.A., Luyten, R.M.J.M., Bakx-Schotman, T.J.M.T. and van Damme, J.M.M. (1997) The dynamics of gynodioecy in *Plantago lanceolata* L. I. Frequencies of male-steriles and their cytoplasmic male sterility types. *Heredity* 79, 453–462.

Ellstrand, N.C. (2009) Evolution of invasiveness in plants following hybridization. *Biological Invasions* 11, 1089–1091.

Ellstrand, N.C. and Schierenbeck, K.A. (2000). Hybridization as a stimulus for the evolution of invasiveness in plants? *Proceedings of the National Academy of Sciences of the USA* 97, 7043–7050.

Escarre, J. and Thompson, J.D. (1991) The effects of successional habitat variation and time of flowering on seed production in *Rumex acetosella*. *Journal of Ecology* 79, 1099–1112.

Franklin-Tong, V.E. and Franklin, F.C.H. (2003) The different mechanisms of gametophytic self-incompatibility. *Philosophical Transactions of the Royal Society of London B* 358, 1025–1032.

Freas, K.E. and Murphy, D.D. (1988) Taxonomy and the conservation of the critically endangered Bakersfield saltbush, *Atriplex tularensis*. *Biological Conservation* 46, 317–324.

Freeman, D.C. and McArthur, E.D. (1984) The relative influences of mortality, nonflowering, and sex change on the sex ratios of six *Atriplex* species. *Botanical Gazette* 145, 385–394.

Freeman, D.C., Harper, K.T. and Charnov, E.L. (1980) Sex change in plants: old and new observations and new hypotheses. *Oecologia* 47, 222–232.

Gross, K.L. and Werner, P.A. (1978) The biology of Canadian weeds. 28. *Verbascum thapsus* L. and *V. blattaria* L. *Canadian Journal of Plant Science* 58, 401–413.

Guttieri, M.J., Eberlein, C.V. and Souza, E.J. (1998) Inbreeding coefficients of field populations of *Kochia scoparia* using chlorsulfuron resistance as a phenotypic marker. *Weed Science* 46, 521–525.

Harder, L.D. and Barrett, S.C.H. (1995) Mating cost of large floral displays in hermaphrodite plants. *Nature* 373, 512–515.

Hegde, S.G., Nason, J.D., Clegg, J.M. and Ellstrand, N.C. (2006) The evolution of California's wild radish has resulted in the extinction of its progenitors. *Evolution* 60, 1187–1197.

Hiscock, S.J. and Tabah, D.A. (2003) The different mechanisms of sporophytic self-incompatibility. *Philosophical Transactions of the Royal Society of London B* 358, 1037–1045.

Howe, H.F. and Westley, L.C. (1997) Ecology of pollination and seed dispersal. In: Crawley, M. (ed.) *Plant Ecology*, 2nd edn. Blackwell Scientific, London, pp. 262–283.

Hutchinson, I., Colosi, J. and Lewin, R.A. (1984) The biology of Canadian weeds. 63. *Sonchus asper* (L.) Hill and *S. oleraceus. Canadian Journal of Plant Science* 64, 731–744.

Huxel, G.R. (1999) Rapid displacement of native species by invasive species: effects of hybridization. *Biological Conservation* 89, 143–152.

Johnson, S.D. and Steiner, K.E. (2000) Generalization versus specialization in plant pollination systems. *Trends in Ecology and Evolution* 15, 140–143.

Klinkhamer, P.G.L., de Jong, T.J. and Neil, H.W. (1994) Limiting factors for seed production and phenotypic gender in the gynodioecious species *Echium vulgare* (Boraginaceae). *Oikos* 71, 469–478.

Knight, T.M., Steets, J.A., Vamosi, J.C., Mazer, S.J., Burd, M., Campbell, D.R., Dudash, M.R., Johnston, M.O., Mitchell, R.J. and Ashman, T.-L. (2005) Pollen limitation of plant reproduction: pattern and process. *Annual Review of Ecology, Evolution, and Systematics* 36, 467–497.

Lalonde, R.G. and Roitberg, B.D. (1994) Mating system, life-history, and reproduction in Canada thistle (*Cirsium arvense*; Asteraceae). *American Journal of Botany* 81, 21–28.

Lefol, E., Danielou, V., Fleury, A. and Darmency, H. (1996) Gene flow within a population of the outbreeding *Sinapsis arvensis*: isozyme analysis of half-sib families. *Weed Research* 36, 189–195.

Les, D.H. (1988) Breeding systems, population structure and evolution in hydophilous angiosperms. *Annals of the Missouri Botanical Garden* 75, 819–835.

Madden, A.D. (1995) An assessment, using a modelling approach, of inbreeding as a possible cause of reduced competitiveness in triazine-resistant weeds. *Weed Research* 35, 289–294.

Mallet, J. (2005) Hybridization as an invasion of the genome. *Trends in Ecology and Evolution* 20, 229–237.

Motten, A.F. and Antonovics, J. (1992) Determinants of outcrossing rate in a predominantly self-fertilizing weed, *Datura stramonium* (Solanaceae). *American Journal of Botany* 79, 419–427.

Mulligan, G.A. and Findlay, J.N. (1970) Reproductive systems and colonization in Canadian weeds. *Canadian Journal of Botany* 48, 859–860.

Mulligan, G.A. and Junkins, B.E. (1977) The biology of Canadian weeds. 23. *Rhus radicans* L. *Canadian Journal of Plant Science* 57, 515–523.

Murphy, S.D. (1999) Pollen allelopathy. In: Inderjit, Dakshini, K.M.M. and Foy, C.L. (eds) *Principles and Practices in Plant Ecology: Allelochemical Interactions.* CRC Press, Boca Raton, Florida, pp 129–148.

Novak, S.J., Soltis, D.E. and Soltis, P.S. (1991) Ownbey's *Tragopogons*: 40 years later. *American Journal of Botany* 78, 1586–1600.

Nunez-Farfan, J., Cabrales-Vargas, R.A. and Dirzo, R. (1996) Mating system consequences on resistance to herbivory and life history traits in *Datura stramonium. American Journal of Botany* 83, 1041–1049.

Pannell, J. (1997) Mixed genetic and environmental sex determination in an androdioecious population of *Mercurialis annua. Heredity* 78, 50–56.

Parker, M.A. (1992) Outbreeding depression in a selfing annual. *Evolution* 46, 837–841.

Rhymer, J.M. and Simberloff, D. (1996) Extinction by hybridization and introgression. *Annual Review of Ecology and Systematics* 27, 83–109.

Rieseberg, L.H., Raymond, O., Rosenthal, D.M., Lai, Z., Livingstone, K., Nakazato, T., Durphy, J.L., Schwarzbach, A.E., Donovan, L.A. and Lexer, C. (2003) Major ecological transitions in wild sunflowers facilitated by hybridization. *Science* 301, 1211–1216.

Ross, C.A. and Auge, H. (2008) Invasive *Mahonia* plants outgrow their native relatives. *Plant Ecology* 199, 21–31.

Schierenbeck, K.A. and Aïnouche, M.L. (2006) The role of evolutionary genetics in studies of plant invasions. In: Cadotte, M.W., McMahon, S.M. and Fukami, T. (eds) *Conceptual Ecology and Invasion Biology.* Springer, Dordrecht, The Netherlands, pp. 193–221.

Schierenbeck, K.A. and Ellstrand, N.C. (2009) Hybridization and the evolution of invasiveness in plants and other organisms. *Biological Invasions* 11, 1093–1105.

Sims, T.L. (1993) Genetic regulation of self-incompatibility. *CRC Reviews in Plant Sciences* 12, 129–167.

Southwick, E.E. (1984) Photosynthate allocation to floral nectar: a neglected energy investment. *Ecology* 65, 1775–1779.

Stanton, M.L. (1994) Male–male competition during pollination in plant populations. *American Naturalist* 144, S40–S68.

Stephens, P.A., Sutherland, W.J. and Freckleton, R.P. (1999) What is the Allee effect? *Oikos* 87, 185–190.

Steven, J.C., Peroni, P.A. and Rowell, E. (1999) The effects of pollen addition on fruit set and sex expression in the andromonoecious herb horsenettle *(Solanum carolinense)*. *American Midland Naturalist* 141, 247–252.

Sun, M. and Ritland, K. (1998) Mating system of yellow starthistle *(Centaurea solstitalis)*, a successful colonizer in North America. *Heredity* 80, 225–232.

Taylor, C.M. and Hastings, A. (2005) Allee effects in biological invasions. *Ecological Letters* 8, 895–908.

Waser, N.M. and Price, M.V. (1989) Optimal outcrossing in *Ipomopsis aggregata*: seed set and offspring fitness. *Evolution* 43, 1097–1109.

Waser, N.M. and Price, M.V. (1993) Crossing distance effects on prezygotic performance in plants: an argument for female choice. *Oikos* 68, 303–308.

Whitehead, D.R. (1983) Wind pollination: some ecological and evolutionary perspectives. In: Real, L. (ed.) *Pollination Biology.* Academic Press, Toronto, Ontario, Canada, pp. 97–108.

Willson, M.F. (1994) Sexual selection in plants: perspective and overview. *American Naturalist* 144, S13–S39.

Wilson, H.D. and Payne, J.S. (1994) Crop/weed microgametophyte competition in *Cucurbita pepo* (Cucurbitaceae). *American Journal of Botany* 81, 1531–1537.

Wyatt, R. (1983) Pollinator–plant interactions and the evolution of breeding systems. In: Real, L. (ed.) *Pollination Biology.* Academic Press, Toronto, Ontario, Canada, pp. 51–95.

5 Asexual Reproduction

5.1 Introduction

Plants do not need to rely on sexual reproduction to pass on their genes to the next generation. Asexual reproduction is the creation of new individuals without involving genetic recombination. As a result, asexual offspring are genetically identical to a single parent plant. The ability to reproduce without sex can be a great advantage to plants which cannot move to find a mate or avoid an inhospitable environment. Asexual reproduction is of interest to invasion ecology because it allows a single individual to invade a new habitat and become established as a population without requiring a mate.

Like many areas of biology, there is the usual confusion of terminology. We use the term 'apomixis' to mean all types of asexual reproduction, but it is sometimes used synonymously with a specific form known as 'agamospermy' (Mogie, 1992; de Meeûs *et al.*, 2007). Agamospermy is a type of asexual reproduction in which seeds are produced from the mitosis of maternal ovules and fertilization does not occur. The most common type of asexual reproduction is vegetative growth, which is sometimes called clonal growth.

5.2 Vegetative Reproduction

Vegetative reproduction is the production of ramets (genetically identical shoots) from somatic (non-reproductive) cells. Ramets often appear to be independent but may still be physically connected, usually by roots or some other tissue, to the original 'genet' (which produced the ramets). A genet is the entire genetic individual and is composed of ramets (Fig. 5.1). Vegetative reproduction differs from the production of branches or leaves, which do not usually persist independently. Ramets may become physically separated from the originating plant, but they will still be part of one single genet – a single, large, genetically identical cluster, or clone. Common forms of vegetative reproduction are the stolons or runners produced by strawberries (*Fragaria* spp.) and invasive quack (or couch) grass (*E. repens*), and root sprouting or suckering by aspens (*Populus* spp.) and buckthorns (*Rhamnus* spp.).

This type of reproduction is a successful plant strategy and has evolved independently many times in different taxa, i.e. it is called a polyphyletic trait. About 28% of dicotyledons have some sort of vegetative reproduction (Leakey, 1981). Of these,

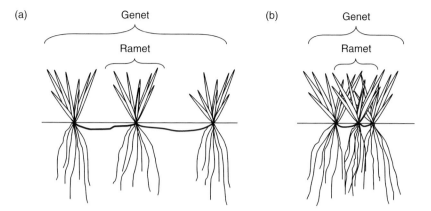

Fig. 5.1. Illustration of ramets of a genet in both (a) guerrilla (loosely packed) and (b) phalanx (densely packed) growth forms.

about 40% reproduce predominantly through vegetative reproduction. In North America, over 70% of monocotyledon families reproduce predominantly through vegetative reproduction. When considered on an area-covered basis, vegetative reproduction is important. For example, the ten most widespread species in Britain reproduce vegetatively, covering 19% of the land mass (Callaghan *et al.*, 1992). While individual ramets may have shorter lifespans, genets may survive for thousands of years and cover thousands of square metres (Cook, 1985). Many of the most successful invasive plants use vegetative reproduction. Aarssen (2008) suggested that vegetative reproduction is especially beneficial for species that are small in stature because it promotes longevity and allows offspring to be produced without the costs of fertilization but also without requiring the individual to reach full reproductive size.

Vegetative reproduction is a successful strategy in both favourable environments and in harsh, but still relatively low-disturbance, habitats such as in the Arctic (van Groenendael *et al.*, 1996; Peterson and Jones, 1997). While vegetative reproduction allows an individual to persist even in a poor habitat, an alternative strategy is to use sexual reproduction to escape. However, if the scale of disturbance is large and prolonged then once again vegetative reproduction can be useful. For example, in Central Europe, plants that could use vegetative reproduction during various historical periods of centuries of climatic stress were able to persist in colder, wetter, nutrient-poor habitats (Klimeš *et al.*, 1997), whereas those that relied on sexual reproduction experienced repeated failures

or low success and were selected against during such periods.

Mechanisms of vegetative reproduction

For herbaceous plants, the classification of structures is based on the tissue of origin (stem or root), the position of the growing tip (above or below ground), the structure of storage organs (e.g. bulbs or tubers), and the length and longevity of the connections between ramets (Klimeš *et al.*, 1997) (Fig. 5.2; Table 5.1). Flowering rush (*Butomus umbellatus*) is invasive in North America and succeeded because it relies on underground vegetative bulblets. Aerial bulblets – more properly called 'bulbils' – also exist; these are vegetative shoots produced where the flower normally exists, e.g. as in invasive wild onion (*Allium vineale*). Elephant ear (*Colocasia esculenta*) invaded Florida from the tropics because it uses corms (underground swollen stems) to reproduce. Goutweed (*Aegopodium podagraria*) invades because it produces long networks of fragile rhizomes (underground stems) that, if detached, each become new independent ramets. This is the same basic mechanism used by aquatic green cabomba (*Cabomba caroliniana*), which has invaded Australia from North America. Giant salvinia (*Salvinia molesta*) may in fact be a nearly worldwide invasive that is almost entirely one genet. It is reliant on asexual reproduction via plantlets that break off as floating leaves with rootlets.

Vegetative reproduction in woody plants has long been exploited by nursery workers. It allows growers to bypass the stages of seed production, seed germination and seedling establishment,

Root-derived organs of clonal growth

1. *Trifolium pratense* (disintegrating primary root)

2. *Alliaria petiolata* (main root with adventitious buds)

3. *Rumex acetosella* (lateral roots with adventitious buds)

4. *Ramunculus ficaria* (root tubers)

Stem-derived organs of clonal growth

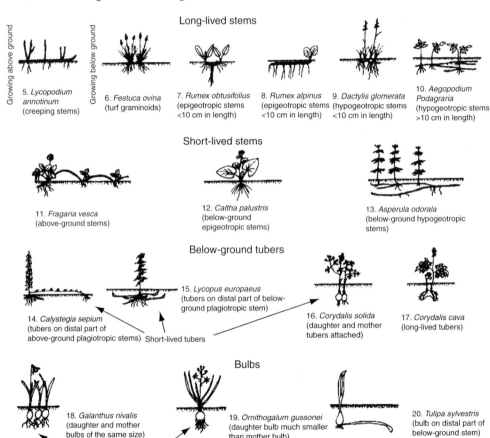

Growing above ground   Growing below ground

Long-lived stems

5. *Lycopodium annotinum* (creeping stems)

6. *Festuca ovina* (turf graminoids)

7. *Rumex obtusifolius* (epigeotropic stems <10 cm in length)

8. *Rumex alpinus* (epigeotropic stems <10 cm in length)

9. *Dactylis glomerata* (hypogeotropic stems <10 cm in length)

10. *Aegopodium Podagraria* (hypogeotropic stems >10 cm in length)

Short-lived stems

11. *Fragaria vesca* (above-ground stems)

12. *Caltha palustris* (below-ground epigeotropic stems)

13. *Asperula odorala* (below-ground hypogeotropic stems)

Below-ground tubers

14. *Calystegia sepium* (tubers on distal part of above-ground plagiotropic stems)   Short-lived tubers

15. *Lycopus europaeus* (tubers on distal part of below-ground plagiotropic stem)

16. *Corydalis solida* (daughter and mother tubers attached)

17. *Corydalis cava* (long-lived tubers)

Bulbs

18. *Galanthus nivalis* (daughter and mother bulbs of the same size)

Daughter bulb formed inside the mother bulb

19. *Ornithogalum gussonei* (daughter bulb much smaller than mother bulb)

20. *Tulipa sylvestris* (bulb on distal part of below-ground stem)

Special adaptations

21. *Dentaria bulbifera* (adventitious and axillary buds, dormant apices, turions, plant fragments, budding plants)

Fig. 5.2. Classification of clonal growth types based on Central European vegetation (Klimeš *et al.*, 1997).

Table 5.1. Definitions of asexual reproduction structures and examples of weeds using them.

Term	Definition	Examples
Creeping stems		
Rhizome	A horizontal, underground structure connecting ramets. It may bear roots and leaves and it may be cordlike or fleshy.	Bermuda grass (*Cynodon dactylon*) Quack grass (*Elymus repens*) Kentucky bluegrass (*Poa pratensis*) Field horsetail (*Equisetum arvense*)
Stolon and runner	An above-ground, horizontal branch (stolon) or stem (runner) connecting ramets or plantlets. Roots and shoots develop from nodes.	Bermuda grass (*C. dactylon*) Creeping bent grass (*Agrostis stolonifera*) Strawberries (*Fragaria* spp.) Crabgrass (*Dactylis glomerata*)
Tuber	An underground storage organ formed from the stem or root and lasting only 1 year. New tubers are formed each year from different tissue.	Yellow nutsedge (*Cyperus esculentus*) Purple nutsedge (*Cyperus rotundus*) Field horsetail (*E. arvense*)
Shoot bases		
Bulb	A fleshy underground storage organ composed of leaf bases and swollen scale leaves.	Wild onion (*Allium vineale*) Lilies (*Lilium* spp.) Wild garlic (*Allium sativum*)
Bulbil	A small bulb developing from an above-ground shoot either in place of a flower (vivipory) or on a lateral shoot.	Wild onion (*A. vineale*) Wild garlic (*A. sativum*)
Corm	A non-fleshy underground storage organ formed from the swollen base of the stem.	Buttercup (*Ranunculus bulbosus*) Oat grass (*Arrhenatherum etatius*)
Root suckers		
	Above-ground shoots that emerge from creeping roots, tap roots or root tubers.	Canada thistle (*Cirsium arvense*) Field bindweed (*Convolvulus arvensis*)

decreasing both the time and mortality rate inherent in these stages of development. In woody plants, new ramets develop either when shoots (trunk, branches and twigs) bear root primordia, or when roots bear shoot buds (Fig. 5.3). The most common types of vegetative reproduction in woody plants are through the sprouting of roots (root suckers) and from the layering of branches and stems when they come in contact with the soil. Root suckering is exhibited almost exclusively by angiosperms (e.g. buckthorn, as noted previously).

The guerrilla versus phalanx strategies

The two general types of vegetative reproduction are guerrilla and phalanx, but these are really two ends of a continuum (Lovett Doust, 1981) (Fig. 5.1). Guerrilla-type growth forms are loosely packed, often linear patches. This type of growth is a foraging strategy that maximizes movement of

a species into new habitats. The guerrilla habit results in a community of mixed species and genotypes (Honnay and Jacquemyn, 2008) because individual shoots develop in established populations. Such species are likely to invade new habitats and vacate other ones over the course of a season (Hutchings and Mogie, 1990). For example, at your local golf course or in your own lawn, red-top grass (*Agrostis stolonifera*) will spread by sending ground-level stems (stolons) into an area that is occupied by other plants. Once there, the stolons produce new shoots that, in turn, produce more stolons for further colonization. At first, only a few stems of red-top grass appear, as if by stealth, but eventually the habitat area is taken over as more stolons and shoots are produced.

Phalanx-type growth is the result of slow-growing, branched clones which form dense patches. This type of growth exploits space by maximizing the occupation of a site and deterring

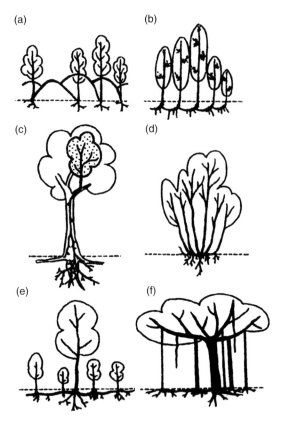

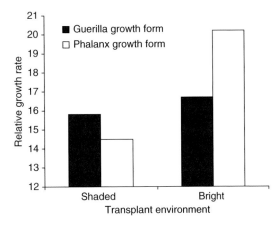

Fig. 5.3. Examples of how clonal growth occurs in woody plants showing (a) layering from drooping branches, (b) sprouting rhizomes, (c) reiteration by aerial shoots or from within roots, (d) basal rooting of coppice shoots, (e) suckering from root buds, and (f) rooting of free-hanging roots (from Jeník, 1994).

in high-quality environments, whereas the guerrilla habit is more likely in poor-quality environments (Lovett Doust, 1981). Sally-my-handsome (*Carpobrotus* aff. *actinaciformis*), an invasive species on dunes and cliffs of Mediterranean islands, is a guerrilla species that changes to a phalanx growth form in optimum environmental conditions (Traveset *et al.*, 2008). Quack grass is another example of this (Amiaud *et al.*, 2008), among others such as salt marsh cordgrass (*Spartina alterniflora*) and phragmites (*Phragmites australis*).

Within species there may be different strategies. The clonal red seaweed (*Asparagopsis armata*) has guerrilla and phalanx genotypes that respond differently to changes in light quality and quantity (Fig. 5.4) (Monro and Poore, 2009). Although both genotypes have increased growth rate under better light conditions, the phalanx genotype is able to increase its growth more than the guerrilla genotype. Johansen (2009) found both phalanx and guerrilla growth strategies within white clover (*Trifolium repens*) patches. In the centre of patches stolons were smaller, grew slowly, had more branches and grew in many directions but stayed within the patch (phalanx strategy), whereas in the peripheral areas of the patch, stolons were longer, grew faster, had fewer branches and grew out of the patch (guerrilla strategy). These two strategies allowed clones to both increase in size and persist in their existing habitat.

invasion from other species (Honnay and Jacquemyn, 2008). Such species form dense monocultures of the same genotype with individuals of approximately equal size. Patches have little movement over the course of a season (Hutchings and Mogie, 1990). Often, peripheral (younger) ramets are dependent on interior (older) ramets for resources, while interior ramets flower and set fruit (Waller, 1988). There are a few invasive plants that exhibit almost total phalanx strategies, e.g. ox-eye daisy (*L. vulgare*).

Many species may use both guerrilla and phalanx strategies. de Kroon and Schieving (1990) described a consolidating strategy in which plants combine phalanx and guerrilla strategies by altering internode length and the number of branches. Usually, phalanx growth is more typical

Fig. 5.4. Effect of the initial growth form (guerrilla or phalynx) on relative growth rate (percentage increase in size per day) (based on Monro and Poore, 2009).

Table 5.2. Costs and benefits of vegetative reproduction.

Costs	Benefits
Cost of maintaining rhizomes, stolons and other connecting tissue	Supports new ramets
	Buffering of environmental heterogeneity and stress
Higher risk of genet mortality and extinction	Resource sharing, division of labour among ramets
Resource dilution	Regulation of competition among ramets through the control of ramet production
	Recycling of resources

Costs and benefits of vegetative reproduction

As with agamospermy, there are costs and benefits associated with vegetative reproduction (Table 5.2). One of the main benefits of clonal growth is that it allows the individual to bypass the juvenile stage of growth necessary for individuals that reproduce by seeds. The seedling stage is often where the highest mortality occurs for plants. Thus, new clonal individuals have lower mortality and can take up biological space that might otherwise become occupied by competitors. A second general benefit of vegetative reproduction is that new ramets can move into other habitats, thereby allowing the genet to invade new space or to enter into a better environment while still maintaining a presence in the old habitat.

The main cost of vegetative reproduction is that there is no new genetic recombination. This reduces a species' ability to adapt to new environments. In addition, if ramets remain attached, it is more likely that an entire genet will be killed by disturbance, disease or management.

Ecological aspects of vegetative reproduction

The ecological importance of asexual reproduction varies with climate and habitat type. Harsh environmental conditions and a lack of mates (pollen donors) favours individuals that can reproduce asexually. Therefore, the distribution of species that have asexual capabilities increases towards the North and South Poles (Pyšek, 1997; Richards, 2003) (Fig. 5.5).

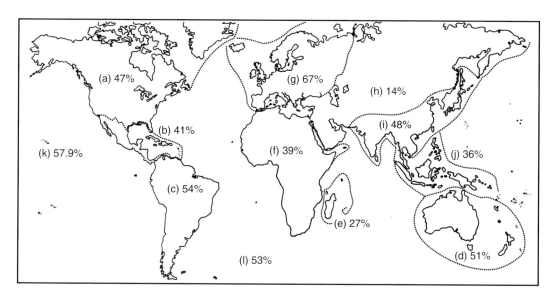

Fig. 5.5. Proportion of the most aggressive non-native species in natural habitats that are capable of vegetative reproduction. Regions are: (a) North America; (b) Central America; (c) South America; (d) Australasia; (e) Malagassia (islands of the south-west Indian Ocean); (f) Africa; (g) Europe; (h) North Asia; (i) South Asia; (j) Malesia; (k) Pacific; (l) Oceanic Islands (adapted from Pyšek, 1997).

Species persistence

Vegetatively reproducing species often have the ability to persist at the edges of their distribution because they are not dependent on sexual reproduction. This pattern is thought to occur because seed production requires higher temperatures than vegetative growth, and because fewer appropriate pollinators are found in stressful habitats (Abrahamson, 1980). Vegetatively reproducing species may form remnant populations if environmental conditions become unsuitable for seed production. Arctic dwarf birch (*Betula glandulosa*), for example, forms vegetatively reproducing stands at the northern edge of its distribution in sites where the species was once widely distributed (Hermanutz *et al.*, 1989). Populations that appear to reproduce entirely vegetatively may revert to sexual seed production when conditions improve. Isolated clonal stands can act as seed sources for recolonization when conditions improve. For example, whorled wood aster (*Oclemena acuminatus*) remains in small clonal populations under a forest canopy, but produces seed when a canopy gap opens (Hughes *et al.*, 1988). Recent studies indicate that this also occurs in the invasive periwinkle (*Vinca minor*) which usually spreads by stolons, although some cultivars spread more rapidly by setting seed in forest canopy gaps (Murphy, personal communication).

Woody plants that reproduce vegetatively benefit from better physical stability and protection from most risks (e.g. fire, wind and herbivores) (Peterson and Jones, 1997). Peripheral ramets may protect inner ones by buffering them from damage. In addition, if a disturbance removes the above-ground biomass, then sprouts from an existing rootstock will provide a 'sprout bank' (Ohkubo *et al.*, 1996). Species with root sprouts have a better chance of re-establishing than species that are reliant on seeds. Invasive buckthorns and northern populations of tree of heaven (*Ailanthus altissima*) are examples of this (Meloche and Murphy, 2006)

Physical and physiological integration of ramets

Ramets may remain physically attached through connectors, such as stolons or rhizomes, or they may fragment into independent parts. Ramets separate naturally when specialized tissues (called 'plantlets') are abscised or when connecting tissues

decay. Fragmentation is also caused by disturbance. In agricultural systems, for example, tillage will fragment rhizomes. In China, those who might try to manage invasive Canada goldenrod (*S. canadensis*) by cutting the rhizomes will simply create a few dozen smaller but now physiologically independent invasive individuals. There are, of course, both costs and benefits to remaining integrated (Table 5.3).

Integration benefits the entire genet because it increases the longevity of the clone and prolongs the occupation of the site. Ramets of woody species may remain integrated for decades. This results in long-lived woody clones that spread extensively. In fact, the largest plant is argued to be a clonal patch of trembling aspen (*Populus tremuloides*) found in Utah, USA. A single male clone contains approximately 47,000 trees (ramets) and covers 43 ha (Grant, 1993).

When ramets remain integrated, they continue to share resources. As a result, older ramets may send resources to younger ones, thus helping them to get established. Reallocation of resources allows genets to exploit nutrient-rich sites by rapidly increasing ramet density in a favourable microhabitat (Hutchings and Mogie, 1990) or by modifying root and shoot structure to optimize resource use within their environment (de Kroon and Hutchings, 1995; van Groenendael *et al.*, 1996). This creates a 'division of labour' (Alpert and Stuefer, 1997).

The benefits of remaining integrated increase in heterogeneous environments (Wijesinghe and Handel, 1994). Genets with integrated ramets effectively live in two or more places at once, because every shoot is anchored in a different microhabitat (Alpert and Stuefer, 1997). Having multiple rooting sites reduces the risk to the intact genet, because resources can be shared between ramets and, therefore, ramets in poorer sites are supported by ones in better sites. Sharing resources may result in less biomass accumulation of individual ramets, but the biomass of the genet will increase. The integration of smooth brome (*Bromus inermis*) clones helped it to proliferate into resource-poor native prairies (Otfinowski and Kenkel, 2008).

In heterogeneous environments, genets may fragment but the fragments are much smaller because the microhabitat is smaller than in consistently nutrient-rich environments. Tall goldenrod (*Solidago altissima*), Canada goldenrod, eastern lined aster (*Symphyotrichum lanceolatum*), and New York aster (*Symphyotrichum novi-belgii*) are examples of

Table 5.3. Benefits and costs associated with maintaining physiological integration among ramets (based on text in Jónsdóttir and Watson, 1997).

Benefits and costs	Description
Benefits	
Movement to better environment	Creation of new ramets allows the genet to move spatially and thus invade new, possibly better environments. This buffers the genet from spatial variability.
Sequestering of biological space	The occupation of space and increased potential to capture resources will decrease the chance of invasion by other species.
Lower mortality	New ramets have lower mortality than seedlings.
Invasion potential	Movement of large genets increases potential to invade and displace competitors. Ramets that remain attached can draw resources from a wide patch thus supporting the invasion front.
Increased resource acquisition	Spreading plants have a high potential to invade nutrient-rich environments. This may be of benefit in spatially or temporally heterogeneous habitats.
Buffering of temporal variability	Storage organs increase survival during stressful periods and changing environments.
Risk aversion to the genet	The risk to a genet is spread among the ramets.
No 'cost of sex'	Creation of ramets does not incur the costs associated with sexual reproduction.
Persistence	Some clones are extremely long lived.
Costs	
Loss of genetic recombination	Lack of genetic recombination through sexual reproduction means the benefits from novel genotypes are lost.
Vulnerable to disturbance	Spatial integrity of clones can make them more vulnerable to large-scale disturbances such as floods, fire and frost heave.
Mortality of individual ramets	Nutrients that are shared among ramets, and therefore survivorship of an individual ramet, may be decreased in favourable habitat.
Transmission of disease	A disease may be able to spread throughout the portions of the genet that remain connected.
Decreased sexual reproduction	The creation of new clones decreases the allocation of resources to sexual reproduction.

species that form small patches in moderately disturbed, shaded environments. In nutrient-rich environments, genets are more likely to fragment, allowing them to colonize and monopolize large tracts of land. Examples of fragmenting species include bracken fern (*Pteridium aquilinum*), Kentucky bluegrass (*Poa pratensis*), quack grass and white clover; these all form larger patches in open habitats with adequate moisture (Jónsdóttir and Watson, 1997). While fragmented genets allow for greater colonization, the trade-off is that such species are less likely to spread to new, favourable habitats because they concentrate their resources in one place (Hutchings and Mogie, 1990).

5.3 Sexual Reproduction in Vegetatively Reproducing Species

Vegetative reproduction rarely occurs to the total exclusion of sexual reproduction. A trade-off occurs between sexual and vegetative reproduction because only a finite amount of resources is available to allocate to reproduction (Abrahamson, 1980). The allocation of resources to sexual versus vegetative reproduction will change over the life of the genet.

Sexually and vegetatively produced offspring may have different genetic and ecological characteristics (Table 5.4). For example, offspring produced through sexual reproduction will differ genetically from their parents and have the ability to disperse, but suffer a high mortality rate in the seedling stage. Offspring produced vegetatively will develop immediately and have a low mortality rate, but less dispersal potential.

As usual, plants express a range of reproductive options even within one species or a population of a species. Those expecting consistency in reproductive strategies will be surprised. For example, wild garlic (*Allium sativum*) was referred to in Table 5.1

Table 5.4. Expected differences between asexually and sexually produced offspring (adapted from Williams, 1975; and Abrahamson, 1980).

Asexual offspring	Sexual offspring
Mitotically standardized	Meiotically diversified
Produced continuously	Produced seasonally
Develop close to parent	Can be widely dispersed
Develop immediately	Can be dormant
Develop more directly to reproductive stage	Develop more slowly from seedling stage to reproductive stage
Environment and optimum genotype predictable from those of parent as they are genetically the same as the parent	Environment and optimum genotype unpredictable because genetic recombination has occurred
Low mortality rate	High mortality rate – especially during seedling stage

as an example of an asexually reproducing plant; it actually can produce various combinations of two types of bulbs (larger bulbs and bulbils) and three types of reproductive shoots (sexually reproducing flowers versus asexual bulbils versus both flowers and bulbils). The relative allocation of resources to bulbs, bulbils and flowers is under strong genetic control (Ronsheim and Bever, 2000).

Initial and repeated seedling recruitment

Seedling recruitment may occur only at some times during the life of an otherwise vegetatively reproducing species. For example, seedling recruitment of Canada goldenrod usually occurs within the first year after colonizing a site, although it occurs for up to 6 years (Hartnett and Bazzaz, 1983). Following the initial establishment, the population size increases through vegetative reproduction. This pattern of recruitment is called 'initial seedling recruitment' (ISR) (Eriksson, 1993). ISR results in a population with an even-aged structure because new individuals are recruited at approximately the same time, and ISR genets may be long-lived because once established they can be virtually immortal unless a disturbance kills the entire genet. In Canada goldenrod, the combined strategies of ISR followed by clonal growth have different functions in the invasion process. Seed production, along with self-incompatible obligate outcrossing, means that the genetic diversity of this species is maintained (Dong *et al.*, 2006). Long-distance dispersal of seeds (Chapter 6) is important for the spread of the species to new habitats, whereas vegetative growth allows the species to maintain and expand the occupied space (Dong *et al.*, 2006).

White clover is an example of the opposite type of recruitment pattern, where there is continual recruitment of new genets into the population via seed production (Barrett and Silander, 1992). This type of recruitment is called 'repeated seedling recruitment' (RSR). Such populations have an uneven-aged structure. Following a disturbance, some genets die, thus making room for new genet recruitment. RSR genets have shorter lifespans because they are continually being replaced (Eriksson, 1993). Of course, many species are located along a continuum between ISR and RSR and some species may undergo both strategies, depending on their environment. For example, mouse-ear hawkweed (*Hieracium pilosella*) follows an IRS strategy in a mesic environment, but it follows a RSR strategy in a drought-stressed environment (Bruun *et al.*, 2007).

5.4 Agamospermy

Agamospermy is the production of seeds without fertilization – superficially the seeds look like almost any other seed produced through sexual reproduction. Facultative agamospermy is the production of asexual seeds if pollination fails. It is present in some cinquefoils (*Potentilla* spp.). This trait is particularly useful to invasive species because they can produce seeds both with and without pollen, and this can aid the spread of a species when pollinators are absent in the new habitat. For example, the dioecious species screw pine (*Pandanus tectorius*) was able to invade islands because it could produce agamospermic seeds, and therefore male plants were not necessary for it to colonize (Cox, 1985).

Obligate agamosperms are only able to produce seeds asexually. Species with obligate agamospermy are often triploids or pentaploids (contain three or five sets of chromosomes) and therefore cannot reproduce via pollen. Obligate agamospermy rarely occurs to the total exclusion of sexual reproduction, although many raspberries (*Rubus* spp.) and most dandelions (*Taraxacum* spp.) are obligate agamosperms.

Costs and benefits of agamospermy

Agamospermic reproduction has many of the benefits of sexual reproduction (Table 5.5). For example, agamospermic seeds disperse (in time and space) like sexually produced seed. However, while agamospermy may avoid the cost of sex, the lack of recombination means that deleterious mutations may accumulate and novel genotypes are not formed. In jubata grass (*Cortaderia jubata*), an invasive agamospermic species from the Andes Mountains, there was a great loss of genetic diversity following its colonization (the founder effect) (Okada *et al.*, 2009). Molecular studies (Chapter 13) revealed that this species consists of one genetic individual, probably originally from Ecuador, that was introduced for horticultural use in California, Maui (Hawaii) and New Zealand. A number of invasive species have undergone similar genetic bottlenecks following introduction (Poulin *et al.*, 2005).

Not every cost and benefit will apply to all agamospermic species. For example, some agamosperms require pollen chemicals to help form the endosperm (the nutritive tissue around the embryo), although the gametes of the pollen are not used in the creation of the new individual (e.g. blackberry, *Rubus fruticosus*). Therefore, pollen may still be needed in the reproduction process even though it does not contribute genetically.

Ecological aspects of agamospermy

Agamospermy is very common among ferns. It is not present in gymnosperms and it is present in only about 10–15% of angiosperm families (Richards, 1997). In angiosperms, approximately 75% of agamospermic taxa are in the daisy (Asteraceae), grass (Poaceae) and rose (Rosaceae) families. A high proportion of the species in dandelions, hawkweeds (*Hieracium* spp.) and raspberries are also agamospermic.

The occurrence of agamospermy is often associated with the following traits or conditions: polyploidy, phenotypic plasticity (Chapter 7),

Table 5.5. Costs and benefits associated with agamospermy (based on text in Richards, 1997).

Costs and benefits	Description
Costs	
Accumulate detrimental mutations	Non-lethal detrimental mutations will remain in the population because there is no recombination and selection to remove them.
No recombination	Agamospecies lack genetic recombination, which can create novel advantageous genotypes that may be more fit, especially in cases of habitat or climatic changes.
Narrow niche	Outcrossing creates genetic variation among individuals of a population, which will lead to an increased likelihood of inhabiting more niches. This is lacking in agamospermic populations, although there is some evidence of high levels of somatic mutations in asexual lineages.
Lack 'fine tuning'	Recombination can create genotypes that are better adapted to local environments. Agamospermic populations are more likely to be generalists (weedy).
Benefits	
Assured reproduction	In the absence of pollination, seed production is assured (although some agamosperms still require the 'cue' from pollination to create asexual seeds).
Advantages of seeds	Obtain dispersal and dormancy but maintain advantages of vegetative reproduction.
Avoids 'cost of meiosis'	No 'unfit' zygotes created through recombination that may disrupt co-adapted genotypes. Offspring have the same fitness as the maternal parent.
Avoids 'cost of males'	Energy does not go towards the creation of pollen (although many agamosperms do produce pollen).
Benefits from 'extremely fit genotype'	Many agamosperms are highly heterozygous and thus have high fitness. Less fit genotypes will decrease through natural selection.

Table 5.6. Traits and conditions associated with agamospermy (based on information from Asker and Jerling, 1992; Richards, 1997).

Trait or character	Description
Hybridization	Hybridization is thought to bring about the conditions necessary for agamospermy. Hybrids may be more vigorous, long-lived and partially sterile.
Polyploidy (multiple sets of base chromosomes)	Polyploidy 'may buffer against the effects of deleterious mutations'. Polyploidy is also associated with other changes, such as altered secondary metabolism, increased seed size and seedling vigour, and a switch from annual to perennial habit.
Phenotypic plasticity	As in inbreeding populations, agamospermic species tend to have higher phenotypic plasticity. Selection is more likely to encourage phenotypic plasticity in populations with less genetic variability.
Polycarpic perennials, often rosette forming	Very few annuals, biennials and monocarpic species are agamospermic.
Pollen limitation	When seed production is limited by the lack of pollen, seeds produced by individuals carrying an agamospermic mutation are more likely to persist in higher numbers.

perennial habit, hybridization (Chapter 4), and pollen limitation (Chapter 4) (Table 5.6). These traits do not necessarily cause agamospermy to develop; they may either be conducive to its development or occur as a result of it. In some cases, the association of agamospermy with these traits is not fully understood. For example, some perennial invasive species are agamospermous whereas others are not, and it is not always clear why a particular species has developed this trait. It need not be adaptive; it may reflect 'phylogenetic constraint' because a descendant species cannot always re-develop what was lost in ancestors, or must function with structures or processes imposed by its ancestral lineage. It appears that this may apply to dandelions, hawkweeds and some cinquefoils.

We have said that the ability to reproduce via agamospermy can improve the chances of colonization success, and gave screw pine as an example earlier in this chapter. Not all agamospermic species are equally good colonizers. While agamospermy increases the chance of colonization, other traits are also required. For example, two closely related species of agamospermous dandelion, which co-occur in sand dunes of Northumberland, UK, have different life history strategies in spite of their similar morphologies. Rock dandelion (*Taraxacum lacistophyllum*) is a more successful colonizer than smooth-bracted dandelion (*Taraxacum brachyglossum*) because it has a faster growth rate, shorter lifespan, and lighter and more dispersible seeds, and also reproduces earlier (Ford, 1985).

Populations (or clones) of agamospermic species with one or a few genotypes may cover large areas, up to several hundred square kilometres (van Dijk, 2003). Single genotype clones may be more susceptible to disease or to pests. This is discussed more in Chapter 13 on molecular aspects of invasion. When several genotypes co-occur they may be ecologically differentiated (van Dijk, 2003). For example, dandelion clones may differ in competitive ability (Chapter 8), reproductive capacity or response to disturbance (van Dijk, 2003).

5.5 Summary

Asexual reproduction increases a species' invasiveness because it means that only one individual is needed to establish a population and no pollinators are necessary. The lack of genetic diversity of asexually produced individuals has costs and benefits depending on the environment. Generally, asexual reproduction can be a highly successful strategy for invasive species. The trade-off between agamospermy and vegetative reproduction is that agamosperms produce seeds that allow long-dispersal colonization, while cloning increases colonizing ability and persistence because it eliminates seedlings – the stage of growth when the risk of mortality is highest. These early stages of growth are considered in the next chapter.

5.6 Questions

1. Why does vegetative reproduction become important in northern environments or in other harsh environments?

2. Explore the available literature and list some examples of invasive species that have a phalanx growth form or a guerrilla growth form.

3. How would asexual reproduction in invasive species influence their management?

4. Considering the management regime for a lawn, design a perfect weed for colonization and persistence.

Further Reading

de Kroon, H. and van Groenendael, J. (1997) *The Ecology and Evolution of Clonal Plants*. Backhuys Publishers, Leiden, The Netherlands.

de Meeûs, T., Prugnolle, F. and Agnew, P. (2007) Asexual reproduction: genetics and evolutionary aspects. *Cellular and Molecular Life Sciences* 64, 1355–1372.

Oborny, B. and Bartha, S. (1995) Clonality in plant communities – an overview. *Abstracta Botanica* 19, 115–127.

Richards, A.J. (2003) Apomixis in flowering plants: an overview. *Philosophical Transactions of the Royal Society of London B* 358, 1085–1093.

References

Aarssen, L.W. (2008) Death without sex–the 'problem of the small' and selection for reproductive economy in flowering plants. *Evolutionary Ecology* 22, 279–298.

Abrahamson, W.G. (1980) Demography and vegetative reproduction. In: Solbrig, O.T. (ed.) *Demography and Evolution in Plant Populations*. Blackwell Scientific, Oxford, UK, pp. 89–106.

Alpert, P. and Stuefer, J.E. (1997) Division of labour in clonal plants. In: de Kroon, H. and van Groenendael, J. (eds) *The Ecology and Evolution of Clonal Plants*. Backhuys Publishers, Leiden, The Netherlands, pp. 137–154.

Amiaud, B., Touzard, B., Bonis, A. and Bouzillé, J.-B. (2008) After grazing exclusion, is there any modification of strategy for two guerrilla species: *Elymus repens* (L.) Gould and *Agrostis stolonifera* (L.)? *Plant Ecology* 197, 107–117.

Asker, S.E. and Jerling, L. (1992) *Apomixis in Plants*. CRC Press, Boca Raton, Florida.

Barrett, J.P. and Silander, J.A. Jr (1992) Seedling recruitment limitation in white clover (*Trifolium repens*; Leguminosae). *American Journal of Botany* 79, 643–649.

Bruun, H.H., Scheepens, J.F. and Tyler, T. (2007) An allozyme study of sexual and vegetative regeneration in *Hieracium pilosella*. *Canadian Journal of Botany* 85, 10–15.

Callaghan, T.V., Carlsson, B.A., Jónsdóttir, I.S., Svensson, B.M. and Jonasson, S. (1992) Clonal plants and environmental change: introduction to the proceedings and summary. *Oikos* 63, 341–347.

Cook, R.E. (1985) Growth and development in clonal plant populations. In: Jackson, J.B.C., Buss, L.W. and Cook, R.E. (eds) *Population Biology and Evolution of Clonal Organisms*. Yale University Press, New Haven, Connecticut, pp. 259–296.

Cox, P.A. (1985) Islands and dioecism: insights from the reproductive ecology of *Pandanus tectorius* in Polynesia. In: White, J. (ed.) *Studies on Plant Demography*. Academic Press, London, pp. 359–372.

de Kroon, H. and Hutchings, M.J. (1995) Morphological plasticity in clonal plants: the foraging concept reconsidered. *Journal of Ecology* 83, 143–152.

de Kroon, H. and Schieving, F. (1990) Resource partitioning in relation to clonal strategy. In: van Groenendael, J. and de Kroon, H. (eds) *Clonal Growth in Plants: Regulation and Function*. SPB Academic Publishing, The Hague, The Netherlands, pp. 113–130.

de Meeûs, T., Prugnolle, F. and Agnew, P. (2007) Asexual reproduction: genetics and evolutionary aspects. *Cellular and Molecular Life Sciences* 64, 1355–1372.

Dong, M., Lu, B.-R., Zhang, H.-B., Chen, J.-K. and Li, B. (2006) Role of sexual reproduction in the spread of an invasive clonal plant *Solidago canadensis* revealed using intersimple sequence repeat markers. *Plant Species Biology* 21, 13–18.

Eriksson, O. (1993) Dynamics of genets in clonal plants. *Trends in Ecology and Evolution* 8, 313–316.

Ford, H. (1985) Life history strategies in two coexisting agamospecies of dandelion. *Biological Journal of the Linnean Society* 25, 169–186.

Grant, M.C. (1993) The trembling aspen. *Discover* 14, 82–89.

Hartnett, D.C. and Bazzaz, F.A. (1983) Physiological integration among intraclonal ramets of *Solidago canadensis*. *Ecology* 64, 779–788.

Hermanutz, L.A., Innes, D.J. and Weis, I.M. (1989) Clonal structure of arctic dwarf birch (*Betula glandulosa*) at its northern limit. *American Journal of Botany* 76, 755–761.

Honnay, O. and Jacquemyn, H. (2008) A meta-analysis of the relationship between mating system, growth form and genotypic diversity in clonal plant species. *Evolutionary Ecology* 22, 299–312.

Hughes, J.W., Fahey, T.J. and Bormann, F.H. (1988) Population persistence and reproductive ecology of a forest herb: *Aster acuminatus*. *American Journal of Botany* 75, 1057–1064.

Hutchings, M.J. and Mogie, M. (1990) The spatial structure of clonal plants: control and consequences. In: van Groenendael, J. and de Kroon, H. (eds) *Clonal Growth in Plants: Regulation and Function*. SPB Academic Publishing, The Hague, The Netherlands, pp. 57–76.

Jeník, J. (1994) Clonal growth in woody plants: a review. *Folia Geobotanica* 29, 291–306.

Johansen, L. (2009) Clonal growth strategies in simultaneously persistent and expanding *Trifolium repens* patches. *Plant Ecology* 201, 435–444.

Jónsdóttir, I.S. and Watson, M.A. (1997) Extensive physiological integration: an adaptive trait in resource-poor environments. In: de Kroon, H. and van Groenendael, J. (eds) *The Ecology and Evolution of Clonal Plants*. Backhuys Publishers, Leiden, The Netherlands, pp. 109–136.

Klimeš, L., Klimešová, J., Hendrik, R. and van Groenendael, J. (1997) Clonal plant architecture: a comparative analysis of form and function. In: de Kroon, H. and van Groenendael, J. (eds) *The Ecology and Evolution of Clonal Plants*. Backhuys Publishers, Leiden, The Netherlands, pp. 1–29.

Leakey, R.R. (1981) Adaptive biology of vegetatively regenerating weeds. *Advances in Applied Biology* 6, 57–90.

Lovett Doust, L. (1981) Population dynamics and local specialization in a clonal perennial (*Ranunculus repens*). I. The dynamics of ramets in contrasting habitats. *Journal of Ecology* 69, 743–755.

Meloche, C. and Murphy, S.D. (2006) Management of invasive tree-of-heaven (*Ailanthus altissima*) in parks and protected areas: a case study of Rondeau Provincial Park (Ontario, Canada). *Environmental Management* 37, 764–772.

Mogie, M. (1992) *The Evolution of Asexual Reproduction in Plants*. Chapman and Hall, London.

Monro, K. and Poore, A.G.B. (2009) Performance benefits of growth-form plasticity in a clonal red seaweed. *Biological Journal of the Linnean Society* 97, 80–89.

Ohkubo, T., Tanimoto, T. and Peters, R. (1996) Response of Japanese beech (*Fagus japonica* Maxim.) sprouts to canopy gaps. *Vegetatio* 124, 1–8.

Okada, M., Lyle, M. and Jasieniuk, M. (2009) Inferring the introduction history of the invasive apomictic grass *Cortaderia jubata* using microsatellite markers. *Diversity and Distribution* 15, 148–157.

Otfinowski, R. and Kenkel, N.C. (2008) Clonal integration facilitates the proliferation of smooth brome clones invading northern fescue prairies. *Plant Ecology* 199, 235–242.

Peterson, C.J. and Jones, R.H. (1997) Clonality in woody plants: a review and comparison with clonal herbs. In: de Kroon, H. and van Groenendael, J. (eds) *The Ecology and Evolution of Clonal Plants*. Backhuys Publishers, Leiden, The Netherlands, pp. 263–289.

Poulin, J., Weller, S.G. and Sakai, A.K. (2005) Genetic diversity does not affect the invasiveness of fountain grass (*Pennisetum setaceum*) in Arizona, California and Hawaii. *Diversity and Distribution* 11, 241–247.

Pyšek, P. (1997) Clonality and plant invasions: Can a trait make a difference? In: de Kroon, H. and van Groenendael, J. (eds) *The Ecology and Evolution of Clonal Plants*. Backhuys Publishers, Leiden, The Netherlands, pp. 405–427.

Richards, A.J. (1997) *Plant Breeding Systems*, 2nd edn. Chapman and Hall, London.

Richards, A.J. (2003) Apomixis in flowering plants: an overview. *Philosophical Transactions of the Royal Society of London B* 358, 1085–1093.

Ronsheim, M.L. and Bever, J.D. (2000) Genetic variation and evolutionary trade-offs for sexual and asexual reproductive modes in *Allium vineale* (Liliaceae). *American Journal of Botany* 87, 1769–1777.

Traveset, A., Moragues, E. and Valladares, F. (2008) Spreading of the invasive *Carpobrotus* aff. *acinaciformis* in Mediterranean ecosystems: the advantage of performing in different light environments. *Applied Vegetation Science* 11, 45–54.

van Dijk, P.J. (2003) Ecological and evolutionary opportunities of apomixis: insights from *Taraxacum* and *Chondrilla*. *Philosophical Transactions of the Royal Society of London B* 358, 1113–1121.

van Groenendael, J.M., Klimeš, L., Klimešová, J. and Hendriks, R.J.J. (1996) Comparative ecology of clonal plants. *Philosophical Transactions of the Royal Society of London, Series B* 351, 1331–1339.

Waller, D.W. (1988) Plant morphology and reproduction. In: Lovett Doust, J. and Lovett Doust, L. (eds) *Plant Reproductive Ecology: Patterns and Strategies*. Oxford University Press, New York, pp. 203–227.

Williams, G.C. (1975) *Sex and Evolution*. Princeton University Press, Princeton, New Jersey.

Wijesinghe, D.K. and Handel, S.N. (1994) Advantages of clonal growth in heterogeneous habitats: an experiment with *Potentilla simplex*. *Journal of Ecology* 82, 495–502.

6 From Seed to Seedling

Concepts

- The number of seeds that a plant produces will depend genetic constraints, germination date, plant size and environmental conditions. There is a trade-off between seed size and seed number.
- Seed dispersal moves seeds away from the parent plant, which may lead to higher seedling survivorship. Seeds are dispersed by animals, water, wind and human activities.
- Humans are highly effective long-distance seed dispersers.
- Seeds are dispersed over time by remaining dormant in the seed bank. Dormancy is controlled by many interacting factors.
- To germinate, seeds require specific environmental conditions such as light, temperature and moisture.
- The seedling emergence and establishment stages have the highest mortality rate in a plant's life cycle.

6.1 Introduction

Reproduction is but one part of the larger life cycle of plants and there is no start or finish (Fig. 6.1). We started our discussion of plant life cycles with the process of seed production, although we could very well have started with seed dispersal or with germination. In this chapter, we will focus on seed production, seed dispersal, germination and seedling emergence. We discuss how biotic and abiotic processes affect each of these stages, and how events during one stage will influence the progress of subsequent stages.

Seeds are the primary mobile stage of the life cycle. Like pollen dispersal, the fate of seeds is dependent on the wind, water or animals that disperse them. Furthermore, because they often contain high levels of nutrients, seeds are a good food source for many animals. In some cases, consumption results in their dispersal to favourable habitats, while in other cases seeds are destroyed or end up in hostile environments. Essentially, the seed must find its 'safe site', a set of favourable environmental conditions in which to survive, to germinate and produce a seedling. Even if seeds survive, the resultant seedlings have high rates of mortality because

they are small and lack long-term nutritional reserves. The odds are against the survival of seeds and seedlings, and many species have strategies that increase their chances of survival.

6.2 Fruit and Seed Production

A seed develops from the fertilized ovule (with the exception of agamospermic seed, see Chapter 5) and contains an embryonic plant surrounded by a protective seedcoat. It also contains nutritional reserves in the form of either endosperm or cotyledons. In angiosperms, seeds may be dispersed within a fruit formed from the flower ovary or receptacle. A fruit may contain one or many seeds. Fruits are divided into two basic types: fleshy and dry. Fleshy fruits include peaches (*Prunus persica*), tomatoes (*Solanum lycopersicum*) and buckthorn (*Rhamnus* spp.), while acorns (oak, *Quercus* spp.), rice (*O. sativa*), garlic mustard (*A. petiolata*) and proso millet (*Panicum miliaceum*) are dry fruits.

Seed set

As a plant has a limited amount of resources for reproduction, there is a trade-off between the

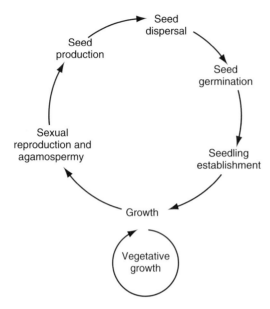

Fig. 6.1. The life cycle of plants has no real beginning or end. One chooses an arbitrary 'start'.

number of seeds and the size of seeds that an individual plant can produce (Coomes and Grubb, 2003). The number of successful seedlings produced by a plant depends on the total number of seeds it produces and on seed size.

Seed number

The number of seeds produced by an individual plant will depend on the number of ovules produced, their rate of fertilization, and on how many fertilized ovules survive to become mature seeds. What determines the actual number of seeds produced? First, there are the genetic constraints over the number (and size) of seeds that a species can produce. Orchids produce thousands of dust-sized seeds but cannot produce coconut-sized seeds; coconut trees cannot produce as many seeds as orchids do.

Within these constraints, seed number is influenced by the availability of resources and by the environmental conditions during pollination and seed development. For example, the number of seeds produced by red-root pigweed (*Amaranthus retroflexus*) decreases as light level decreases (McLachlan *et al.*, 1995). The benefit of producing many seeds is that there are more opportunities for colonization because of the sheer numbers of seeds produced. In

addition, seed-eating organisms will not be able to find and consume all the seeds (Chapter 9).

Seed size

Seed size has many repercussions for dispersal and seedling establishment. The main benefit of having large seeds is that the resultant seedlings are more likely to survive because they have more nutrient reserves and can endure harsher conditions for longer periods of time. Seedlings from large seeds are better able to withstand drought, defoliation, shade, (plant) litter and competition from established vegetation or seedlings emerging concurrently from relatively small seeds (Westoby *et al.*, 1996; Leishman, 2001). Larger seeds, however, require more energy to produce and are more likely to be consumed by herbivores in search of an easy-to-find and nutritious meal (Reader, 1993; Thompson *et al.*, 1993; Rees, 1996).

There are advantages to having small seeds. Individuals with small seeds produce many more of them than individuals with large seeds, and can do so because small seeds require less energy per seed to produce (Leishman, 2001). Because so many seeds are produced, it is less likely that all of them will be destroyed before they have a chance to germinate.

While seed size was once thought to be a genetically stable trait within a species (Harper, 1977), we now know that environmental variation often causes seed size to vary greatly within species, populations and individuals (Michaels *et al.*, 1988; Westoby *et al.*, 1992). For example, the seed size of the annual weedy cucumber (*Sicyos deppei*) is dependent on the environment in which the fruit develops (Orozco-Segovia *et al.*, 2000). This weed is a vine that climbs up the stems and trunks of the other vegetation in the fields and disturbed forests of Mexico. As a result, some fruits develop in full sunlight while others develop in the shade. Seeds that develop in full sunlight are larger and heavier than shaded seed, but seed viability is the same.

Because seed size is genetically controlled and environmentally influenced, selection pressures could lead to a change in seed size. For example, seed size of the weedy gold of pleasure (*Camelina sativa*) has diverged over time depending on the type of flax crop (*Linum* spp.) the weed grows in. Weed seeds became similar to crop seeds because both pass through the winnowing machine at the same time, and so weed seeds similar in size and weight to flax are selected for. In fibre flax, gold of

pleasure seeds are flat; in flax grown for oil production, the weed seeds are smaller and plumper.

Seed size can also vary within a species. Two
populations of lantana (*Lantana camara*) differed
in mean seed mass; seed mass also varied within
populations (Vivian-Smith and Panetta, 2009).
Seed dormancy and survival and seedling emergence of the two populations differed in response
to irrigation and natural rainfall.

Trade-off between seed number and seed size

Producing a very few, nutrient-rich seeds can be disastrous if all of the seeds die because they end up in
unsuitable habitats or are destroyed (e.g. eaten).
Likewise, producing many nutrient-poor seeds can
be equally disastrous if the seedlings are not able to
survive the biotic and abiotic stresses of their environment. So how will a plant resolve this trade-off?
Eriksson (2000) proposed a model based on seed size
to explain the dispersal and colonizing ability of a
species (Fig. 6.2). In this, seed size decreases with
increased seed production. The number of seedlings
that survive to become adults (i.e. are 'recruited')
increases as seed size increases, but this reaches a
limit at the maximum recruitment threshold. As a
result, the maximum combined dispersal and colonizing ability is at an intermediate level of seed size.
Eriksson (2000) recognized that this relationship is
dependent on habitat and community type. The peak
of the dispersal and colonizing ability curve moves
leftward in disturbed sites because the benefits of
increasing seed number outweigh those of increasing

seed size. Additionally, these maxima will vary with
short- and long-term environmental variation.

In a global survey of seed mass and seed production per individual, Mason *et al.* (2008) concluded
that invasive species tend to produce a greater
number of seeds and smaller seeds than native species, resulting in a greater reproductive output for
invasive species. This increase contributes to a species' invasiveness because a greater reproductive
output increases the chance that some seeds will be
dispersed longer distances to a new habitat. Of
course, seed size and number are just two traits of
many that influence a species' invasiveness.

Fruit and seed polymorphisms

Sometimes, some populations of some species may
produce two or more types of seeds that differ in
morphology. Species with two types of seeds are
dimorphic, and those with more are polymorphic.
Seed morphs may have different sets of germination
requirements or different dispersal mechanisms
associated with them. Having two or more morphs
is a form of bet-hedging. By producing seeds with
different germination and requirements, the plant is
likely to have at least *some* seeds germinate. This
strategy is advantageous in environmentally variable habitats. The trade-off with bet-hedging is that
while it increases the chance of at least some seeds
germinating in most conditions, fewer seeds will
germinate in optimal conditions.

Seeds of common lambsquarters (*Chenopodium
album*) are dimorphic for two characters: seed wall
and seed texture. There are thin-walled brown seeds
that germinate immediately, and thick-walled black
seeds that are dormant (in a resting state, unable to
germinate; see Section 6.4). In addition, both brown
and black seeds have smooth coat and textured coat
morphs. The proportion of each of the four seed
types varies among populations (Harper *et al.*,
1970). Seed polymorphisms are common in the
daisy (Asteraceae), goosefoot (Chenopodiaceae),
grass (Poaceae), and mustard (Brassicaceae) families. These families, not coincidentally, also contain
many invasive species (Harper, 1977).

Wingpetal (*Heterosperma pinnatum*) is a summer annual that has different types of achenes (a
dry, single-seeded fruit) within each flowering
head. The polymorphisms ensure that some of
these seeds will germinate each year. Venable *et al.*
(1995) divided these into three morphological
types (central, intermediate and peripheral) based

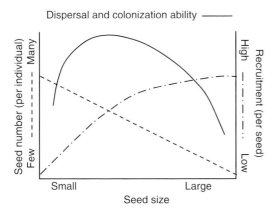

Fig. 6.2. A model of the relationship between seed size,
seed number, recruitment and dispersal, and colonizing
ability (redrawn from Eriksson, 2000).

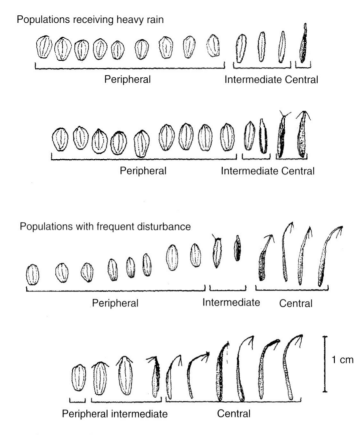

Populations receiving heavy rain

Peripheral   Intermediate Central

Peripheral   Intermediate Central

Populations with frequent disturbance

Peripheral   Intermediate   Central

1 cm

Peripheral intermediate   Central

Fig. 6.3. Three morphs of wingpetal (*Heterosperma pinnatum*) achenes from two populations (Venable *et al.*, 1995).

on length to width ratio, and the presence of a beak and/or wing (Fig. 6.3). Central achenes are awned (winged) and tend to disperse further. They lose dormancy earlier than other morphs and germinate in the spring. Peripheral achenes do not disperse as far but tend to have higher germination under harsh conditions. The relative proportion of morphs differs among populations; this is a result of selection under different environments. For example, populations in disturbed habitats have a higher proportion of central achenes as these are more likely to disperse away from a habitat that may be eliminated. In populations receiving heavy early rains, there are more peripheral achenes that germinate late and can withstand this environment.

6.3 Seed Dispersal (Dispersal in Space)

Seeds are dispersed away from parents for several reasons. First, it avoids seedlings competing with their parents or siblings. Being dispersed away from

the maternal plant also decreases the likelihood of mating with a sibling, which could lead to inbreeding depression. Second, dispersal increases colonization opportunities. If all seeds fall directly around their parent plant, then the species has little chance of colonizing new habitats or expanding its range. Seedling establishment is sometimes higher in habitats away from the parent plant, for example, where a parent plant creates too much shade for the seedling. Finally, dispersal also reduces the chance of all seeds from an individual being eaten by herbivores or destroyed by parasites or pathogens. If there was no dispersal, the intensity of attack by parasites or pathogens attracted to specific genotypes of seeds would increase.

There is a trade-off between dispersal and seed production. The cost of dispersal is that energy is allocated to dispersal structures (e.g. wings or fleshy fruits) and away from seed production itself. The benefits of seed dispersal must outweigh this cost (Howe and Smallwood, 1982; Willson, 1992;

Eriksson and Kiviniemi, 1999). Plants have developed different dispersal mechanisms or 'syndromes' to maximize dispersal and minimize energy losses.

Dispersal vectors and structures

Seeds can be dispersed via several types of vectors, including wind, water and animals – primarily vertebrates and ants. These vectors are often associated with specific dispersal structures. For example, wind-dispersed seeds, such as dandelion (*Taraxacum officinale*) and bull thistle (*Cirsium vulgare*), are usually small and light, and tend to have wings or plumes that increase their dispersal distance. Seeds such as bluebur (*Lappula echinata*) and common burdock (*Arctium minus*) that are dispersed on the outside of animals have structures such as barbs or hooks, or are sticky, causing them to attach to fur, hair or feathers. Seeds such as European buckthorn (*Rhamnus cathartica*) and common barberry (*Berberis vulgare*) that are dispersed internally by animals tend to have fleshy fruit that attracts animals, but also hard seedcoats to protect the seed while it passes through the animal's gut. The behaviour of an animal determines the fate and the distribution of a seed; for example, hoarding animals may produce seed caches – clumps of seeds (and seedlings) (Howe and Westley, 1997). A special type of animal dispersal is by ants. These seeds offer external structures containing fat tissues (elaiosomes) to entice ants. Bull thistle and mile-a-minute weed (*Polygonum perfoliatum*) are ant dispersed. Ballistically dispersed seeds, such as bur cucumber (*Sicyos angulatus*) and touch-me-not (*Impatiens glandulifera*), are usually housed in fruit that has a trigger mechanism to propel the seed away from the parent plant. Some species, such as lambsquarters, have seeds with no special dispersal structures (van der Pijl, 1982; Willson *et al.*, 1990).

Primary and secondary dispersal

Seeds will not be evenly dispersed throughout a habitat because they are subject to myriad abiotic and biotic factors. These may be categorized as primary dispersal, in which the seeds are dispersed away from the parent plant, and secondary dispersal, in which the seeds are subsequently moved to other sites by a different vector (Chambers and MacMahon, 1994; Vander Wall and Longland, 2004) (Fig. 6.4).

The distance that seeds are dispersed through primary dispersal is determined by the morphological characteristics of the seed. For example, small winged or plumed seeds (e.g. fleabanes, *Erigeron* spp.) travel further in wind than unadorned and relatively heavy seeds (e.g. pigweeds, *Amaranthus* spp.). Seeds enclosed

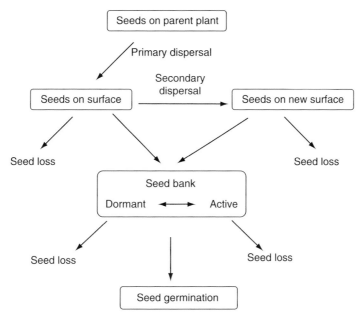

Fig. 6.4. Movements and fates of seeds (based on Chambers and MacMahon, 1994).

in a fruit may be consumed by animals and then deposited elsewhere. This first step moves the seeds away from the maternal plant to places where they are widely dispersed from rather than clumped around the parent plant, but not necessarily to sites that are ideal for germination (Vander Wall and Longland, 2004).

The distance that seeds move through secondary dispersal is dependent on seed characteristics and how these interact with abiotic factors (gravity, wind, rain and frost heaving) and the physical structure of the community (topography, vegetation and soil) (Chambers and MacMahon, 1994). Even small differences in the physical structure can change secondary dispersal patterns. A slight change in soil particle size can determine whether a seed moves in response to wind. Secondary dispersal is also influenced by biotic factors. Many types of seed movement are the result of animals (digging, scatter hoarding). The presence of earthworms, for example, increases the depth of annual bluegrass (*Poa annua*) seeds. Animals show specific preferences for seed types, but their level of consumption is dependent on seed density (Cromar *et al.*, 1999). Secondary dispersal may reduce the likelihood of seed predation (Chapter 9) and may move seeds to sites that are more favourable to germination and establishment (Vander Wall and Longland, 2004).

Humans as dispersal agents

Humans are an excellent agent of seed dispersal. While seeds may stick to our hair, skin, clothing and the vehicles we drive, we are an important seed-dispersal agent more because of our technology and mobility. Much of the dispersal is rather passive, or at least accidental in terms of the motives of the dispersal agent (McCanny and Cavers, 1988;

Schmidt, 1989; Mack, 1991; Lonsdale and Lane, 1994). Deliberate dispersal of invasive species occurs when seeds or plants are introduced before it is realized that they are invasive. We have intentionally introduced seeds for landscaping, farming or erosion control. Kudzu (*P. lobata*), for example, is a highly invasive species introduced from China to the southeastern USA as an ornamental species and for erosion control. Many species, including dame's rocket (*Hesperis matronalis*), have been introduced as part of commercial seed mixes for agriculture and ecological restoration purposes.

Humans are, by far, the most efficient seed dispersers, moving seeds faster than non-human dispersal methods, and they have assisted seed dispersal for millennia. However, since the 1950s, technological advancement and the push towards globalization of trade have greatly increased the magnitude of invasions (Hulme, 2009).

Long-distance dispersal

So far we have mostly discussed dispersal of seeds within their natural range. In most species, most seeds are dispersed within metres of the parent plant. As a result, changes in a species' range will occur gradually over time through edge expansion (Wilson *et al.*, 2009). In this type of dispersal, there is continuous gene flow among individuals in the population. Occasionally, however, seeds are dispersed well beyond their usual range. This is called long-distance dispersal. In long-distance dispersal there is no physical connection between the original and new habitats, and there may or may not continue to be gene flow among populations (Wilson *et al.*, 2009). While long-distance dispersal occurs mainly through human activities, there are other vectors and mechanisms (Table 6.1) (Nathan *et al.*, 2008).

Table 6.1. Major types of long-distance dispersal of seeds (based on Nathan *et al.*, 2008).

Type	Description
Open terrestrial landscapes	Seeds will disperse further over open landscape owing to a lack of obstacles for the seed and vector.
Large animals	Large animals tend to transport seeds over longer distances as they have larger home ranges and travel faster than small animals.
Migratory animals	Migratory birds and mammals travel long distances relatively quickly, often beyond the plant's native range.
Extreme meteorological events	Meteorological events, such as storms and floods, can rapidly disperse seeds.
Ocean currents	Floating or rafting seeds can move between continents by ocean currents.
Human transportation	The movement of seeds through global trade and transportation is virtually limitless.

6.4 Seed Banks and Seed Dormancy (Dispersal in Time)

When a seed is dispersed away from its maternal parent, it may be either dormant (in a resting state and unable to germinate) or non-dormant (able to germinate) (Fig. 6.5). Both dormant and non-dormant seeds eventually become incorporated into the soil as part of the 'seed bank' – think of this as a historical repository of seeds that will be withdrawn over time. Seeds that remain in the seed bank for long periods of time can do so because they are dormant.

Seed banks

The seed bank is referred to as 'dispersal in time' because it provides the same essential benefit as dispersal through space – it increases the chance that at least some seeds will survive to germinate under suitable environmental conditions. Unfortunately for plants, the seed bank is not a benign place, and seeds cannot survive indefinitely. Over time, the seed bank decreases because seeds die as a result of failed germination, physiological death, disease, herbivory or pathogens. Seed germination is also delayed, sometimes for a long time, by adverse soil conditions (pH and moisture) and deep burial (Simpson *et al.*, 1989). Seeds in the seed bank are continually redistributed via worms, water and other forms of dispersal. For example, human disturbance such as tillage can alter the distribution (Fig. 6.6) and density of seeds in the soil (Clements *et al.*, 1996).

A species' seed bank can be classified based on seasonal variation in germinable seeds (Thompson and Grime, 1979). The two main types are transient seed banks and persistent seed banks. These two main types can be further subdivided: transient seed banks into those that have seeds present only in the summer (Type I) or only in the winter (Type II); and persistent seed banks into those with (Type III) or without (Type IV) a seasonal peak (Fig. 6.7).

Transient seed banks contain seeds that rarely last for more than one year; they may contain either autumn-germinating seeds (Type I, e.g. perennial ryegrass, *Lolium perenne*) or spring-germinating seeds (Type II, e.g. Himalayan balsam, *I. glandulifera*). Giant hogweed (*H. mantegazzianum*) is considered to have a transient seed bank because very few seeds survive in the top 5 cm of soil from autumn to the following summer (Krinke *et al.*, 2005).

Persistent seed banks contain seeds that remain viable for more than 1 year. They either contain many seeds that germinate in autumn but maintain a small seed bank throughout the year (Type III, e.g. Rhode Island bent grass, *Agrostis tenuis*) or they may have a large persistent seed bank year round

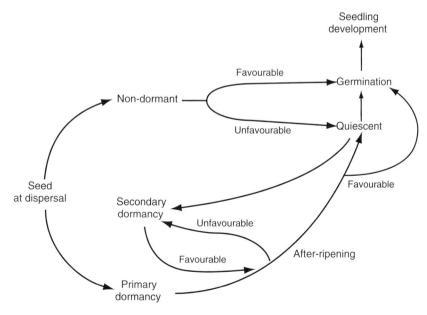

Fig. 6.5. The fate of a seed from when it disperses to when it germinates (Foley, 2001).

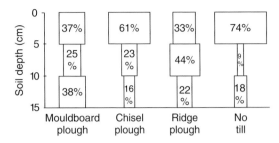

Fig. 6.6. Effect of different types of tillage (or no till) on percentage seed distribution by soil depth (based on Clements *et al.*, 1996).

(Type IV, e.g. common chickweed, *Stellaria media*). Most of the world's worst weeds have large persistent seed banks (Holm *et al.*, 1977). Species that are invasive are more likely to have a persistent seed bank than species that are decreasing in abundance.

Once buried, seeds are more likely to persist because they have lower predation rates and are less likely to encounter germination-inducing environmental conditions, such as light. For example, Harrison *et al.* (2007) found that seed survival of giant ragweed (*Ambrosia trifida*) was higher in surface seeds after 1 year, but that survival was greater in buried seeds after 2–4 years. Interestingly, while seeds buried at 20 cm depth had the highest seed bank survival, no seedlings emerged from these depths. Therefore, some disturbance would be

required for these seeds to emerge. Giant ragweed seeds are polymorphic, with individuals producing larger or smaller seeds (Sako *et al.*, 2001). For smaller seeds, emergence was higher when seeds were buried at 5 cm depth or on the surface, whereas for larger seeds emergence was highest at 5 cm depth but was lower for seeds on the surface.

Seed dormancy

Seeds that remain viable in the seed bank are usually dormant (Table 6.2). Dormant seeds cannot germinate until a specific set of environmental and physiological conditions are met. During dormancy the seed exhibits little growth or development and respiration is reduced (Rees, 1997; Benech-Arnold *et al.*, 2000). This allows the seed to persist but expend few resources on maintenance. Dormancy both prevents germination while the seed is still on the parent plant and ensures temporal dispersal into environments favourable to seedling survival (Murdoch and Ellis, 1992). Long-lived perennials tend to have short-lived seed banks, whereas short-lived species are more dependent on dormancy.

Primary and secondary dormancy

Seeds that are unable to germinate when they first mature have primary dormancy. When dormancy

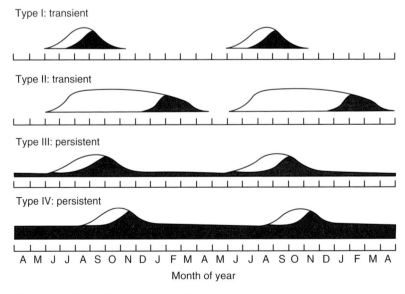

Fig. 6.7. Types of transient and persistent seed banks. The curves show seasonal seed abundance of immediately germinable seeds (shaded areas) and of viable but dormant seeds (unshaded areas) (Thompson and Grime, 1979).

Table 6.2. Definitions of terms associated with seed dormancy (see also Fig. 6.5).

Term	Definition
After-ripening	A physiological process whereby an embryo gradually matures and is able to germinate over a broader range of conditions.
Dormant	Seeds unable to germinate even though they have imbibed water and are under favourable environmental conditions.
Non-dormant	Being able to germinate under favourable environmental conditions.
Primary dormancy	Seeds that are unable to germinate when they mature and are either dispersed or still attached to the maternal parent plant.
Quiescent	Seeds that are unable to germinate owing to unfavourable environmental conditions.
Secondary dormancy	Dormancy that is imposed on the seed after being dispersed.

is imposed after seeds have dispersed, this is called secondary dormancy. Dormancy is usually imposed when environmental conditions are unfavourable for prolonged periods of time, and can be adaptive because it prevents seeds from germinating during seasons when environmental conditions are unsuitable.

Seeds may cycle in and out of dormancy, changing from dormant to conditionally dormant (where they germinate under a smaller range of conditions) to non-dormant; this cycle repeats and can result in the annual dormancy cycles observed in many invasive species such as barnyard grass (*E. crus-gallii*) (Honek *et al.*, 1999) (Fig. 6.8). Dormancy of summer annuals such as common ragweed (*A. artemisiifolia*) and lady's thumb (*Polygonum persicaria*) is released in the spring by low winter temperatures and reinduced by early summer high temperatures. Winter annuals such as ivy-leaved speedwell (*Veronica hederifolia*) and henbit (*Lamium amplexicaule*) require higher summer temperatures to release dormancy for autumn germination. Dormancy cycles ensure that seeds remain viable

over seasons (by not expending resources) but are able to germinate when conditions are appropriate for them.

Secondary dormancy is maintained by several mechanisms. Physiological mechanisms in the seed embryo may prevent it from germinating, such as in wild oat (*Avena fatua*) and annual sunflower (*H. annuus*), which have embryo dormancy. Physical mechanisms can also enforce dormancy (Foley, 2001). A hard seedcoat, for example, prevents water and/or gases from entering the seed. Seeds with a hard seedcoat usually require physical or chemical abrasion to break dormancy. Alternatively, chemical inhibitors within the seedcoat can maintain dormancy. These chemicals must be removed, for example by leaching, before the seed can germinate. Velvetleaf (*A. theophrasti*) and field bindweed (*Convolvulus arvensis*) are examples of weeds with seedcoat-imposed dormancy.

Dormant seeds often cannot germinate until the embryo fully matures. This process of physiological and physical maturation is called 'after-ripening'

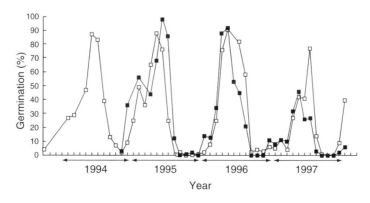

Fig. 6.8. Patterns of cyclic dormancy in barnyard grass (*E. crus-gallii*). Seeds buried in 1993 (□) and 1994 (■) were periodically retrieved and germinated in light at 25 °C (Honek *et al.*, 1999).

(Table 6.2, Fig. 6.5). During after-ripening, seeds are gradually able to germinate over a broader range of conditions (Baskin and Baskin, 1989). The environmental conditions required for after-ripening to occur are specific to individuals (and often broadly to populations and species). For example, common ragweed and giant ragweed require cool, moist conditions (Bazzaz, 1970; Ballard *et al.*, 1996), whereas common cocklebur (*X. strumarium*) requires warm, dry conditions (Esashi *et al.*, 1993) to after-ripen.

Breaking dormancy

The conditions required to break dormancy differ not only among species, but also within species or populations. This is especially true if a species' geographical range or habitat has a high degree of environmental variation (Allen and Meyer, 1998). In species that normally experience wide environmental variation, such as beardlip penstemon (*Penstemon barbatus*) and blue flax (*Linum perenne*), only some seeds respond to dormancy-breaking conditions and they maintain a long-term seed bank as a hedge against sudden environmental change within a growing season. Within a given species, populations that experience colder winters have more dormant seeds and require longer periods of cold to break dormancy than populations with milder winters. Intermediate populations often have variable dormancy, with differences occurring either within or among individual plants (Allen and Meyer, 1998).

Even seeds produced from one individual may have different dormancy-breaking requirements. This often occurs in species with polymorphic seeds. For example, seeds from the peripheral flowers on tansy ragwort (*Senecio jacobaea*) are larger, heavier and require longer periods of time to break dormancy, whereas seeds from central flowers are small, lighter, dispersed further and less dormant (McEvoy, 1984). Similarly, common lambsquarters produces mainly dormant black seeds but a few are non-dormant brown seeds (Roman *et al.*, 2000).

The cues for breaking dormancy and germination differ

When dormancy is broken, this does not necessarily lead to seed germination because the conditions required to break dormancy are not necessarily the same as those required for germination (Benech-Arnold *et al.*, 2000). A seed that breaks dormancy may either germinate, become dormant again or die. Breaking dormancy and seed germination generally are considered as two sequential processes.

6.5   Seed Germination

Non-dormant seeds will only germinate when they experience favourable environmental conditions. These are termed 'quiescent' seeds (Foley, 2001); they are 'seeds in waiting'. Quiescent seeds are able to germinate immediately once they encounter favourable environmental conditions, but they may enter secondary dormancy (Fig. 6.5).

The critical factor for seeds is to be able to germinate at an appropriate time, a daunting task as environmental conditions vary on small spatial scales and also are rarely constant from day to day and year to year. For example, in temperate regions, seeds that germinate late often experience intense competition from other individuals and have a shorter growing season to complete their life cycle. Conversely, seeds that germinate early may experience high mortality from unfavourable environmental conditions (e.g. frost). However, the risk may be worth it as the ones that survive are better competitors for light and other resources and have higher fitness (Ross and Harper, 1972; Marks and Prince, 1981; Gross, 1984).

The timing of seed germination is triggered by environmental cues. The most common cues are light quality and quantity, temperature, moisture and gases (O_2 and CO_2). These generally vary on large scales (latitudinal scales) but they also vary locally. On a local scale, seeds of different species are sensitive to different aspects of their physical micro-environment. For example, Sheldon (1974) found that annual sow thistle (*Sonchus oleraceus*) and dandelion seeds that had the attachment end (the end that was formerly attached to the seed head) in closest contact with a moist substrate had the highest percentage germination (Fig. 6.9).

Requirements differ among species but species with small seeds tend to require light for germination more than large-seeded species (Milberg *et al.*, 2000). Alternating temperatures are required for the seed germination of many invasive species, including common lambsquarters, large crabgrass (*Digitaria sanguinalis*), field bindweed, orchard grass (*Dactylis glomerata*) and some species of dock (*Rumex* spp.). The effect of temperature fluctuations depends on the amplitude (difference between maximum and minimum temperatures),

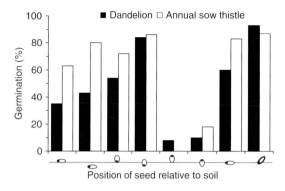

Fig. 6.9. Effect of seed position (in relation to the soil surface) on the germination of dandelion (*Taraxacum officinale*) and annual sow thistle (*Sonchus oleraceus*). The dark end of the seed is the attachment end (adapted from Sheldon, 1974).

mean temperature and thermoperiod (time above mean temperature each day) (Probert, 1992).

6.6 Seedling Emergence and Establishment

The term 'seedling' is simply another way of saying 'young plant' and implies no specific age or stage. Trees are often referred to as seedlings into their second or third year, simply because this stage is small compared with their adult form, whereas a fast-growing annual may be called a seedling only for a matter of days.

The distinction between seed germination, seedling emergence and establishment is not always clear and the terminology can be confusing. Germination normally means that the seed is physiologically active and the embryo is undergoing mitosis to produce a shoot and root which emerges from the seedcoat. Emergence usually refers to the appearance of a shoot above the soil. Establishment is generally considered to have occurred once a seedling no longer depends on seed reserves (endosperm and cotyledons) and is physiologically independent.

Factors affecting seedling emergence and establishment

The seedling stage often has the highest mortality rate of the different stages of a plant's life cycle because seedlings are vulnerable to environmental stress (Harper, 1977). As with germination, the timing of seedling emergence is important because it determines whether an individual will be able to compete with its neighbours, or be subjected to herbivory or disease; it is also important in relation to the timing of other life history events (Forcella *et al.*, 2000).

The timing of seedling emergence is determined by the interaction of abiotic factors (e.g. soil temperature, temperature fluctuations, soil moisture, depth of burial, light) with seed size, dormancy, germination and the rate of stem and root elongation (Allen and Meyer, 1998; Forcella *et al.*, 2000; Roman *et al.*, 2000). For example, Benvenuti *et al.* (2001) examined the effect of burial depth on 20 invasive species and concluded that seeds emerged faster when they were closer to the soil surface and that larger seeds were able to emerge from a greater depth.

6.7 Seeds and Seedlings Must Find Safe Sites

A seed must land in a site that provides all the conditions necessary for the seed to germinate and emerge from the soil. This is called a safe site. The environmental conditions needed for a safe site vary among individuals, populations and species. Conditions that are important include the environmental conditions necessary to break dormancy and allow germination to proceed, as well as to protect the seed and seedling from hazards such as herbivores, competitors and disease. The problem is that most seeds do not end up in a safe site. A seed is not guaranteed to find a safe site because it cannot control its own dispersal. Dispersal vectors do not guarantee safe passage and delivery to a good place to germinate and grow. Even if the seed finds a safe site and then germinates, the seedling may not survive because safe germination sites may not promote seedling emergence. A seedling may require different environmental conditions, or the environmental conditions may change by the time a seedling emerges.

6.8 Summary

There are many risks associated with the seed, seed germination, seedling emergence and seedling establishment stages. The risks of each stage are different for each species and these will change over time. The odds against a seed becoming successfully established are enormous. We will discuss environmental risks (or filters) in a community context in Chapter 11, on community assembly. Plants have many adaptations that improve the likelihood of survival. Adaptations include producing many seeds, increasing resources allocated to seeds so that more germinate, and remaining dormant in the seed bank in wait until ideal growing conditions occur. In the next chapter we look at the subsequent stages of growth.

Box 6.1. Invasive species case study: seeds and seedlings.

- Consider the seed and seedling stages of your chosen invasive species.
- How are seeds and/or fruits dispersed in your selected species?
- What type of seed bank does the species have?
- Describe a safe site for the species.

6.9 Questions

1. When trying to reduce the seed bank of an invasive species, how would the type of dormancy affect control strategies?

2. Explain why in the Eriksson (2000) model (Fig. 6.2), the hump of the dispersal and colonization ability curve moves to the left in disturbed sites.

3. Explain how invasive species benefit from being dispersed in time and space.

4. What variables limit the ability of a species to disperse?

Further Reading

Benech-Arnold, R.L., Sánchez, R.A., Forcella, F., Kruk, B.C. and Ghersa, C.M. (2000) Environmental control of dormancy in weed seed banks in soil. *Field Crops Research* 67, 105–122.

Coomes, D.A. and Grubb, P.J. (2003) Colonization, tolerance, competition and seed-size variation within functional groups. *Trends in Ecology and Evolution* 18, 283–291.

Howe, H.F. and Miriti, M.N. (2000) No question: seed dispersal matters. *Trends in Ecology and Evolution* 15, 434–436.

Wilson, J.R.U., Dormontt, E.E., Prentis, P.J., Lowe, A.J. and Richardson, D.M. (2009) Something in the way you move: dispersal pathways affect invasion success. *Trends in Ecology and Evolution* 24, 136–144.

References

Allen, P.S. and Meyer, S.E. (1998) Ecological aspects of seed dormancy loss. *Seed Science Research* 8, 183–191.

Ballard, T.O., Bauman, T.T. and Foley, M.E. (1996) Germination, viability, and protein changes during cold stratification of giant ragweed (*Ambrosia trifida* L.) seed. *Journal of Plant Physiology* 149, 229–232.

Baskin, C.C. and Baskin, J.M. (1989) Physiology of dormancy and germination in relation to seed bank ecology. In: Leck, M.A., Parker, V.T. and Simpson, R.L. (eds) *Ecology of Soil Seed Banks*. Academic Press, San Diego, California, pp. 53–66.

Bazzaz, F.A. (1970) Secondary dormancy in the seeds of the common ragweed *Ambrosia artemisiifolia*. *Bulletin of the Torrey Botanical Club* 97, 302–305.

Benech-Arnold, R.L., Sánchez, R.A., Forcella, F., Kruk, B.C. and Ghersa, C.M. (2000) Environmental control of dormancy in weed seed banks in soil. *Field Crops Research* 67, 105–122.

Benvenuti, S., Macchia, M. and Miele, S. (2001) Quantitative analysis of emergence of seedlings from buried weed seeds with increasing soil depth. *Weed Science* 49, 528–535.

Chambers, J.C. and MacMahon, J.A. (1994) A day in the life of a seed: movements and fates of seeds and their implications for natural and managed systems. *Annual Review of Ecology and Systematics* 25, 263–292.

Clements, D.R., Benoit, D.L., Murphy, S.D. and Swanton, C.J. (1996) Tillage effects on weed seed return and seedbank composition. *Weed Science* 44, 314–322.

Coomes, D.A. and Grubb, P.J. (2003) Colonization, tolerance, competition and seed-size variation within functional groups. *Trends in Ecology and Evolution* 18, 283–291.

Cromar, H.E., Murphy, S.D. and Swanton, C.J. (1999) Influence of tillage and crop residue on post-dispersal predation of weed seeds. *Weed Science* 47, 184–194.

Eriksson, O. (2000) Seed dispersal and colonization ability of plants – assessment and implications for conservation. *Folia Geobotanica* 35, 115–123.

Eriksson, O. and Kiviniemi, K. (1999) Evolution of plant dispersal. In: Vuorisalo, T.O. and Mutikainen, P.K. (eds) *Life History Evolution in Plants*. Kluwer, Dordrecht, The Netherlands, pp. 215–238.

Esashi, Y., Ogasawara, M., Gorecki, R. and Leopold, A.C. (1993) Possible mechanisms of afterripening in *Xanthium* seeds. *Physiologia Plantarum* 87, 359–364.

Foley, M.E. (2001) Seed dormancy: an update on terminology, physiological genetics, and quantitative trait loci regulating germinability. *Weed Science* 49, 305–317.

Forcella, F., Benech-Arnold, R.L., Sánchez, R. and Ghersa, C.M. (2000) Modeling seedling emergence. *Field Crops Research* 67, 123–139.

Gross, K.L. (1984) Effects of seed size and growth form on seedling establishment of six monocarpic perennial plants. *Journal of Ecology* 72, 369–388.

Harper, J.L. (1977) *Population Biology of Plants*. Academic Press, London.

Harper, J.L., Lovell, P.H. and Moore, K.G. (1970) The shapes and sizes of seeds. *Annual Review of Ecology and Systematics* 1, 327–356.

Harrison, S.K., Regnier, E.E., Schmoll, T.J. and Harrison, J.M. (2007) Seed size and burial effects on giant ragweed (*Ambrosia trifida*) emergence and seed demise. *Weed Science* 55, 16–22.

Holm, L.G., Plucknett, D.L., Pancho, J.V. and Herberger, J.P. (1977) *The World's Worst Weeds: Distribution and Biology*. University Press of Hawaii, Honolulu, Hawaii.

Honek, A., Martinkova, Z. and Jarosik, V. (1999) Annual cycles of germinability and differences between primary and secondary dormancy in buried seeds of *Echinochloa crus-galli*. *Weed Research* 39, 69–79.

Howe, H.F. and Smallwood, J. (1982) Ecology of seed dispersal. *Annual Review of Ecology and Systematics* 13, 201–228.

Howe, H.F. and Westley, L.C. (1997) Ecology of pollination and seed dispersal. In: Crawley, M.J. (ed.) *Plant Ecology*, 2nd edn. Blackwell Scientific, Oxford, UK, pp. 262–283.

Hulme, P.E. (2009) Trade, transport and trouble: managing invasive species pathways in an era of globalization. *Journal of Applied Ecology* 46, 10–18.

Krinke, L., Moravcová, L., Pyšek, P., Jarošík, V., Pergl, J. and Perglová, I. (2005) Seed bank of an invasive alien *Heracleum mantegazzianum* and its seasonal dynamics. *Seed Science Research* 15, 239–248.

Leishman, M.R. (2001) Does the seed size/number trade-off model determine plant community structure? An assessment of the model mechanisms and their generality. *Oikos* 93, 294–302.

Lonsdale, W.M. and Lane, A.M. (1994) Tourist vehicles as vectors of weed seeds in Kakadu National Park, Northern Australia. *Biological Conservation* 69, 277–283.

Mack, R.N. (1991) The commercial seed trade: an early dispersers of weeds in the United States. *Economic Botany* 45, 257–273.

Marks, M. and Prince, S. (1981) Influence of germination date on survival and fecundity in wild lettuce *Lactuca serriola*. *Oikos* 36, 326–330.

Mason, R.A.B., Cooke, J., Moles, A.T. and Leishman, M.R. (2008) Reproductive output of invasive versus native plants. *Global Ecology and Biogeography* 17, 633–640.

McCanny, S.J. and Cavers, P.B. (1988) Spread of proso millet (*Panicum miliaceum* L.) in Ontario, Canada. II. Dispersal by combines. *Weed Research* 28, 67–72.

McEvoy, P.B. (1984) Dormancy and dispersal in dimorphic achenes of tansy ragwort, *Senecio jacobaea* L. (Compositae). *Oecologia* 61, 160–168.

McLachlan, S.M., Murphy, S.D., Tollenaar, M., Weise, S.F. and Swanton, C.J. (1995) Light limitation of reproduction and variation in the allometric relationship between reproductive and vegetative biomass in *Amaranthus retroflexus* (redroot pigweed). *Journal of Applied Ecology* 32, 157–165.

Michaels, H.J., Benner, B., Hartgerink, A.P., Lee, T.D., Rice, S., Willson, M.F. and Bertin, R.I. (1988) Seed size variation: magnitude, distribution and ecological correlates. *Evolutionary Ecology* 2, 157–166.

Milberg, P.J., Andersson, L. and Thompson, K. (2000) Large-seeded species are less dependent on light for germination than small-seeded ones. *Seed Science Research* 10, 99–104.

Murdoch, A.J. and Ellis, R.H. (1992) Longevity, viability and dormancy. In: Fenner, M. (ed.) *Seeds: the Ecology of Regeneration in Plant Communities*. CAB International, Wallingford, UK, pp. 193–229.

Nathan, R., Schurr, F.M., Spiegel, O., Steinitz, O., Trakhtenbot, A. and Tsoar, A. (2008) Mechanisms of long-distance seed dispersal. *Trends in Ecology and Evolution* 23, 638–647.

Orozco-Segovia, A., Brechú-Franco, A.E., Zambrano-Polano, L., Osuna-Fernández, R., Languna-Hernández, G. and Sánchez-Coronado, M.E. (2000) Effects of maternal light environment on germination and morphological characteristics of *Sicyos deppei* seeds. *Weed Research* 40, 495–506.

Probert, R.J. (1992) The role of temperature in germination ecophysiology. In: Fenner, M. (ed.) *Seeds: the Ecology of Regeneration in Plant Communities*. CAB International, Wallingford, UK, pp. 285–325.

Reader, R.J. (1993) Control of seedling emergence by ground cover and seed predation in relation to seed size for some old-field species. *Journal of Ecology* 81, 169–175.

Rees, M. (1996) Evolutionary ecology of seed dormancy and seed size. *Philosophical Translations of the Royal Society of London B* 351 1299–1308.

Rees, M. (1997) Seed dormancy. In: Crawley, M.J. (ed.) *Plant Ecology*, 2nd end. Blackwell Science, Oxford, UK, pp. 214–238.

Roman, E.S., Murphy, S.D. and Swanton, C.J. (2000) Simulation of *Chenopodium album* seedling emergence. *Weed Science* 48, 217–224.

Ross, M.A. and Harper, J.A. (1972) Occupation of bio-logical space during seedling establishment. *Journal of Ecology* 60, 77–88.

Sako, Y., Regnier, E.E., Daoust, T., Fujimura, K., Harrison, S.K. and McDonald, M.B. (2001) Computer image analysis and classification of giant ragweed seeds. *Weed Science* 49, 738–745.

Schmidt, W. (1989) Plant dispersal by motor cars. *Vegetatio* 80, 147–152.

Sheldon, J.C. (1974) The behaviour of seeds in soil. III. The influence of seed morphology and the behav-iour of seedlings on the establishment of plants from surface-lying seeds. *Journal of Ecology* 62, 47–66.

Simpson, L.R., Leck, M.A. and Parker, V.T. (1989) Seed banks: general concepts and methodological issues. In: Leck, M.A., Parker, V.T. and Simpson, R.L. (eds) *Ecology of Soil Seed Banks.* Academic Press, San Diego, California, pp. 3–8.

Thompson, K. and Grime, J.P. (1979) Seasonal variation in the seed banks of herbaceous species in ten contrasting habitats. *Journal of Ecology* 67, 893–921.

Thompson, K., Band, S.R. and Hodgson, J.G. (1993) Seed size and shape predict persistence in soil. *Functional Ecology* 7, 236–241.

van der Pijl, L. (1982) *Principles of Dispersal in Higher Plants,* 3rd edn. Springer-Verlag, Berlin.

Vander Wall, S.B. and Longland, W.S. (2004) Diplochory: are two seed dispersers better than one? *Trends in Ecology and Evolution* 19, 155–161.

Venable, D.L., Dyreson, E. and Morales, E. (1995) Popula-tion dynamic consequences and evolution of seed traits of *Heterosperma pinnatum* (Asteraceae). *American Journal of Botany* 82, 410–420.

Vivian-Smith, G. and Panetta, F.D. (2009) Lantana (*Lan-tana camara*) seed bank dynamics: seedling emer-gence and seed survival. *Invasive Plant Science and Management* 2, 141–150.

Westoby, M., Jurado, E. and Leishman, M. (1992) Com-parative evolutionary ecology of seed size. *Trends in Ecology and Evolution* 7, 368–372.

Westoby, M., Leishman, M. and Lord, J. (1996) Comparative ecology of seed size and dispersal. *Philosophical Transla-tions of the Royal Society of London B* 351, 1309–1318.

Willson, M.F. (1992) The ecology of seed dispersal. In: Fenner, M. (ed.) *Seeds: the Ecology of Regeneration in Plant Communities.* CAB International, Wallingford, UK, pp. 61–86.

Willson, M.F., Rice, B.L. and Westoby, M. (1990) Seed dis-persal spectra: a comparison of temperate plant com-munities. *Journal of Vegetation Science* 1, 547–562.

Wilson, J.R.U., Dormontt, E.E., Prentis, P.J., Lowe, A.J. and Richardson, D.M. (2009) Something in the way you move: dispersal pathways affect invasion suc-cess. *Trends in Ecology and Evolution* 24, 136–144.

7 Growing Up, Getting Old and Dying

Concepts

- Phenology, the study of the timing of regularly occurring life-cycle events, influences components of fitness such as reproduction and maturity.
- A plant's phenology is determined by the interaction of its genetic make-up and its biotic and abiotic environment.
- Plants allocate limited resources to growth, maintenance and reproduction. Allocation patterns vary with species and change with the environment.
- Phenotypic plasticity allows species to adapt to challenging environmental conditions. Many invasive species are thought to be phenotypically plastic.
- Senescence is the natural deterioration of plants leading to death. Mechanisms of senescence differ between monocarpic and polycarpic species.

7.1 Introduction

In the previous chapter, we discussed the importance of the seed and early seedling stages of a plant's life cycle. In this chapter we continue the theme of plant growth and development from seedling establishment through to death. We focus on how abiotic factors influence these life-cycle events. We also look at how plants allocate resources to their growth, maintenance and reproduction, and how allocation changes over the life of a plant. Finally, we discuss the natural process of plant death.

7.2 Phenology

A seedling becomes independent when it has used up all the nutrients in the endosperm and cotyledons. Following this, a plant will continue to grow, producing stems, leaves and roots by obtaining energy from the sun (via photosynthesis) and nutrients from the roots. Eventually, most plants will flower and produce seeds, and then either die (if they are monocarpic), or have repeated periods of flowering throughout their life (if they are polycarpic). The timing of these regular life-cycle events

is called phenology. Phenology is an important aspect of a plant's life cycle because it influences major components of fitness such as how the plant acquires resources, grows and reproduces (Nord and Lynch, 2009). For example, the timing of flowering is an important indicator of physiological stress or climate change (Cleland *et al.*, 2007). Phenological development is influenced by the interaction of genetics with environment. For example, the biennial common mullein (*V. thapsus*) does not usually flower in its first year and will only flower in its second year if it has reached a certain biomass (Gross and Werner, 1978). The rate of biomass accumulation by the rosette is influenced by the environmental conditions occurring at the time of growth. Most of the critical life-history events of a plant will be influenced by several interacting abiotic variables such as temperature, light, water and nutrient stress.

Effect of abiotic factors on phenology

The abiotic environment affects phenology both directly and indirectly. Directly, an environmental cue may trigger a specific phenological event. For example, a species may not flower until a certain day length is reached. Indirectly, the environment can influence phenology because it can affect the rates of physiological processes, such as photosynthesis and cell division. There is usually a range of environmental factors under which the response of a plant is optimal (Fig. 7.1). Above and below this range, the rate of the physiological processes will decrease. For example, most plants will not survive in either very cold or very hot temperatures, and will have optimum growth at intermediate temperatures. Every species (population or individual) has a different set of responses to a factor.

Abiotic factors that influence phenology include light, temperature, water, nutrients, gases (oxygen and carbon dioxide) and soil characteristics (pH and texture). In the next sections, we look at two specific factors (temperature and light), and then discuss how abiotic factors, in general, interact to influence phenology.

Temperature

Temperature has a strong influence on plant growth rate. We see this when garden plants appear to stop growing on cold spring days, and then 'grow before our eyes' on the next warm day. Below a temperature threshold, plants will die from freezing damage (or from chilling in warmer climates).

As the temperature increases, the metabolic rate increases and growth rate increases; however, a few degrees beyond the optimum, proteins begin to be denatured (destroyed) and the plant dies. The critical thresholds will change depending on the life stage of the plant and environment. For example, a seed will have different temperature limits from a seedling or mature plant. Furthermore, individuals growing in colder climates may be adapted or acclimatized to lower temperatures than those in warmer climates.

Other processes, such as dormancy, germination (Chapter 6), and bud and flower initiation, are also influenced by temperature. Canada thistle (*Cirsium arvense*), for example, produces new shoots from overwintering roots only when temperatures reach 5 °C (Sheley and Petroff, 1999). Some species require a cold period to promote flowering; if kept in warm temperatures, they will continue to grow vegetatively but will not flower. Dalmatian toadflax (*Linaria dalmatica*), for example, requires a winter dormancy period and then exposure to temperatures between 10 and 20 °C to produce floral stems (Sheley and Petroff, 1999).

The critical temperatures that are required will differ among the various processes in the plant. For example, seedling emergence of wild mustard (*Sinapis arvensis*) was fastest under day/night temperatures of 29/19 °C and 35/25 °C, while plants grew taller at 23/13 °C (Huang *et al.*, 2001) (Fig. 7.2). However,

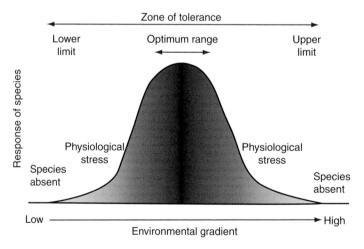

Fig. 7.1. Effect of environmental factors (e.g. temperature and light) on the rate of important physiological processes (e.g. photosynthesis, cell division). Generally, the response of a plant increases as the level of the factor concerned increases until an optimal level is reached and then begins to decrease. Beyond an upper and lower limit of the factor, the plant will not survive.

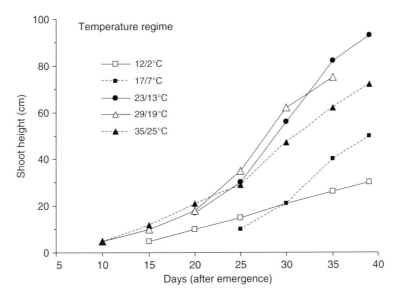

Fig. 7.2. Change in shoot height of wild mustard (*Sinapsis arvensis*) over time at five temperature regimes (redrawn from Huang *et al.*, 2001).

for shoot elongation, the optimum temperature (and high and low thresholds) was 24.5 °C (5.5–40.0 °C) whereas for leaf appearance it was 27 °C (1.5–48 °C) (Huang *et al.*, 2001). While specific temperatures are required for some processes, temperature fluctuations are required for others, such as seed germination (see Chapter 6). Because temperature influences so many plant processes, it is an important factor controlling the overall distribution of many species (Cleland *et al.*, 2007; Post *et al.*, 2008).

Light

Light (irradiance) has a direct effect on many plant processes, such as growth, flowering, stem elongation, seed dormancy, formation of storage organs and leaf abscission, and on many plant morphological characteristics, such as leaf number and shape (Salisbury and Ross, 1985; Lambers *et al.*, 1998). Plants respond to changes in light quality (spectral composition), quantity (intensity) and photoperiod (periodicity of light and dark cycles). It is sometimes difficult to determine which type of light effect is occurring because they interact; for example, as light quantity changes the spectral composition changes.

Reduced light quantity will have species-specific effects. In Canada thistle, for example, shading reduces shoot and root production, and the number

of inflorescences produced (Zimdahl *et al.*, 1991) (Fig. 7.3). Conversely, itchgrass (*Rottboellia exaltata*) plants growing in shade are taller than those growing in full sunlight. This is likely to be why itchgrass is competitive with maize (Patterson, 1985). Pattison *et al.* (1998) compared the relative growth rates of native and non-native species grown under conditions representing the full sun, partial sun and full shade environments found in Hawaiian rainforests. At low light levels, all species had lower relative growth rates, but native species were less affected (Fig. 7.4). This is why non-native species are sometimes more successful in open, disturbed habitats: they are able to take advantage of high light levels whereas the native species cannot.

As light passes through the leaf canopy the quality of light is also altered. The spectrum of light wavelengths changes because plants reflect green and far-red wavelengths while absorbing blue and red wavelengths. This causes the ratio of red to far-red (R:FR) light to decrease in shaded conditions (Fig. 7.5). Many species respond to this change in the R:FR ratio. For example, higher levels of FR light trigger internode extension in white mustard (*Sinapis alba*), Chinese datura (*Datura ferox*) and common lambsquarters (*C. album*), causing plants to grow taller into better (higher) light situations (Alm *et al.*, 1991; Mahoney and Swanton, 2008). Lambsquarters grew taller, produced more leaf area

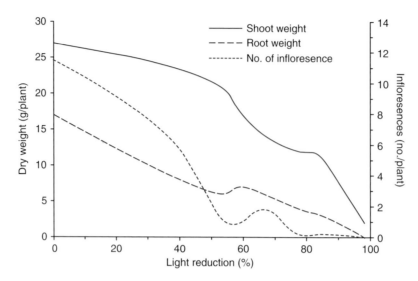

Fig. 7.3. Effect of light intensity on shoot and root dry weight and on inflorescence number of Canada thistle (*Cirsium arvense*) (adapted and redrawn from Zimdahl *et al.*, 1991).

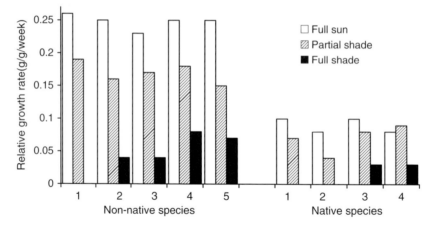

Fig. 7.4. Effect of light conditions (shading) on the relative growth rate of four native and five non-native species of Hawaiian rainforest (redrawn from Pattison *et al.*, 1998).

and had delayed flowering when grown in a high FR environment (Mahoney and Swanton, 2008).

Photoperiod is what synchronizes many of the seasonal events observed in nature. Alternating light and dark cycles give an accurate cue as to the time of year. While temperature can vary unpredictably from day to day, and light quantity and quality is altered by the surrounding vegetation, photoperiod is not changed by the environment. In wild mustard, for example, photoperiod influences many reproductive events (Huang *et al.*, 2001). As photoperiod shortens, the duration of the period

from seedling emergence to the production of floral primordia and from the flower primordia stage to flowering both increase, but the duration of the period from flowering to initiation of seed set decreases (Fig. 7.6).

The flowering response to photoperiod can take many forms (Table 7.1). A response to photoperiod typically requires several cycles, although some species require only one cycle to trigger a response. For example, a single short-day cycle will induce flowering in red-root pigweed (*A. retroflexus*). While we refer to the importance of day length, it is

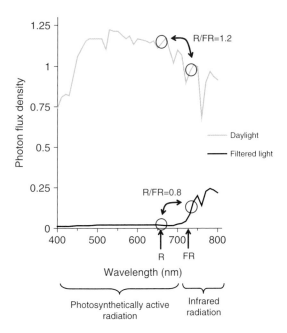

Fig. 7.5. Spectral photon distribution in the 400–800nm wavelength region on an overcast day in full daylight and filtered through a canopy of ash (*Fagus*) (redrawn and adapted from Pons, 2000).

actually the length of the dark period that usually triggers a response. For example, a 1 h interruption of florescent light during the dark period will inhibit

the flowering of red-root pigweed (Gutterman, 1985). Plants that respond to photoperiod generally go through three stages, of sensitivity: a pre-inductive stage, where photoperiod has no effect; an inductive phase, where photoperiod triggers reproductive response; and a post-inductive stage, where reproduction will continue irrespective of photoperiod (Patterson, 1995).

Because photoperiod changes with latitude, there are often 'biotypes' within species that respond to abiotic factors differently. A biotype is a group of individuals which have similar genetic structure and respond to their environment in different ways from other biotypes of the same species. For example, both common lambsquarters and common cockle-bur (*X. strumarium*) have biotypes that respond differently to photoperiod. Northern biotypes usually require shorter nights to initiate flowering. For example, in short days (12h light), Minnesota populations of velvetleaf (*A. theophrasti*) produce more vegetative growth than the Mississippi populations, but the reverse was true for long days (16h light) (Patterson, 1993). In the longest photoperiods, Minnesota populations allocate resources to reproduction, thereby limiting further vegetative growth. Similarly, northern populations of side-oats grama (*Bouteloua curtipendula*) are long-day biotypes while southern populations are short-day biotypes (Olmsted, 1944).

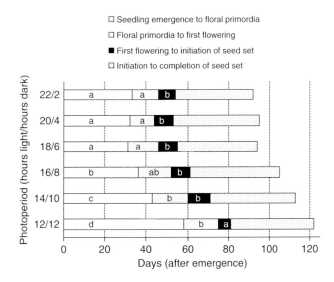

Fig. 7.6. The effect of photoperiod on the duration of the periods from emergence to floral primordia, to first flowering, to seed set initiation and to completion of seed set in wild mustard (*A. petiolata*) (from data in Huang *et al.*, 2001). Different letters within each developmental stage indicate significant differences; the lengths of the final developmental stages were not significantly different.

Table 7.1. Types of photoperiod responses and examples of representative species (adapted from Salisbury and Ross, 1985).

Photoperiod response	Examples
Short day	Red goosefoot (*Chenopodium rubrum*)
	Goosefoot (*Chenopodium polyspermum*)
	Common cocklebur (*X. strumarium*)
	Red-root pigweed (*A. retroflexus*)
Long day	Scarlet pimpernel (*Anagallis arvensis*)
	White mustard (*Sinapis alba*)
	Henbane (*Hyoscyamus niger*)
Short-long day	White clover (*Trifolium repens*)
	Kentucky bluegrass (*Poa pratensis*)
Intermediate day	Purple nutsedge (*Cyperus rotundus*)
Day neutral	Onion (*Allium cepa*)
	Wild carrot (*D. carota*)
	Barnyard grass (*E. crus-galli*)
	Indian goosegrass (*Eleusine indica*)
	Portulaca (*Portulaca oleracea*)
	Itchgrass (*Rottboellia exaltata*)

Interaction of abiotic factors

We have looked at how plants respond to changes in temperature and light separately. Data of this type are usually derived from controlled experiments where one variable is changed while all others remain constant. Such results do not necessarily reflect real situations for several reasons. First, environmental variables tend to fluctuate in tandem (when the sun comes out, light and temperature increase), therefore manipulation of one variable is not realistic. Second, plants respond in a complex fashion to the array of environmental factors they face and this may not be predictable by looking at one factor at a time. The response to one environmental factor will affect how an individual responds to another factor.

Reproduction is often determined by an interaction between photoperiod and temperature. For example, at low temperatures, poinsettia (*Euphorbia pulcherrima*) and morning glory (*Ipomoea purpurea*) are long-day plants, whereas at high temperatures they are triggered by short days. At intermediate temperatures they are day neutral. This interaction can lead to somewhat surprising outcomes. Tropical Kikuyu grass

(*Pennisetum clandestinum*) was bred as turfgrass for milder temperate areas, where it quickly became invasive (Wilen and Holt, 1996). Given its tropical origins, there was the realized expectation that photosynthesis is maximized in hotter temperatures during summer, or in ecozones such as Florida where the spring and autumn are still relatively warm (25–40 °C). What was not predicted was that photosynthesis can also be maintained during somewhat cooler spring and autumn seasons in areas such as Mediterranean ecozones. This is because photosynthesis is also influenced by light–dark cycles and in these ecozones, the day lengths are still long enough and the temperate climate may not be hot, but it is still warm enough. The case of Kikuyu grass illustrates the general principle that many successful invasive species can accommodate and maximize growth over a wide range of light and temperature interactions (Plowman and Richards, 1997; Roche *et al.*, 1997; Kibbler and Bahinsch, 1999).

Water stress can also affect a plant's ability to respond to other environmental triggers. For example, water stress can limit or prevent flowering in single-cycle photoperiod species such as common cocklebur and ryegrass (*Lolium temulentum*) (Chiariello and Gulmon, 1991).

Before a plant can respond to an environmental cue, it may have to reach a morphological stage at which it is able to sense the cue. The classic example is cocklebur, which must reach a certain size before it responds to photoperiod cues. In this species, leaves must be at least 1 cm long before the plant will respond to light. Other plants are less sensitive to morphological stage and can take advantage of favourable conditions when they occur earlier than expected, e.g. purple loosestrife (*L. salicaria*) and garlic mustard (*A. petiolata*) (Dech and Nosko, 2004; Murphy, 2005). With the occurrence of anthropogenic climate change this may become more common (Cleland *et al.*, 2007; Post *et al.*, 2008).

Plants can have severe constraints placed on them when they are introduced into a new habitat because the environmental cues that trigger important phenological events may be different there. For example, the day length in a new habitat may not be long enough to trigger flowering, or the cool temperatures may reduce the growth rate of an annual species so that it does not accumulate enough biomass to flower. For this reason, many successful invasive species are flexible in their response to the environment and are able to grow and reproduce under a wide range of environmental conditions.

7.3   Resource Allocation

The 'principle of allocation' states that plants have a limited supply of resources and that this is allocated to various physical structures and biochemical processes in a way that maximizes lifetime fitness (Bazzaz, 1996; Barbour *et al.*, 1999). Obviously, plants do not make conscious decisions of where to allocate resources. This is determined by the interaction of their genotype and their environment. How a plant allocates resources is important because if too much is spent on one function, then other functions may suffer. For example, if a perennial species allocates too many resources to reproduction and not enough to

storage, then it may not be able to survive a harsh winter.

The amount of resources allocated to growth, reproduction and maintenance will vary with the plant's life history strategy and will change over the course of a plant's life cycle (Fig. 7.7). Early on, plants accumulate biomass and nutrients in roots, shoots and leaves. In annual species, reproduction events require expenditures of resources towards the production of reproductive structures and towards the care of maturing embryos (Willson, 1983); therefore, as the season progresses, more resources will be devoted to reproduction, and less to vegetative structures; Fig. 7.8a shows the example of corn marigold (*Chrysanthemum segetum*). If resources are limiting, then the individual may not be able to reproduce, or may reproduce, but at the cost of future fitness or survival.

In perennials, allocation patterns differ primarily because fewer resources are allocated to reproduction.

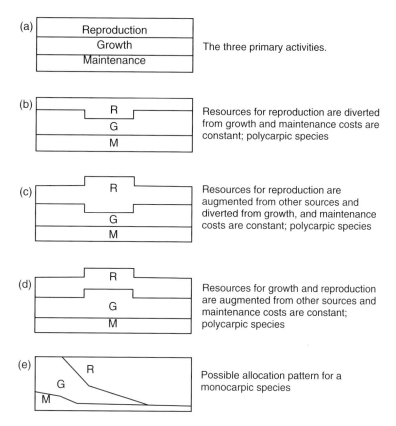

Fig. 7.7. Theoretical resource budgets of plants over one growing season. Resources are allocated to maintenance (M), growth (G) and reproduction (R) (redrawn from Willson, 1983).

For example, the Jerusalem artichoke (*Helianthus tuberosus*) is a herbaceous perennial that is well adapted to invading open areas, particularly cultivated fields (Swanton and Cavers, 1989). Here, a relatively large proportion of biomass is allocated to structural organs such as stems, leaves and branches (Fig. 7.8b).

Over the season, the allocation to storage organs such as roots, rhizomes and tubers increases and is much larger than the biomass allocated to flowers and seeds. This pattern of allocation ensures long-term survival through clonal structures as well as through seed production.

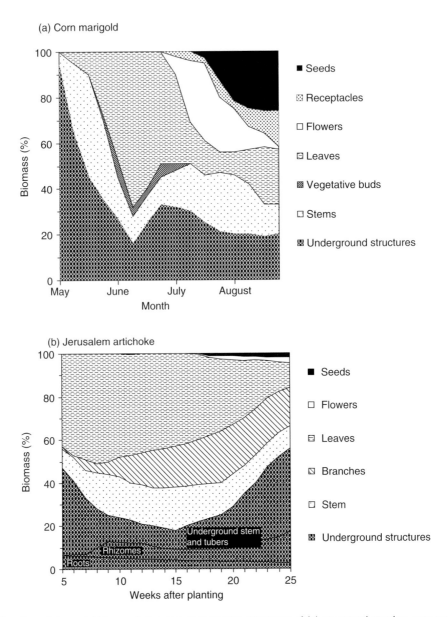

Fig. 7.8. Allocation of dry weight to vegetative and reproductive structures of (a) an annual species, corn marigold (*Chrysanthemum segetum*) and (b) a perennial species, Jerusalem artichoke (*Helianthus tuberosus*) (adapted and redrawn from Harper (1977) and Swanton and Cavers (1989)).

Chapter 7

Each reproductive event comes at a cost that must be compensated for either through the accumulation of new resources or through a trade-off within the plant, often between fecundity and survival (Matsuyama and Sakimoto, 2008). In polycarpic species, for example, the plant diverts only some of its resources towards each reproductive event (Fig. 7.7b,c,d). In some cases, the cost occurs at the expense of growth (Fig. 7.7b), while at other times the cost may be covered by the uptake of additional resources which will partially or totally compensate (Fig. 7.7c,d). This increase in resource supply occurs when reproductive structures take up more resources, for example when flowers and fruits are photosynthetic, or when they enhance the uptake of resources through vegetative structures. During reproduction, for example, leaf photosynthesis of quack (or couch) grass (*E. repens*) may increase or decrease depending on the plant's genotype and nutrient status (Reekie and Bazzaz, 1987). Where the change in leaf photosynthesis is positive, the cost of reproduction is offset. In monocarpic species, the plant diverts most of its stored resources towards reproduction at the end of its life cycle.

Allocation patterns differ among species. Common groundsel (*Senecio vulgaris*), for example, allocates proportionally more resources to stems and less to flowers than corn marigold, and yet both are annuals (Harper, 1977). Within species, allocation patterns will vary with the environment. In common groundsel, for example, more resources are allocated to roots in stressful environments (Harper, 1977). Similarly, bitter-cress (*Cardamine leucantha*) and Japanese green briar (*Smilacina japonica*), both rhizomatous perennials found in deciduous forests, have different allocation patterns in shade and light gaps (Ida and Kudo, 2009).

7.4 Phenotypic Plasticity

Phenotypic plasticity is the ability of an individual to change its appearance in response to biotic and abiotic conditions. It is the ability of the phenotype to change in response to the environment's effect on the genotype. Therefore, one genotype will produce many phenotypes, depending on the environment (Miner *et al.*, 2005). Within a species, genotypes will vary in their ability to respond phenotypically.

Phenotypic plasticity can be used to take advantage of a sudden, temporary improvement in environmental conditions or to avoid stressful environments (Meerts, 1995; Sans and Masalles, 1997; Wulff *et al.*, 1999; Weinig, 2000). Plasticity can increase biodiversity within a habitat, help to stabilize a population, modify an environment and, as a result, possibly influence ecosystem functioning (Miner *et al.*, 2005).

Invasive species are thought to have a high degree of phenotypic plasticity, although there are relatively few studies that clearly demonstrate this (Clements *et al.*, 2004). Phenotypically plastic invasive species are able to mature and reproduce under a broad range of environmental conditions. For example, showy crotalaria (*Crotalaria spectablis*) can reproduce in heavy shade even though it is substantially smaller in the shade (Patterson, 1982); this species can also produce seed under a range of temperatures in spite of decreased size and biomass. Barnyard grass (*E. crus-galli*) showed extreme plasticity among six cohorts planted from March to September (Bazzaz *et al.*, 2000); while vegetative biomass was over 3000 g in early cohorts and less than 25 g in later cohorts, late cohorts still flowered, although the number of flowers/plant was reduced from 10,000 to less than 100.

Other weeds exhibiting phenotypic variation are dandelion (*T. officinale*), Jimson weed (*D. stramonium*) and velvetleaf (Clements *et al.*, 2004), Sally-my-handsome (*C. aff. actinaciformis*), a Mediterranean invasive species (Traveset *et al.*, 2008), and the clonal red seaweed (*A. armata*), which switches from a guerrilla to a phalanx growth form, depending on the light regime (Chapter 5) (Monro and Poore, 2009). Two similar weeds, wild oat (*A. fatua*) and slender wild oat (*A. barbata*), rely on different mechanisms to respond to environmental heterogeneity (Marshall and Jain, 1968). Wild oat is more genetically variable, and this allows it to persist in many types of environments because populations contain many genotypes, some of which will be able to survive and reproduce in the ambient environment. Slender wild oat is more genetically uniform but is able to persist in a variety of habitats because it is phenotypically plastic. Thus, genotypic variation and phenotypic plasticity are two mechanisms through which plants deal with environmental stress and heterogeneity.

7.5 Senescence

It is easy to understand why plants die when they are eaten, trampled, or run out of water. It is not as obvious why plants die naturally, unthreatened by external forces, at the end of their life cycle. Senescence is the process of deterioration that leads to the natural death of a plant (or plant part). It is an internally controlled process that determines the lifespan of a whole organism or its parts; fundamentally, it is related to how long the short strands of DNA ('telomeres') at the ends of chromosomes continue to function as cells age – past a threshold too many cells die and so does the organism (this work won a 2009 Nobel Prize for Elizabeth H. Blackburn, Carol W. Greider and Jack W. Szostak). Here, we discuss one type of senescence: whole plant senescence. For reviews of senescence, see Noodén *et al.* (1997), Chandlee (2001) and Thomas *et al.* (2009).

The mechanisms of senescence may differ between monocarpic and polycarpic species. Monocarpic plant senescence involves the deterioration of existing vegetative tissue with an associated decline in meristematic cell activity. This decline in the production of somatic cells will prevent the production of new photosynthetic tissue (Noodén, 1988; Hensel *et al.*, 1993). Senescence generally occurs during or shortly after reproductive development and results in the remobilization of nutrients to the seeds and eventual death of the whole plant. Wilson (1997) reviewed several hypothesized mechanisms for the senescence of monocarpic species. The 'death hormone' hypothesis suggests that the rapid death of monocarpic plants is preprogrammed and likely to be hormonally controlled; however, this has not been adequately tested. The alternative hypothesis is that senescence in monocarps is caused by nutrient starvation following flowering. Here it is suggested that the act of reproduction diverts stored nutrients away from vegetative tissue, causing it to die. For example, senescence is delayed in the annual species beggar's tick (*Bidens pilosa*) when flowers are removed (Zobolo and van Staden, 1999). Wilson (1997) points out that while this may occur in some species, the biological evidence does not generally support this alternative hypothesis.

The death of polycarpic perennials is somewhat harder to explain, partly because it is difficult to separate the effects of age and size. Evidence can be found of senescence at the cell or tissue level, but whole-plant senescence is rare, if not improbable (Munné-Bosch, 2008), As a plant gets larger, it declines physiologically because it has more tissue to support, it must transport water and nutrients further, and it is more susceptible to herbivores and pathogens (Watkinson, 1992). Thus, the effects of size are similar to those of age. However, increasing size does not fully explain senescence of old plants. Senescence of woody plants could be caused by a number of factors, including the collapse of structural tissue and toxins accumulating in cells, and deleterious mutations that may accumulate over time. Some clonal plants appear to have escaped senescence altogether, by evidence of their extremely old age. The oldest living clone (King's lomatia, *Lomatia tasmanica*) has been reported to be 43,600 years old (Lynch *et al.*, 1998).

7.6 Summary

We have now completed our look at the dynamics of individual plant populations. Until now, we have looked at population dynamics as being separate from the dynamics of other species. Common sense tells us that this does not reflect real life. Even a monoculture maize field has hundreds (thousands?) of other species, including weeds, insects, soil fungi and bacteria, mammals and birds, many of which will influence the population dynamics of the maize. To begin our examination of how populations interact, we look first at competition in Chapter 8, and then at herbivory and other types of interactions in Chapter 9. From there, we have a chapter on how populations and their interactions are studied (Chapter 10).

Box 7.1. Invasive species case study: growing up, getting old and dying.

- Describe the phenology and resource allocation patterns of your selected invasive species.
- What environmental factors trigger important phenological events?

7.7 Questions

1. Why would knowledge of phenology be important for predicting the range of a species?
2. Explain why phenotypic plasticity is an adaptive trait for an invasive species.
3. Does phenotypic plasticity help to explain a species range?
4. Plants inevitably senesce. Why then should we worry about invasive species?

Further Reading

Elzinga, J.A., Atlan, A., Biere, A., Gigord, L., Weis, A.E. and Bernasconi, G. (2007) Time after time: flowering phenology and biotic interactions. *Trends in Ecology and Evolution* 22, 432–439.

Sultan, S.E. (2001) Phenotypic plasticity for components of *Polygonum* species of contrasting ecological breadth. *Ecology* 82, 328–343.

Thomas, H., Huang, L., Young, M. and Ougham, H. (2009) Evolution of plant senescence. *BMC Evolutionary Biology* 9, 163.

References

Alm, D.M., McGiffen, M.E. Jr and Hesketh, J.D. (1991) Weed phenology. In: Hodges, T. (ed.) *Predicting Crop Phenology*. CRC Press, Boca Raton, Florida, pp. 191–218.

Barbour, M.G., Burk, J.H., Pitts, W.D., Gilliam, F.S. and Schwartz, M.W. (1999) *Terrestrial Plant Ecology*, 3rd edn. Addison Wesley Longman, Menlo Park, California.

Bazzaz, F.A. (1996) *Plants in Changing Environments: Linking Physiological, Population and Community Ecology*. Cambridge University Press, New York.

Bazzaz, F.A., Ackerly, D.D. and Reekie, E.G. (2000) Reproductive allocation in plants. In: Fenner, M. (ed.) *Seeds: the Ecology of Regeneration in Plant Communities*, 2nd edn. CAB International, Wallingford, UK, pp. 1–29.

Chandlee, J.M. (2001) Current molecular understanding of the genetically programmed process of leaf senescence. *Physiologia Plantarum* 113, 1–8.

Chiariello, N.R. and Gulmon, S.L. (1991) Stress effects on plant reproduction. In: Mooney, H.A., Winner, W.E. and Pell, E.J. (eds) *Response of Plants to Multiple Stresses*. Academic Press, San Diego, California, pp. 161–188.

Cleland, E.E., Chuine, I., Menzel, A., Mooney, H.A. and Schwartz, M.D. (2007) Shifting plant phenology in response to global change. *Trends in Ecology and Evolution* 22, 357–365.

Clements, D.R., DiTommaso, A., Jordan, N., Booth, B.D., Cardina, J., Doohan, D., Mohler, C.L., Murphy, S.P.

and Swanton, C.J. (2004) Adaptability of plants invading North American cropland. *Agriculture, Ecosystems and Environment* 104, 379–398.

Dech, J.P. and Nosko, P. (2004) Rapid growth and early flowering in an invasive plant, purple loosestrife (*Lythrum salicaria* L.) during an El Niño spring. *International Journal of Biometeorology* 49, 26–31.

Gross, K.L. and Werner, P.A. (1978) The biology of Canadian weeds. 28. *Verbascum thapsus* L. and *V. blattaria* L. *Canadian Journal of Plant Science* 58, 401–413.

Gutterman, Y. (1985) Flowering, seed development, and the influences during seed maturation on seed germination of annual weeds. In: Duke, S.O. (ed.) *Weed Physiology. Volume I, Reproduction and Ecophysiology*. CRC Press, Boca Raton, Florida, pp. 1–26.

Harper, J.L. (1977) *Population Biology of Plants*. Academic Press, Toronto, Ontario.

Hensel, L.L., Grbi, V., Baumgarten, D.A. and Bleecker, A.B. (1993). Developmental and age-related processes that influence the longevity and senescence of photosynthetic tissues in *Arabidopsis*. *The Plant Cell* 5, 553–564.

Huang, J.Z., Shrestha, A., Tollenaar, M., Deen, W., Rajcan, I., Rahimian, H. and Swanton, C.J. (2001) Effect of temperature and photoperiod on the phenological development of wild mustard (*Sinapis arvensis* L.). *Field Crops Research* 70, 75–86.

Ida, T.Y. and Kudo, G. (2009) Comparison of light harvesting and resource allocation strategies between two rhizomatous herbaceous species inhabiting deciduous forest. *Journal of Plant Research* 122, 171–181.

Kibbler, H. and Bahnisch, L.M. (1999) Distribution of *Hymenachne acutigluma* (Steudel) Guilliland in ponded pasture is limited by photosynthetic response to temperature. *Australian Journal of Experimental Agriculture* 39, 437–443.

Lambers, H., Chapin, F.S. III and Pons, T.L. (1998) *Plant Physiological Ecology*. Springer-Verlag, New York.

Lynch, A.J.J., Barnes, R.W., Cambecèdes, J. and Vaillancourt, R.E. (1998) Genetic evidence that *Lomatia tasmanica* (Proteaceae) is an ancient clone. *Australian Journal of Botany* 46, 25–33.

Mahoney, K.J. and Swanton, C.J. (2008) Exploring *Chenopodium album* adaptive traits in response to light and temperature stress. *Weed Research* 48, 552–560.

Marshall, D.R. and Jain, S.K. (1968) Phenotypic plasticity of *Avena fatua* and *A. barbata*. *American Naturalist* 102, 457–467.

Matsuyama, S. and Sakimoto, M. (2008) Allocation to reproduction and relative reproductive costs in two species of dioecious Anacardiaceae with contrasting phenology. *Annals of Botany* 101, 1391–1400.

Meerts, P. (1995) Phenotypic plasticity in the annual weed *Polygonum aviculare*. *Botanica Acta* 108, 414–424.

Miner, G.B., Sultan, S.E., Morgan, S.G., Padilla, D.K. and Relyea, R.A. (2005) Ecological consequences of phenotypic plasticity. *Trends in Ecology and Evolution* 20, 685–692.

Monro, K. and Poore, A.G.B. (2009) Performance benefits of growth-form plasticity in a clonal red seaweed. *Biological Journal of the Linnean Society* 97, 80–89.

Munné-Bosch, S. (2008) Do perennials really senesce? *Trends in Plant Science* 13, 216–220.

Murphy, S.D. (2005) Concurrent management of an exotic species and initial restoration efforts in forests. *Restoration Ecology* 13, 584–593.

Noodén, L.D. (1998) The phenomena of senescence and aging. In: Noodén, L.D. and Leopold, A.C. (eds) *Senescence and Aging in Plants.* Academic Press, San Diego, California, pp. 1–50.

Noodén, L.D., Guiamét, J.J. and John, I. (1997) Senescence mechanisms. *Physiologia Plantarum* 101, 746–753.

Nord, E.A. and Lynch, J.P. (2009) Plant phenology: a critical controller of soil resources. *Journal of Experimental Botany* 60 1927–1937.

Olmsted, C.E. (1944) Growth and development in range grasses. IV. Photoperiodic responses in twelve geographic strains of side-oats grama. *Botanical Gazette* 106, 46–74.

Patterson, D.T. (1982) Effects of shading and temperature on showy crotalaria (*Crotalaria spectabilis*). *Weed Science* 30, 692–697.

Patterson, D.T. (1985) Comparative ecophysiology of weeds and crops. In: Duke, S.O. (ed.) *Weed Physiology. Volume I, Reproduction and Ecophysiology.* CRC Press, Boca Raton, Florida. pp. 101–129.

Patterson, D.T. (1993) Effects of temperature and photoperiod on growth and development of sicklepod (*Cassia obtusifolia*). *Weed Science* 41, 574–582.

Patterson, D.T. (1995) Effects of photoperiod on reproductive development in velvetleaf (*Abutilon theophrasti*). *Weed Science* 43, 627–633.

Pattison, R.R., Goldstein, G. and Ares, A. (1998) Growth, biomass allocation and photosynthesis of invasive and native Hawaiian rainforest species. *Oecologia* 117, 449–459.

Plowman, A.B. and Richards, A.J. (1997) The effect of light and temperature on competition between atrazine susceptible and resistant *Brassica rapa. Annals of Botany* 80, 583–590.

Pons, T.L. (2000) Seed responses to light. In: Fenner, M. (ed.) *Seeds: the Ecology and Regeneration of Plant Communities*, 2nd edn. CAB International, Wallingford, UK, pp. 237–260.

Post, E., Pedersen, C., Wimers, C.C. and Forchhammer, M.C. (2008) Warming, plant phenology and the spatial dimension of trophic mismatch for large herbivores. *Proceedings of the Royal Society B* 275, 2005–2013.

Reekie, E.G. and Bazzaz, F.A. (1987) Reproductive effort in plants. I. Carbon allocation to reproduction. *American Naturalist* 129, 876–896.

Roche, C.T., Thill, D.C. and Shafi, B. (1997) Prediction of flowering in common crupina (*Crupina vulgaris*). *Weed Science* 45, 519–528.

Salisbury, F.B. and Ross, C.W. (1985) *Plant Physiology*, 3rd edn. Wadsworth Publishing, Belmont, California.

Sans, F.X. and Masalles, R.M. (1997) Demography of the arable weed *Diplotaxis erucoides* in central Catalonia, Spain. *Canadian Journal of Botany* 75, 86–95.

Sheley, R.L. and Petroff, J.K. (1999) *Biology and Management of Noxious Rangeland Weeds.* Oregon State University Press, Corvallis, Oregon.

Swanton, C.J. and Cavers, P.B. (1989) Biomass and nutrient allocation patterns in Jerusalem artichoke (*Helianthus tuberosus*). *Canadian Journal of Botany* 67, 2880–2887.

Thomas, H., Huang, L., Young, M. and Ougham, H. (2009) Evolution of plant senescence. *BMC Evolutionary Biology* 9, 163. doi:10.1186/1471-2148-9-163.

Traveset, A., Moragues, E. and Valladares, F. (2008) Spreading of the invasive *Carpobrotus* aff. *acinaciformis* in Mediterranean ecosystems: the advantage of performing in different light conditions. *Applied Vegetation Science* 11, 45–54.

Watkinson, A. (1992) Plant senescence. *Trends in Ecology and Evolution* 7, 417–420.

Weinig, C. (2000) Limits to adaptive plasticity: temperature and photoperiod influence shade-avoidance responses. *American Journal of Botany* 87, 1660–1668.

Wilen, C.A. and Holt, J.S. (1996) Physiological mechanisms for the rapid growth of *Pennisetum clandestinum* in Mediterranean climates. *Weed Research* 36, 213–225.

Willson, M.F. (1983) *Plant Reproductive Ecology.* John Wiley and Sons, New York.

Wilson, J.B. (1997) An evolutionary perspective on the 'death hormone' hypothesis in plants. *Physiologia Plantarum* 99, 511–516.

Wulff, R.D., Causin, H.F., Benitez, O. and Bacalini, P.A. (1999) Intraspecific variability and maternal effects in the response to nutrient addition in *Chenopodium album. Canadian Journal of Botany* 77, 1150–1158.

Zimdahl, R.L., Lin, J. and Dall'Armellina, A.A. (1991) Effect of light, watering frequency, and chlorsulfuron on Canada thistle (*Cirsium arvense*). *Weed Science* 39, 590–594.

Zobolo, A.M. and van Staden, J. (1999) The effects of deflowering and defruiting on growth and senescence of *Bidens pilosa* L. *South African Journal of Botany* 65, 86–88.

8 Competition

8.1 Interactions in Populations and Communities

Invasive species live in contact and interact with individuals of their own species and other species. We have already discussed this interaction in previous chapters where we mentioned pollination and the influence of organisms on each other. Perhaps the most important implication of interactions is that the fate of an individual cannot be disentangled from its interactions with others. Interactions are complex as they can benefit, harm or have no effect, and will vary among the many individuals found in populations and communities. The structure and dynamics of populations and communities are based on environmental variation and individual characteristics; however, the sheer number of possible interactions and outcomes complicates our understanding of populations and communities. In reality, many individuals are interacting at once, but it is easier for us to visualize and discuss interactions between a pair of individuals (or species) (see Table 8.1). In this chapter, we focus on competition, an interaction in which at least one individual is negatively affected.

8.2 Competition Basics

Competition is a common, mutually harmful interaction that occurs when two or more individuals vie for the same limited resources. These limited resources include light, water, nutrients and space. Competition can cause changes to the physiology and morphology of individuals which, in turn, lead to changes in the growth, reproduction and fitness in individuals. Ultimately, competition is a key process that influences population and community dynamics. There is no evidence, however, that competition has caused global extinctions (Gurevitch and Padilla, 2004; Sax *et al.*, 2007). Therefore while competition from invasive species may lead to the extirpation of a native species, it is unlikely to cause a species extinction.

While competition is said to occur when resources are limited, individuals may interact before this actually occurs. For example, plants may detect neighbours through changes in light quality, specifically the red to far-red ratio (R:FR), caused when light is reflected from the leaf surfaces of neighbouring plants (Smith *et al.*, 1990; Ballare and Casal, 2000). Physiological and morphological changes,

Table 8.1. Summary of interactions that might occur between two species.

| Interaction | Species | | Explanation |
	A	B	
Neutralism	0	0	Neither species affected
Competition	0/–	–	Both species inhibited, or one species affected, the other not
Allelopathy	0	–	Species A releases a chemical than inhibits species B
Herbivory	+	–	Species A (animal) consumes part of species B (plant)
Mutualism	+	+	Both species benefit
Commensalism	+	0	Species A benefits while species B is not affected
Parasitism	+	–	Species A (parasite) exploits species B (host) by living on or in it

such as increases in plant height, leaf area, stem elongation, delayed phenological development and a reduction in root biomass, occur in response to changes in R:FR (Rajcan and Swanton, 2001; Rajcan *et al.*, 2004; Liu *et al.*, 2009; Page *et al.*, 2009). These physiological and morphological responses to changes in the R:FR ratio are a first response in determining the outcome of competition. The subtle issue is that these changes occur before the onset of competition for resources of light, water, nutrients and space. Also, these changes in growth come at a physiological cost, a cost that reduces a plant's ability to respond to the eventuality of competition caused by limited resources.

Ideally, it would be better for all organisms to avoid competition because it harms all involved. Being a better competitor just means that you suffer less and have a greater relative fitness. If you could avoid allocating resources to characteristics that only help you to compete and, for example, allocate resources instead to producing more seeds, then your absolute fitness would increase. Because most plants require the same types of resources, avoiding competition altogether is rarely possible. Thus, natural selection often favours traits that allow organisms to outcompete others in order to insure their relative fitness.

The competitive ability of a plant can be measured in two ways (Aarssen, 1989; Goldberg, 1990; Goldberg and Landa, 1991; Navas and Violle, 2009). The competitive effect is the ability of an individual to obtain resources and is characterized by traits that allow the plant to capture resources quickly and efficiently. The competitive response is the ability of an individual to survive the depletion of resources and is characterized by traits that allow the plant to tolerate low levels of resources or acquire resources in a different way from competitors.

Competition involving plants has been analysed by many authors (Grace and Tilman, 1990;

Connolly *et al.*, 2001; Berger *et al.*, 2008; Freckleton *et al.*, 2009). We focus on the following questions related to competition as these are especially relevant to the ecology of invasive species:

- What is the importance of competition?
- What evidence do we have that competition occurs?
- What do plants compete for?
- What factors determine the outcome of competition?

It is generally agreed that in resource-rich environments, plants are less likely to compete for nutrients because nutrients will not be limited. However, plants will still compete for light because more individuals will survive and develop stems and leaves that reduce the available light (Goldberg, 1990) or change the quality of light. The discussion on whether plants compete in resource-poor environments is more polarized. Grime (1979) maintained that competition was not important in resource-poor environments because resources cannot be depleted further. He argued that species that are adapted to tolerate stress will dominate in these environments, whereas competitive species will be favoured in resource-rich environments. Tilman (1988, 1990) argued that competition does occur in resource-poor environments. While some plants do adapt to tolerate these stressful environments, some individuals may deplete resources more than others by having a high efficiency of nutrient uptake.

Grace (1990) suggested that these two theories were compatible and reflected different aspects of competition. Grime focused on the longer-term competitive effect on a community, while Tilman focused on the shorter-term competitive response of individuals. Furthermore, if the ability to tolerate being denied resources is considered a competitive trait (Aarssen 1989, 1992), then stress tolerance is actually a competitive trait. Indeed, most plants do compete in resource-poor environments; recall that one of the conditions for competition is the existence of limited resources. Like so many concepts in ecology, the issue is generally one of semantics; in this case it is about what we define as resource rich and resource poor. There is no good way to quantify the division between these two. Hence, most studies find evidence for both theories.

Establishing the relative importance of competition in real world habitats requires evidence that competition affects population size or biomass, or that it affects community composition (Goldberg *et al.*, 1995; Goldberg and Barton, 1992). To do this, a series of events must occur (Fig. 8.1). Even if we can establish that there is an effect on a population or community, the effect may be ephemeral and the population or community may return to its original state. Complicating matters further is the fact that even if observed patterns suggest competition, there can be other explanations, such as other species' interactions or environmental changes. Patterns also may be the result of unknown past events or interactions. Attempting to determine the exact cause (i.e. was it caused by competition) is like chasing the ghost of these past events (Connell, 1980).

8.3   What Evidence Do We Have That Competition Occurs?

When you observe a habitat, how can you tell whether individual plants are competing? Competition may be subtle and difficult to observe, but it is easier to observe competition between individual plants than to observe the effects of competition on populations or communities. Thus, ecologists often use controlled experiments to test whether inferences made in the field about competition are real. The easiest way to detect competition experimentally is to grow plants together (in competition) and apart (without competition), and compare their growth, survival and fitness. There are many variations on this basic experimental design.

One example of how competition was assessed is from Wilson and Tilman (1995). They planted three species of native grasses into prairie soil and measured their growth in three competition treatments:

- competition for light and nutrients by allowing roots and shoots to interact
- competition for nutrients by excluding the shoots from interactions
- no competition – by not allowing either roots or shoots to interact.

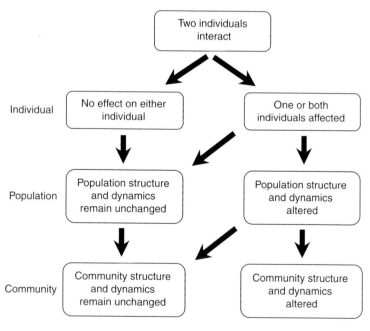

Fig. 8.1. Series of events that must occur before competition between two individuals will influence populations or communities.

They found that excluding the shoots of neighbours from direct competition had no effect on plant growth; i.e. they prevented any possible competition for light but competition for nutrients still was possible. Excluding both shoots and roots from competition caused an increase in growth of all three species; i.e. when the possibility of competition for nutrients was also prevented, the plants benefited. Therefore, the plants were competing for nutrients but not light.

While controlled experiments are useful, they are limited in their broad-scale application because they ignore all other species normally found in a habitat. For example, if we wished to look at the competitive effect of invasive species on a forest ecosystem, we would have to decide:

- what to measure (e.g. survivorship or growth)
- the number and types of species to measure it on (e.g. one invasive species and one native species, or many species)
- what timespan to cover (e.g. 1 year or 10 years).

This approach was used to study the suppression of garlic mustard (*A. petiolata*) by varying densities of selected native species (Murphy, 2005; Murphy *et al.*, 2007). The experimental design involved a range of eight densities of two native species. One species (bloodroot, *Sanguinaria canadensis*) had large leaves and emerged concurrently with garlic mustard; the other (May apple, *Podophyllum peltatum*) was even larger in terms of leaf size and height, but emerged later. This design did not allow for tests of relative importance of mechanisms of competition (e.g. for light or water) although a vast literature on garlic mustard indicates that it is most competitive for light. It did, however, allow tests of the outcome of competition by measuring 14 variables (ranging from leaf area to height to root structure) and, importantly, by finding out which densities of native species would significantly reduce populations of garlic mustard over time. This was more of a management-style design because the objective was to determine how to reduce the impact of this invasive species.

Apparent competition

Even with well-designed experiments, an observer may conclude that competition is occurring between individuals even though some other reason may explain the outcome. Apparent competition is a type of indirect effect where a third species mediates the interaction between two other species (Holt, 1977; Connell, 1990; Sessions and Kelly, 2002; White *et al.*, 2006; Orrock *et al.*, 2008). For example, common sage (*Salvia officinalis*) attracts small herbivores because it provides shelter for them. These herbivores, in turn, consume other vegetation around the sage plant. The ultimate result is that other species do poorly when growing near sage. In an experiment, if you did not look for the herbivores, you might conclude erroneously that sage was outcompeting other species. Apparent competition has other types of mechanisms besides herbivory. For example, hybrid cattail (*Typha* × *glauca*) reduces native species diversity because the cattail litter reduces the establishment and growth of native species (Farrer and Goldberg, 2009).

8.4 What Do Plants Compete For?

The relationship between the amount and type of resources available and competition is not simple because:

- Plants can compete for more than one resource at a time.
- Plants compete both above and below ground.
- The outcome of competition is modified by abiotic factors and other biotic interactions that influence the rate and efficiency at which the resources are consumed.
- The ability of plants to compete for nutrients, water, light and space will be influenced primarily by the time of seedling emergence relative to other species, seedling density and species traits.
- Some species are genetically variable and thus all individuals of a species will not compete in the same way.

Nutrients

Soil fertility is an important factor influencing the success of a biological invasion. Plants compete mainly for nitrogen, phosphorus and potassium. Nitrogen is usually the most limited nutrient in terrestrial habitats, whereas phosphorus is usually the most limited nutrient in aquatic ecosystems. Potassium is often overlooked, but some terrestrial invasive plants (e.g. the invasive dandelion *T. officinale*) might be managed better if potassium-poor fertilizers were used because some species are limited by this nutrient (Tilman *et al.*, 1999). In general, plants compete directly for nutrients; however, species that germinate and emerge before other plants are able to access the available nutrient resource pool first. The establishment of competitive hierarchies within a plant community is dependent upon time of seedling emergence relative to other species and on biomass accumulation (i.e. plant growth rates). Differences in plant biomass often dominate the outcome of competition for nutrients and water (Everard *et al.*, 2010).

Water

Competition for water depends on rooting volume, the ability of roots to extract water, and water-use efficiency (WUE). WUE is the ability to minimize water use for a given amount of nutrient assimilation. When water is abundant, plants with low WUE (and flood tolerance) are more competitive, whereas in arid conditions, plants with high WUE (and drought tolerance) have the advantage (Walch *et al.*, 1999; Schillinger and Young, 2000). Weedy salt cedars (*Tamarix* spp.) and their impact on the south-western USA are particularly illustrative, as they have invaded 600,000 ha, in part because of their superior WUE (Di Tomaso, 1998). Similarly, trees in Tasmanian forest plantations suffer because some species have a high WUE and can outcompete others for water (Hunt and Beadle, 1998). Plants in more temperate climates are less likely to compete for water because rain and snowmelt are usually abundant.

Light

The availability of light will have a strong influence on plant growth rate and biomass accumulation (Gurevitch *et al.*, 2008). Both light quantity (photosynthetic photon flux density or PPFD) and light quality (red to far-red light ratio, R:FR) are important aspects of competition (Novoplansky, 1991; Mahoney and Swanton, 2008). Many studies that explore competition for light do not separate the effects of these two interacting variables. For example, in common lambsquarters (*Chenopodium album*) low R:FR effects were detected early in the life cycle but by seed set, growth was influenced only by low PPFD. Low PPFD reduced seed number per plant but not seed weight (Mahoney and Swanton, 2008). Other invasive species can compensate for poor light conditions in a number of ways. Many such species germinate early and

grow taller at a faster rate to acquire as much light as possible, e.g. self-heal (*Prunella vulgaris*) (Miller *et al.*, 1994). Because the presence of dense leaf canopies reduces the quantity and quality of light available to plants, competition for light is greatest when plant density is highest. Phenotypic plasticity (Chapter 7) of some invasive plants explains why:

- They tend to be taller when grown in high densities (Nagashima *et al.*, 1995).
- The position and orientation of leaves changes to intercept more light (Aphalo *et al.*, 1999).
- Stems may elongate so that leaves are positioned above competing vegetation (Aphalo *et al.*, 1999).

Physical space

It is easier to understand how plants compete for a resource such as nutrients, water or light because we can imagine these being consumed by the plant. It is harder to imagine how space is something that is competed for, and yet the lack of physical space creates consequences for individuals. Root restriction experiments are one way to test the effect of limiting space. Although not all species react equally, restricting a plant's rooting space generally decreases shoot biomass, height or growth rate, even when ample water and nutrients are supplied (Richards and Rowe, 1977; Gurevitch *et al.*, 1990; McConnaughay and Bazzaz, 1991; Matthes-Sears and Larson, 1999).

Exploiting different resources: competition above and below ground

Above- and below-ground competition for limited resources are considered separately because plants use different structures (e.g. roots versus leaves) to compete for different resources. Above ground, plants compete for space and light; below ground they compete for space, nutrients and water (Casper and Jackson, 1997; Schenk, 2006). Below-ground root competition is complex because the soil is heterogeneous, and water and nutrients vary spatially and temporally (Casper and Jackson, 1997; Schenk, 2006). Furthermore, there are both positive and negative interactions between roots, whereby roots of one species may increase the concentration of some resources and decrease the concentration of others (Schenk, 2006; Berger *et al.*, 2008).

From a physiological perspective, roots and shoots are so integrally related that it is practically impossible to separate them. This integration creates problems because plants must trade off allocating resources between tissues involved in above- and below-ground competition. For example, if the invasive rice cockspur (*E. oryzoides*) increases leaf area to compete for light, it reduces allocation of resources to roots and is vulnerable to competition for nutrients (Gibson and Fischer, 2001).

The relationship between above-ground and below-ground competition is not usually additive; that is, the total competitive effect is not simply the above-ground effect plus the below-ground effect, although this is often presumed (Wilson, 1988). Root and shoot competition may have opposing effects, or be subject to complex interactions, and this may not be evident when they are measured together. One species may outcompete another in below-ground interactions, but this benefit may be negated by above-ground competition (Wilson and Tilman, 1995).

Competition for pollen and pollinators

Competition for pollinators occurs when two species share the same pollinators, and it can reduce the reproductive success of an individual. This competition for pollinators can reduce both visit quantity and visit quality (Kandori *et al.*, 2009; Mitchell *et al.*, 2009). The number of visits a flower receives from a pollinator (visit quantity) will influence the amount and species of pollen that lands on the stigma; if the pollen is heterospecific then there may be pre-emption of stigma or style space. Pollen competition is an important process in species invasions if the invasive species and the native species share pollinators. For example, the invasive dandelion has been replacing native Japanese dandelions (*Taraxacum japonicum*) in urban areas and is spreading into the countryside. Kandori *et al.* (2009) found that pollen competition is occurring because the native dandelions have fewer pollinator visits in the presence of the invasive dandelion.

Allelopathy as a competitive interaction

Allelopathy is the effect of one individual on others through the release of chemical compounds from roots, shoots, leaves or flowers (Rice, 1995). It can influence the competitive interactions among species. It is unclear how often allelopathy occurs because many chemicals released by plants only become toxic after being transformed by other species. Such indirect effects may be considered to be allelopathy, but whether the plant that exuded the original chemical benefits at all is unclear (Connell, 1990; Williamson, 1990). Some researchers consider allelopathy to be a type of competition, but because a resource is often pre-empted (not exploited), allelopathy is also termed as a unique interaction by some (Murphy, 1999; Schenk, 2006).

Allelopathy may be common to invasive species but there are few clear examples (Bais *et al.*, 2003; Kaur *et al.*, 2009; Ren and Zhang, 2009). An example is invasive narrow-leaved cattail (*Typha angustifolia*), which reduced the root and shoot biomass of the native river bulrush (*Bolboschoenus fluviatilis*) (Jarchow and Cook, 2009). Allelopathy in non-native species is sometimes an example of a 'novel weapon' because native plants may have had no experience with the chemical.

Allelopathic chemicals can sometimes exist in pollen and this may facilitate invasions, or at least cause otherwise uncompetitive species to persist for long periods of time and stall succession (Murphy and Aarssen, 1995a,b; Murphy *et al.*, 2009a,b). The longevity of such allelopathic effects is not clear as selection and genetic drift may reduce impacts quickly (He *et al.*, 2009; Lankau *et al.*, 2009; Thorpe *et al.*, 2009).

8.5 Factors Determining the Outcome of Competition

One of the reasons that some invasive species are so successful is that they adapt rapidly to new environmental conditions. Plants do not 'know' how competitive other individuals are. If a plant is at a competitive disadvantage but still produces offspring, there should be selection for the offspring to develop better competitive abilities (as long as the genes are available). What complicates the situation is that plants are subject to selection from other types of interactions.

Competitive traits

A species may have traits that will suppress neighbours or traits that allow the species to avoid being suppressed by neighbours. It is rare to find species that do both well. For example, Goldberg (1990) found that the presence of common ragweed (*A. artemisiifolia*) decreased available sunlight to other plants but did not affect soil moisture.

Its competitor, narrow-leaved plantain (*Plantago lanceolata*), responds only to decreases in moisture. It does not respond to decreases in light. Therefore, ragweed is not competitively superior to plantain (Fig. 8.2).

There are general traits associated with competition. Zimdahl (1999) listed traits associated with highly competitive agricultural weeds (Table 8.2). Possessing these traits, however, does not guarantee competitive success. For example, Sydney golden

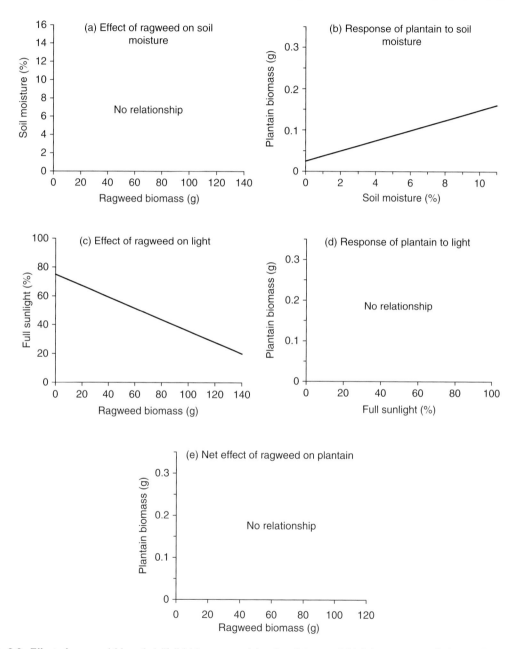

Fig. 8.2. Effect of ragweed (*A. artimisiifolia*) biomass on (a) soil moisture and (b) light; response of plantain (*Plantago lanceolata*) to (c) soil moisture and (d) light; and (e) net effect of ragweed on plantain (redrawn and adapted from Goldberg, 1990).

Table 8.2. List of above- and below-ground characteristics associated with competitive agricultural weeds (adapted from Zimdahl, 1999).

Shoot characteristics	C_4 photosynthetic pathway and low leaf light transmissivity
	Climbing habit
	High allocation of dry matter to build a tall stem
	Horizontal leaves under overcast conditions and obliquely slanted leaves (plagiotropic) under sunny conditions
	Large leaves
	Leaves forming a mosaic leaf arrangement for best light interception
	Rapid expansion of tall, foliar canopy
	Rapid extension in response to shading
Root characteristics	Early and fast root penetration of a large soil area
	High proportion of actively growing roots
	High root density/soil volume
	High root length per root weight
	High root–shoot ratio
	High uptake potential for nutrients and water
	Long and abundant root hairs

wattle (*Acacia longifolia*) has a high photosynthetic rate, but is still outcompeted by the invasive weedy tick berry (*Chrysanthemoides monilifera*). Tick berry has a more efficient leaf arrangement to intercept light and therefore is able to outcompete the native species (Weiss and Noble, 1984).

Size

Size is really not a trait itself; it is more of a general description. We say this because size could mean a species has adapted (or is phenotypically plastic enough) to grow taller, branch out more or produce more roots to capture resources (Goldberg and Werner, 1983; Schoener, 1983; Goldberg, 1987). Hence, size results from interacting traits, such as rates of cell division, leaf expansion and seed germination, and time of seedling emergence.

However, size is not always a determinant of competitive success (Wilson, 1988; Gerry and Wilson, 1995). To test for size advantage, Grace *et al.* (1992) grew six grasses alone and in pairs. During the first 2 years the initial plant size was correlated with competitive success measured as relative yield (a comparison of yield when grown alone and when grown in competition). In the third year, however, the initial size did not confer an advantage. Individuals with higher relative growth rates were at an advantage, rather than those that were initially bigger. Weigelt *et al.* (2002) suggested that size was more important during the seedling stage, whereas species-specific traits, such as biomass allocation patterns, were more important during the adult stages of a plant's life cycle. Size was less likely to be advantageous in situations of low nutrients and high light, where it does not improve an individual's chance of obtaining resources.

The actual impact of size on the outcome of competition can be difficult to quantify. You might expect that an individual that is initially twice the size of a competitor will be twice as competitive. Indeed, this may happen if competition is size-symmetric. In other cases, competition is size-asymmetric, meaning that an individual that is (initially) twice the size of a competitor may be, for example, four times as competitive and therefore has a disproportionate effect on its neighbours. For this reason, many consider competition to be a size-specific interaction rather than a species-specific interaction (Berger *et al.*, 2008).

Below-ground competition is more likely to be size-symmetric whereas above-ground competition is more likely to be size-asymmetric (Casper and Jackson, 1997; Schwinning and Weiner, 1998). This is because a plant that successfully outcompetes others for early-season light often has accelerated growth, leading to faster suppression of competitors, capture of increasingly available light as day length increases, and thus further suppression of competitors. This type of feedback is what leads to size-asymmetric competition. Below ground, individuals that have more roots (or more efficient roots) will capture more resources, but the process is much slower as the water and nutrients are less ubiquitous than light and are harder to find. A lack of accelerated capture of resources means that the competitive advantage of an individual with a large root system is restricted to being closely equivalent to its size advantage.

One of the implications of size-asymmetry is that larger individuals are often those that germinated or emerged first and captured more resources. In agriculture and forestry, early-emerging weeds are the ones that cause crop losses because they compete with the young and vulnerable crops for nutrients and light, depending on the planting conditions used (Forcella, 1993; Van Acker *et al.*, 1993; Knezevic *et al.*, 1994; Chikoye *et al.*, 1995; Weinig, 2000). The same principle applies to competition in non-crop ecosystems (Gerry and Wilson, 1995; Tremmel and Bazzaz, 1995). For example, garlic mustard is likely to be competitive because it germinates in the autumn, overwinters (and perhaps photosynthesizes), and quickly grows tall as soon as the temperature, moisture and light conditions allow, usually before native spring plants. In this manner, garlic mustard captures early-season light, nutrients, moisture and space at the expense of individuals of native species (Murphy, 2005).

Effect of the environment on competition

Selection pressures change such that a trait may be advantageous only in some locations at a given time. We have already emphasized that the existence of spatial and temporal variability in the environment is the reality that plants must survive in. The more unpredictable the environmental variability, the more risk to existence. In terms of competitive traits, a genotype may survive for years with a certain suite of traits, but if the environment changes then that genotype may be placed at a competitive disadvantage. This is actually a principle of any invasive species management: how to outcompete them. The problem, again, is that invasive species tend to adapt more quickly to change and produce a wide variety of genotypes that can be fit in a range of environments.

How adaptable are invasive species to changing environments? Generally, plants can only adapt if they have the genes available. For most invasive species, this is rarely a problem as they reproduce sexually and recombine genes constantly (Chapter 4). When the environment changes, some genotypes will die or at least be disadvantaged, while other genotypes will survive to ensure the population and species survives. In studies conducted in semi-arid grasslands,

Wilson (2007) suggested that less frequent recruitment opportunities rather than competitive intensity would alter community diversity if global warming occurred.

Time of seedling emergence and density

Species that emerge early relative to their neighbours tend to have a competitive advantage (Knezevic *et al.*, 1994; Chikoye *et al.*, 1995; Bosnic and Swanton, 1997). Time of seedling emergence relative to neighbouring plants has a larger influence on competitive outcomes than does density. One would assume that as seedling density increases so does competitive intensity. Density-dependent competition, however, is meaningful when time of seedling emergence relative to neighbouring plants is accounted for. This interaction of timing of seedling emergence and density then becomes a powerful tool to help explain the outcome of competitive interactions. For example, Bosnic and Swanton (1997) reported that barnyard grass (*E. crus-galli*) seedlings reduced maize grain yield when they became established during the first to third-leaf stages of maize. However, when seedlings emerged after the fourth-leaf stage, maximum yield losses were much lower. Similar results were observed by Knezevic *et al.* (1994) in respect of red-root pigweed (*A. retroflexus*) and maize (*Zea mays*) grain yield (Fig. 8.3). Despite the range in pigweed density, time of emergence relative to the crop had the largest influence on the competitive outcome.

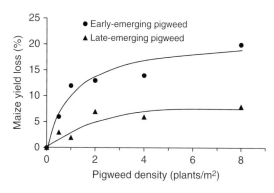

Fig. 8.3. Yield loss in maize (*Zea mays*) as a function of red-root pigweed (*Amaranthus retroflexus*) density and date of pigweed emergence (redrawn from Knezevic *et al.*, 1994).

8.6 Summary

Interactions can have positive, negative, or neutral effects on the individuals involved, and these effects may or may not influence population and community dynamics. When individuals compete, they exploit limited resources, such as nutrients, water, light and space, making them unavailable to competitors. There is no question that competition can play an important role in shaping populations and communities. Individuals can compete for more than one resource at a time and they can compete both above and below ground. The outcome of competition will depend on the interaction of plant traits and the abiotic environment. Allelopathy, a form of competition, will also influence the outcome. In the next chapter we examine types of interactions in which at least one individual benefits.

Box 8.1. Invasive species case study: competition.

- Would you consider the invasive species you selected to be an effective competitor?
- What traits does it have that confer competitive ability?
- Under what situations will your species be competitive and not be competitive?

8.7 Questions

1. What is competition, and why is it so difficult to define?
2. Is competition more important in a resource-rich or a resource-poor environment? Explain your answer.
3. What evidence of competition do you see in a habitat near you?
4. Is allelopathy likely to be more important in agricultural or natural systems? Explain why.
5. What species traits do you think confer the greatest advantage to invasive plants?
6. How does the environment determine competitive outcome?

Further Reading

Murphy, S.D., Sherr, I. and Bullock, C. (2009) Allelopathic pollen in Canadian invasive species: *Alliaria petiolata* and *Hesperis matronalis*. *Allelopathy Journal* 23, 63–70.

Navas, M.-L. and Violle, C. (2009) Plant traits related to competition: how do they shape the functional diversity of communities? *Community Ecology* 10, 131–137.

Schenk, H.J. (2006) Root competition: beyond resource depletion. *Journal of Ecology* 94, 725–739.

Sommer, U. and Worm B. (2002) *Competition and Coexistence* ((Ecological Studies 161). Springer-Verlag, Berlin, Heidelberg.

References

Aarssen, L.W. (1989) Competitive ability and species coexistence: a 'plant's-eye' view. *Oikos* 56, 386–401.

Aarssen, L.W. (1992) Causes and consequences of variation in competitive ability in plant communities. *Journal of Vegetation Science* 3, 165–174.

Aphalo, P.J., Ballaré, C.L. and Scopel, A.L. (1999) Plant–plant signaling, the shade-avoidance response and competition. *Journal of Experimental Botany* 50, 1629–1634.

Bais, H.P., Vepachedu, R., Gilroy, S., Callaway, R.M. and Vivanco, J.M. (2003) Allelopathy and exotic plant invasion: from molecules and genes to species interactions. *Science* 301, 1377–1380.

Ballare, C.L. and Casal, J.J. (2000) Light signals perceived by crop and weed plants. *Field Crops Research* 67, 149–160.

Berger, U., Piou, C., Schiffers, K. and Grimm, V. (2008) Competition among plants: concepts, individual-based modelling approaches, and a proposal for a future research strategy. *Perspectives in Plant Ecology, Evolution and Systematics* 8, 121–135.

Bosnic, A.C. and Swanton, C.J. (1997) Influence of barnyardgrass (*Echinochloa crus-galli*) time of emergence and density on corn (*Zea mays*). *Weed Science* 45, 276–282.

Casper, B. and Jackson, R.B. (1997) Plant competition underground. *Annual Review of Ecology and Systematics* 28, 545–570.

Chikoye, D., Weise, S.F. and Swanton, C.J. (1995) Influence of common ragweed (*Ambrosia artemisiifolia*) time of emergence and density on white bean (*Phaseolus vulgaris*). *Weed Science* 43, 375–380.

Connell, J.H. (1980) Diversity and the coevolution of competitors, or the ghost of competition past. *Oikos* 35,131–138.

Connell, J.H. (1990) Apparent versus "real" competition in plants. In: Grace, J.B. and Tilman, D. (eds) *Perspectives on Plant Competition*. Academic Press, San Diego, California, pp. 9–26.

Connolly, J., Wayne, P. and Bazzaz, F.A. (2001) Interspecific competition in plants: How well do current methods answer fundamental questions? *American Naturalist* 157, 107–125.

Di Tomaso, J.M. (1998) Impact, biology, and ecology of saltcedar (*Tamarix* spp.) in the southwestern United States. *Weed Technology* 12, 326–336.

Everard, K., Seabloom, E.W., Harpole, W.S. and de Mazancourt, C. (2010) Plant water use affects competition for nitrogen: why drought favors invasive species in California. *American Naturalist* 175, 88–97.

Farrer, E.C. and Goldberg, D.E. (2009) Litter drives ecosystem and plant community changes in cattail invasion. *Ecological Applications* 19, 398–412.

Forcella, F. (1993) Seedling emergence model for velvetleaf. *Agronomy Journal* 85, 929–933.

Freckleton, R.P., Watkinson, A.R. and Rees, M. (2009) Measuring the importance of competition in plant communities. *Journal of Ecology* 97, 379–384.

Gerry, A.K. and Wilson, S.D. (1995) The influence of initial size on the competitive responses of six plant species. *Ecology* 76, 272–279.

Gibson, K.D. and Fischer, A.J. (2001) Relative growth and photosynthetic response of water-seeded rice and *Echinochloa oryzoides* (Ard.) Fritsch to shade. *International Journal of Pest Management* 47, 305–309.

Goldberg, D.E. (1987) Neighbourhood competition in an old-field community. *Ecology* 68, 1211–1223.

Goldberg, D.E. (1990) Components of resource competition. In: Grace, J.B. and Tilman, D. (eds) *Perspectives on Plant Competition*. Academic Press, San Diego, California, pp. 27–49.

Goldberg, D.E. and Barton, A.M. (1992) Patterns and consequences of interspecific competition in natural communities: a review of field experiments with plants. *American Naturalist* 139, 771–801.

Goldberg, D.E. and Landa, K. (1991) Competitive effect and response: hierarchies and correlated traits in the early stages of competition. *Journal of Ecology* 79, 1013–1030.

Goldberg, D.E. and Werner, P.A. (1983) Equivalence of competitors in plant communities: a null hypothesis and a field experiment approach. *American Journal of Botany* 70,1098–1104.

Goldberg, D.E., Turkington, R. and Olsvig-Whittaker, L. (1995) Quantifying the community-level consequences of competition. *Folia Geobotanica* 30, 231–242.

Grace, J.B. (1990) On the relationship between plant traits and competitive ability. In: Grace, J.B. and Tilman, D. (eds) *Perspectives on Plant Competition*. Academic Press, San Diego, California, pp. 51–66.

Grace, J.B. and Tilman, D. (eds) (1990) *Perspectives on Plant Competition*. Academic Press, San Diego, California.

Grace, J.B., Keough, J. and Guntenspergen, G.R. (1992) Size bias in traditional analyses of substitutive competition experiments. *Oecologia* 90, 429–434.

Grime, J.P. (1979) *Plant Strategies and Vegetation Processes*. John Wiley, New York.

Gurevitch, J. and Padilla, D.K. (2004) Are invasive species a major cause of extinction? *Trends in Ecology and Evolution* 19, 470–474.

Gurevitch, J., Wilson, P., Stone, J.L., Teese, P. and Stoutenburgh, R.J. (1990) Competition among old-field perennials at different levels of soil fertility and available space. *Journal of Ecology* 78, 727–744.

Gurevitch, J., Howard, T.G., Ashton, I.W., Leger, E.A., Howe, K.M., Woo, E. and Lerdau, M. (2008) Effects of experimental manipulation of light and nutrients on establishment of seedlings of native and invasive woody species in Long Island, NY forests. *Biological Invasions* 10, 821–831.

He, W.-M., Feng, Y., Ridenour, W.M., Thelen, G.C., Pollock, J.L., Diaconu, A. and Callaway, R.M. (2009) Novel weapons and invasion: biogeographic differences in the competitive effects of *Centaurea maculosa* and its root exudate (±)-catechin. *Oecologia* 159, 803–815.

Holt, R.D. (1977) Predation, apparent competition, and the structure of prey communities. *Theoretical Population Ecology* 12, 197–229.

Hunt, M.A. and Beadle, C.L. (1998) Whole-tree transpiration and water-use partitioning between *Eucalyptus nitens* and *Acacia dealbata* weeds in a short-rotation plantation in northeastern Tasmania. *Tree Physiology* 18, 557–563.

Jarchow, M.E. and Cook, B.J. (2009) Allelopathy as a mechanism for the invasion of *Typha angustifolia*. *Plant Ecology* 204, 113–124.

Kandori, I., Hirao, T., Matsunaga, S. and Kurosaki, T. (2009) An invasive dandelion unilaterally reduces the reproduction of a native congener through competition for pollination. *Oecologia* 159, 559–569.

Kaur, H., Kaur, R., Kaur, S., Baldwin, I.T. and Inderjit. (2009) Taking ecological function seriously: soil microbial communities can obviate allelopathic effects of released metabolites. *PLoS ONE* 4, e4700. doi:10.1371/journal.pone.0004700.

Knezevic, S.Z., Weise, S.F. and Swanton, C.J. (1994) Interference of redroot pigweed (*Amaranthus retroflexus*) in corn (*Zea mays*). *Weed Science* 42, 568–573.

Lankau, R.A., Nuzzo, V., Spyreas, G. and Davis, A.S. (2009) Evolutionary limits ameliorate the negative impact of an invasive plant. *Proceedings of the National Academy of Sciences of the USA* 106, 15362–15367.

Liu, J.G., Mahoney, K.J., Sikkema, P.H. and Swanton, C.J. (2009) The importance of light quality in crop–weed competition. *Weed Research* 49, 217–224.

Mahoney, K.J. and Swanton, C.J. (2008) Exploring *Cheonopodium album* adaptive traits in response to light and temperature stresses. *Weed Research* 48, 552–560.

Matthes-Sears, U. and Larson, D.W. (1999) Limitations to seedling growth and survival by the quantity and quality of rooting space: implications for the establishment of *Thuja occidentalis* on cliff faces. *International Journal of Plant Sciences* 160, 122–128.

McConnaughay, K.D.M. and Bazzaz, F.A. (1991) Is physical space a soil resource? *Ecology* 72, 94–103.

Miller, T.E., Winn, A.A. and Schemske, D.W. (1994) The effects of density and spatial distribution on selection for emergence time in *Prunella vulgaris* (Lamiaceae). *American Journal of Botany* 81, 1–6.

Mitchell, R.J., Flanagan, R.J., Brown, B.J., Waser, N.M. and Karron, J.D. (2009) New frontiers in competition for pollination. *Annals of Botany* 103, 1403–1413.

Murphy, S.D. (1999) Pollen allelopathy. In: Inderjit, Dakshini, K.M.M. and Foy, C.L. (eds) *Principles and Practices in Plant Ecology: Allelochemical Interactions*. CRC Press, Boca Raton, Florida, pp. 129–148.

Murphy, S.D. (2005) Concurrent management of an exotic species and initial restoration efforts in forests. *Restoration Ecology* 13, 584–593.

Murphy, S.D. and Aarssen, L.W. (1995a) Allelopathic pollen extract from *Phleum pratense* L. (Poaceae) reduces seed set in sympatric species. *International Journal of Plant Sciences* 156, 435–444.

Murphy, S.D. and Aarssen, L.W. (1995b) Reduced seed set in *Elytrigia repens* caused by allelopathic pollen from *Phleum pratense*. *Canadian Journal of Botany* 73, 1417–1422.

Murphy S.D., Duncan, B., Wilson, D., Noll, K. and Flanagan, J. (2007) Implications for delaying invasive species management in ecological restoration. *Ecological Restoration* 25, 85–93.

Murphy, S.D., Sherr, I. and Bullock, C. (2009a) Allelopathic pollen in Canadian invasive species: *Alliaria petiolata* and *Hesperis matronalis*. *Allelopathy Journal* 23, 63–70.

Murphy, S.D., Flegel, S., Smedes, J., Finney, N., Zhang, B., Walton, K. and Henstra, S. (2009b) Identification of pollen allelochemical in *Hieracium × dutillyanum* Lepage and its ecological impacts on *Conyza canadensis* (L.) Cron. and *Sonchus arvensis* L. dominated community in southern Ontario, Canada. *Allelopathy Journal* 23, 85–94.

Nagashima, H., Terashima, I. and Ketoh, S. (1995) Effects of plant density on frequency distributions of plant height in *Chenopodium album* stands: analysis based on continuous monitoring of height-growth of individual plants. *Annals of Botany* 75, 173–180.

Navas, M.-L. and Violle, C. (2009) Plant traits related to competition: how do they shape the functional diversity of communities? *Community Ecology* 10, 131–137.

Novoplansky, A. (1991) Developmental responses of portulaca seedlings to conflicting spectral signals. *Oecologia* 88, 138–140.

Orrock, J.L., Witter, M.S. and Reichman, O.J. (2008) Apparent competition with an exotic plant reduces native plant establishment. *Ecology* 89, 1168–1174.

Page, E.R., Tollenaar, M., Lee, E.A., Lukens, L. and Swanton, C.J. (2009) Does the shade avoidance response contribute to the critical period for weed control in maize (*Zea mays*)? *Weed Research* 49, 563–571.

Rajcan, I. and Swanton, C.J. (2001) Understanding maize–weed competition: resource competition, light quality and the whole plant. *Field Crops Research* 71, 139–150.

Rajcan, I., Chandler, K. and Swanton, C.J. (2004) Red–far-red ratio of reflected light: a hypothesis of why early-season weed control is important in corn. *Weed Science* 52, 774–778.

Ren, M.X. and Zhang, Q.G. (2009). The relative generality of plant invasion mechanisms and predicting future invasive plants. *Weed Research* 49, 449–460.

Rice, E.L. (1995) *Biological Control of Weeds and Plant Diseases: Advances in Applied Allelopathy*. University of Oklahoma Press, Norman, Oklahoma.

Richards, D. and Rowe, R.N. (1977) Effects of root restriction, root pruning, and 6-benzyl-aminopurine on the growth of peach seedlings. *Annals of Botany* 41, 729–740.

Sax, D.F., Stachowicz, J.J., Brown, J.H., Bruno, J.F., Dawson, M.N., Gaines, S.D., Grosberg, R.K., Hastings, A., Holt, R.D., Mayfield, M.M., O'Connor, M.I. and Rice, W.R. (2007) Ecological evolutionary insights from species invasions. *Trends in Ecology and Evolution* 22, 465–471.

Schenk, H.J. (2006) Root competition: beyond resource depletion. *Journal of Ecology* 94, 725–739.

Schillinger, W.F. and Young, F.L. (2000) Soil water use and growth of Russian thistle after wheat harvest. *Agronomy Journal* 92, 167–172.

Schoener, T.W. (1983) Field experiments on interspecific competition. *American Naturalist* 122, 240–285.

Schwinning, S. and Weiner, J. (1998) Mechanisms determining the degree of size asymmetry in competition among plants. *Oecologia* 113, 447–455.

Sessions, L. and Kelly, D. (2002) Predator-mediated apparent competition between an introduced grass, *Agrostis capillaris*, and a native fern, *Botrychium australe* (Ophioglossaceae), in New Zealand. *Oikos* 96, 102–109.

Smith, H., Casal, J.J. and Jackson, G.M. (1990) Reflection signals and the perception by phytochrome of the proximity of neighbouring vegetation. *Plant, Cell and Environment* 13, 73–78.

Thorpe, A.S., Thelen, G.C., Diaconu, A. and Callaway, R.M. (2009) Root exudate is allelopathic in invaded community but not in native community: field evidence for the novel weapons hypothesis. *Journal of Ecology* 97, 641–645.

Tilman, D. (1988) *Plant Strategies and the Dynamics and Structure of Plant Communities*. Princeton University Press, Princeton, New Jersey.

Tilman, D. (1990) Mechanisms of plant competition for nutrients: the elements of a predictive theory of competition. In: Grace, J.B. and Tilman, D. (eds) *Perspectives on Plant Competition*. Academic Press, San Diego, California, pp. 117–141.

Tilman, E.A., Tilman, D., Crawley, M.J. and Johnston, A.E. (1999) Biological weed control via nutrient competition: potassium limitation of dandelions. *Ecological Applications* 9, 103–111.

Tremmel, D.C. and Bazzaz, F.A. (1995) Plant architecture and allocation for different neighbourhoods: implications for competitive success. *Ecology* 76, 262–271.

Van Acker, R.C., Swanton, C.J. and Weise, S.F. (1993) The critical period of weed control in soybean (*Glycine max* (L.) Merr.). *Weed Science* 41, 194–200.

Walch, J.L., Baskin, J.M. and Baskin, C.C. (1999) Relative competitive abilities and growth characteristics of a narrowly endemic and geographically widespread *Solidago* species (Asteraceae). *American Journal of Botany* 86, 820–828.

Weigelt, A., Steinlein, T. and Beyschlag, W. (2002) Does plant competition intensity rather depend on biomass or on species identity? *Basic and Applied Ecology* 3, 85–94.

Weinig, C. (2000) Differing selection in alternative competitive environments: shade-avoidance strategies and germination timing. *Evolution* 54, 124–136.

Weiss, P.W. and Noble I.R. (1984) Interactions between seedlings of *Chrysanthemoides monilifera* and *Acacia longifolia. Australian Journal of Ecology* 9, 107–115.

White, E.M., Wilson, J.C. and Clarke, A.R. (2006) Biotic indirect effects: a neglected concept in invasion biology. *Diversity and Distribution* 12, 443–455.

Williamson, G.B. (1990) Allelopathy, Koch's postulates, and the neck riddle. In: Grace, J.B. and Tilman, D. (eds) *Perspectives on Plant Competition*. Academic Press, San Diego, California, pp. 143–162.

Wilson, J.B. (1988) Shoot competition and root competition. *Journal of Applied Ecology* 25, 279–296.

Wilson, S.D. (2007) Competition, resources, and vegetation during 10 years in native grassland. *Ecology* 88, 2951–2958.

Wilson, S.D. and Tilman, D. (1995) Competitive responses of eight old-field plant species in four environments. *Ecology* 76, 1169–1180.

Zimdahl, R.L. (1999) *Fundamentals of Weed Science*, 2nd edn. Academic Press, San Diego, California.

9 Herbivory, Parasitism and Mutualism

Concepts

- Herbivory is the consumption of plant tissue (e.g. flowers, roots and seeds) by vertebrate and invertebrate animals.
- Plants have developed a variety of ways to avoid, tolerate or defend against herbivory.
- Herbivory influences population and community dynamics. This knowledge forms the basis for biological control.
- Parasitic plants obtain nutrients, shelter and support from other plants.
- Mutualistic interactions provide benefits to both individuals. Facultative mutualists are more likely to be invasive than obligate mutualists.
- It is often difficult to predict the invasiveness of a species because population processes interact.

9.1 Introduction

There are many types of interactions where one individual gains benefits from the interaction. In herbivory, for example, vertebrate and invertebrate animals benefit from consuming plants. As you will see, the plants may be harmed by or benefit from this interaction. In parasitism, one organism benefits by gaining nutrition, support or shelter from another organism, which may or may not be harmed. Finally, mutualisms are interactions in which both individuals gain benefits. In this chapter we look at herbivory, parasitism and mutualisms as separate interactions, and then consider the net effect of all types of positive and negative interactions on populations.

9.2 Herbivory

Herbivory is the consumption of plant tissue by animals. Plants, in general, are a low-quality food because they are low in nitrogen (needed by animals for protein) and high in complex carbohydrates that are hard for most animals to digest. Herbivores preferentially consume roots, young leaves, flowers, fruits and seeds because these are higher in nutrients and more digestible.

In terrestrial plant communities, about 20% of above-ground plant biomass is consumed by herbivores, but this varies considerably with community type and the number and types of herbivores present (Cyr and Pace, 1993). Herbivores include mammals, such as grazing deer and zebra, sap-sucking insects, seedling-eating molluscs, root-feeding larvae, leaf-eating grasshoppers and seed-eating mice and beetles. Generally, they range from specialists (typically invertebrates) that eat only one or a few types of food to generalists (typically vertebrates) that are unselective feeders (Crawley, 1989). Some herbivores can be both specialist and generalist because they change their feeding behaviour depending on the relative density of the plants.

Herbivory is usually non-lethal unless all of the plant or an essential tissue (e.g. all the roots) is eaten, or if the plant is already stressed. If a plant is not able to compensate for the loss of tissue, such as when a tree is girdled by a deer, it will die. The more immediate effects of herbivory are loss of photosynthetic tissue, stored nutrients and meristems (sites of active growth) (Noy Meir, 1993). These impacts may decrease a plant's ability to photosynthesize, absorb nutrients and water, and grow. There are,

Table 9.1. Mechanisms by which plants respond to herbivory (adapted from Crawley, 1992).

- Increased light intensity for surviving leaf area
- Increase in the rate of carbon fixation at a given light intensity
- Improved water and nutrient availability to the surviving leaf tissue
- Delayed senescence (plus rejuvenation) of leaves
- Increased duration of the growing period
- Redistribution of the photosynthate to the production of new leaves and away from roots, flowers, fruits or storage
- Reduced rate of flower abortion
- Production of new shoots from dormant buds or newly produced epicormic buds
- Ungrazable reserve (e.g. storage in roots and woody stems)
- No regrowth while the herbivore is still around

however, mechanisms that allow plants to compensate for herbivory. Plants respond by mobilizing stored nutrients, increasing their rates of photosynthesis, reactivating dormant meristems, or growing more roots or shoots (depending on what was eaten) (Crawley, 1992) (Table 9.1). The net effect of long-term herbivory will depend on the age of the foliage, the distribution of damage on the plant, and the stage of development and seasonal timing (Crawley, 1992). For example, the loss of young leaves by herbivory may have more long-term repercussions than the loss of older leaves. No plant species is completely immune to herbivory, but many have evolved ways to defend against it or compensate for it.

Plant defences and compensation for herbivory

Plants cannot run away from potential herbivores, but they do have ways to avoid being consumed. The two types of defences are direct (avoidance and tolerance) and indirect (defend themselves by recruiting bodyguards to protect them from herbivores). The types of defences used by plants will change over their lifetime as different herbivore pressures will be different at the seed, seedling, juvenile and adult stages (Boege and Marquis, 2005).

Avoidance

Avoidance defences reduce the chance that a plant will be eaten. Plants have three strategies to avoid herbivory: structural, phenological and chemical (Table 9.2). Structural and phenological defences are 'constituent' traits that are present in the plant whether or not a herbivore is present. Constituent defences deter a herbivore from damaging a plant. Chemical defences, however, may

not exist until they are induced by herbivory. These chemicals can inhibit digestion, deter feeding or intoxicate the herbivore, thereby reducing the damage that it causes. However, some herbivores adapt and can detoxify chemical defences. For example, saliva from the corn earworm caterpillar (*Helicoverpa zea*) suppresses the induction of the toxin nicotine in tobacco (*Nicotiana tabacum*) (Musser *et al.*, 2002). Herbivores may even use the plant chemicals to their own benefit. The classic association is that between the monarch butterfly (*Danaus plexippus*) and common milkweed (*A. syriaca*). Milkweed produces a glycoside that is toxic to the heart and circulatory system in most herbivores. Monarch larvae, however, are able to consume milkweed leaves, and, in turn, the glycoside makes them more unpalatable to their predators.

Tolerance

Tolerance is the ability to minimize damage from herbivory and maintain reproductive fitness (Belsky *et al.*, 1993; Strauss and Agrawal, 1999; Ashton and Lerdau, 2008). The degree to which a plant tolerates herbivory is called compensation. There are several mechanisms used by plants to compensate for and increase their tolerance to herbivory or other types of damage (Table 9.2) (Belsky *et al.*, 1993). Regardless of the mechanism, however, the success of any compensation is related to the cost the individual incurs because it has to allocate resources away from growth or reproduction. Ultimately, this reallocation of resources can reduce the fitness of an individual; for example, Jimson weed (*D. stramonium*) experienced a 15–25% reduction in fitness after defoliation (Foroni and Nunez-Farfan, 2000).

Box 9.1. How to study herbivory.

Herbivory can be tested by using real herbivores or by simulating herbivory by manually removing plant tissue, depending on the types of plants and herbivores examined. Tiffin and Inouye (2000) have reviewed the advantages and disadvantages of using natural versus imposed herbivory. For example, the effect of elephant raids on crop yield cannot be simulated easily in a greenhouse experiment.

Clear Hill and Silvertown (1997) used herbivore addition and removal experiments to examine the interaction of slugs (dominant species grey field slug, *Deroceras reticulatum*) and sheep (*Ovis aries*) on the seedling establishment of several grassland species. Sheep herbivory was managed in two ways. First, some pastures were winter-grazed while others were not. Second, summer grazing was done to maintain a grass height of 3 cm or 9 cm. Thus, there are four sheep grazing treatments (Fig. 9.1). Within each sheep grazing treatment, slug density was manipulated by placing a metal ring that slugs could not cross. Then slugs were trapped in the plots to manipulate plot density so they were either present or absent. Seeds of mouse-eared chickweed (*Cerastium fontanum*) were planted in each plot and their emergence counted.

The authors found that more seedlings emerged in sites that had been intensively grazed by sheep in the summer. This is likely to have occurred because intensive summer grazing created microsites suitable for seed germination. The presence of slugs reduced seedling emergence in all treatments, except that with intense summer grazing but no winter grazing. In this treatment, slugs may have switched from eating chickweed seedlings to other food, such as litter and new growth of established vegetation, which were more available because they had not been removed by winter-grazing sheep.

Clear Hill and Silvertown (1997) were able to do this study because they used organisms whose density was easy to manipulate. Slugs can be trapped and moved, and sheep can be enclosed or excluded using fencing. Experiments like this are much more complicated with more mobile organisms such as birds or mice; while these are easily excluded from treatments, it is harder to envision a way to increase their density but still maintain a natural habitat for them.

As a corollary to this study, we would like to remind you that your actions can influence experimental results. Touching a plant while making observations can induce a change in its growth, survival, resource allocation or many other variables. Cahill *et al.* (2000), for example, found that making weekly visits to a plant and touching it (to simulate the act of taking measurements) increased leaf loss in hemp dogbane (*Apocynum cannabinum*), decreased leaf loss in sulfur cinquefoil (*P. recta*), but did not affect horsenettle (*S. carolinense*), Canada thistle (*C. arvense*) or Kentucky bluegrass (*P. pratensis*). This phenomenon, called the observer effect, applies to all types of experiments in which plants are repeatedly visited or measured, and should be considered when designing experiments.

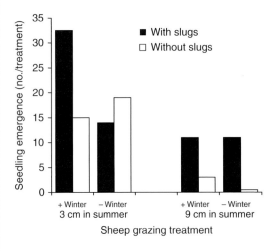

Fig. 9.1. Number of mouse-eared chickweed (*C. fontanum*) seedlings emerging from plots with various combinations of herbivory. Sheep grazing in the summer was managed so that pastures were maintained at either 3 cm or 9 cm heights. In the winter, pastures were either grazed (+Winter) or not grazed (–Winter). In addition, subplots were maintained either with or without slug herbivores (redrawn from Clear Hill and Silvertown, 1997).

Table 9.2. Types of herbivore avoidance and tolerance strategies used by plants to defend against herbivory.

Defence	Type	Example	Explanation
Avoidance	Structural	Hairs, spines, trichomes	Make it harder to consume tissue
		Digestibility reducers	Cellulose, lignin, cutins, tannins affect digestion by blocking digestive enzymes
		Protective coating	Shells of fruits make them difficult to eat
	Chemical	Chemical defences	Reduce the palatability of plants or make them poisonous
	Phenological	Rare or ephemeral	Makes plants hard to find when herbivores are active
		Early growth or reproduction	Plant escapes later-emerging herbivores
Tolerance	Compensation	Increase in net photosynthetic rate	Grazed plants produce more above-ground biomass (after being grazed) than do non-grazed individuals
		High relative growth rates	Plant can gow more quickly following herbivory
		Increased branching or tillering	After apical dominance is released
		High carbon storage	Pre-existing stores of carbons in roots can be reallocated for above-ground production
		Carbon reallocation	Can quickly reallocate more carbon from roots to shoots

Tolerance responses vary with the type of herbivory and the constraints on the plant, i.e. what is morphologically or physiologically possible. For example, cotton (*Gossypium hirsutum*) responds to phloem-sucking aphids by decreasing axillary branching; however, if buds are eaten, branching increases (Sandras, 1996). The weedy leafy spurge (*Euphorbia esula*) seems to be more constrained as it tolerates defoliation by continuously allocating carbon to its large root system, and not by increased allocation to its shoot system (Olson and Wallander, 1999). The internal mechanisms that allow a plant to respond to herbivory are moderated by the timing of herbivory, the availability of nutrients, light and water, and the presence of plant competitors (Strauss and Agrawal, 1999). While tolerance may be an alternative to avoidance, under certain types of herbivory both tolerance and avoidance may be used simultaneously. For example, if slugs start eating wild ginger (*Asarum caudatum*), its palatability decreases (avoidance) and its growth and seed production decrease (tolerance) (Cates, 1975).

A special type of tolerance is 'overcompensation', as found in invasive species such as purple loosestrife (*L. salicaria*) and yellow star-thistle (*C. solstitialis*) (Venecz and Aarssen, 1998; Callaway *et al.*, 2006). This occurs when herbivory actually benefits plants and increases fitness (Aarssen, 1995). The mechanism is relatively simple: if the apical (shoot) meristem is eaten, the lateral meristems (on the branches) are

chemically signalled to grow. As a result, the extra branches and all their leaves may allow the damaged plant to increase photosynthesis and carbohydrate production relative to undamaged plants. If this extra production increases reproductive success, then fitness increases (Venecz and Aarssen, 1998; Agrawal, 2000). Selective feeding of introduced Canada geese (*Branta canadensis*) on exotic grasses over native forbs actually facilitated the invasion of the grasses (Best and Arcese, 2009). The grasses responded to herbivory by forming low, dense mats, giving them a competitive advantage over the native forbs.

Indirect plant defences

Indirect plant defences mean that an individual uses another organism to defend itself against herbivory. For example, a plant may be protected when it grows near an unpalatable species. The unpalatable species can mask features, such as scent, that otherwise would attract herbivores. (Price *et al.*, 1980). Some plants decrease herbivore damage by recruiting body-guards. This means that an individual provides pollen, nectar, habitat or other rewards that increase the population's foraging effectiveness of species that harm herbivores (Price *et al.*, 1980; Sabelis *et al.*, 1999; Elliot *et al.*, 2000). In some cases, bodyguards are recruited only as an induced defence; for example, predatory mites are attracted by chemicals released when herbivorous spider mites damage leaves. It appears that few invasive species use body-guards, though they can have the general character-istic of induced defences (e.g. Jennings *et al.*, 2000).

Seed predation: a special type of herbivory

By convention, when herbivores eat seeds, it is called seed predation. While seed predation can be severe (>90%), the rates of seed loss are temporally and spatially variable because of interactions among plants, herbivores and the environment (Crawley, 1992; Mauchline *et al.*, 2005). Seed predation can occur while seeds are still on the plant (predispersal) or after they have abscised (postdispersal). Predispersal seed predators are usually invertebrates (mainly insects) with a narrow host range or specialized feeding habits. Their life cycles are usually short and are synchronized with the seed matura-tion of the host species (Lewis and Gripenberg, 2008). There are a wider variety of postdispersal seed predators, including vertebrates (birds or rodents), insects (ants and carabid beetles) and mol-luscs (snails and slugs) (Crawley, 1989; Blaney and Kotanen, 2001). These species are usually longer-lived and eat a variety of seed species (and other types of food) (Lewis and Gripenberg, 2008).

Some species have defences against seed preda-tors. For example, velvetleaf (*A. theophrasti*) seeds, which are small and have a hard seedcoat, can often survive intact after being ingested and defaecated. In this case, the seed predator actually serves to disperse seeds. Giant Parramatta grass (*Sporobolus indicus* var. *major*), a major weed of pastures and disturbed areas in Australia, can pass through the guts of cattle (Andrews, 1995). Therefore, to pre-vent invasion of the species into uninfested fields, cattle must be quarantined from the infected field for 7 days after grazing in an infested field.

Producing large seed crops at irregular intervals (masting) is a further method of defence against seed predators because it reduces the chances that all seeds will be destroyed by seed predators. Usually, one to a few years of heavy seed production are fol-lowed by a period of low seed production. During mast years, the number of seeds consumed by preda-tors is high but the probability of any one seed escap-ing predation is also high, therefore new seedlings are produced through the sheer quantity of seeds avail-able (Silvertown, 1980). This appears to be a more effective strategy against invertebrate than vertebrate predators because vertebrates are more likely to migrate towards areas of mast seeding. Vertebrates

also have other food sources to support them during non-mast years (Crawley, 1989). Invasive species that mast include Norway maple (*A. platanoides*) and Monterey pine (*P. radiata*).

It is difficult to isolate the effect of seed predation on population dynamics. While rates of predisper-sal seed predation are usually low, they can occa-sionally be high. Growing evidence suggests that this can affect seedling recruitment and population growth (Maron and Crone, 2006; Kolb *et al.*, 2007). Clark *et al.* (2007) concluded that the degree to which a species is seed-limited will determine the influence of seed predation.

Effect of herbivory on populations

Herbivory can influence the distribution and abun-dance of populations and should be considered by invasion biologists (Gonzales and Arcese, 2008). Herbivory of invasive exotic Monterey pine and Tasmanian blue gum (*Eucalyptus globulus*) trees by exotic vertebrate herbivores reduces seedling estab-lishment and impedes the spread of the trees (Becerra and Bustamante, 2008). Therefore, controlling the exotic herbivores may increase the invasion of these exotic trees. In eastern hemlock (*Tsuga canadensis*) forests, white-tailed deer (*Odocoileus virginianus*) accelerated the invasion of garlic mustard (*A. petio-lata*), Japanese stiltgrass (*Microstegium vimineum*) and Japanese barberry (*Berberis thunbergii*) (Eschtruth and Battles, 2008). The effect of herbivory will generally depend on the type of herbivore involved, the intensity and frequency of herbivory, the plant species, the type and age of tissue consumed and the abiotic environment.

The enemy release hypothesis

We have already discussed how herbivores may con-trol the abundance and distribution of a native spe-cies. In contrast, when a species is introduced to a habitat, there may be few herbivores that consume it and, consequently, the population distribution and abundance of the introduced species may increase. This is termed enemy release because the plant spe-cies is released from the pressure of herbivory (Keane and Crawley, 2002). The enemy release hypothesis states that invasive species have less pressure from herbivores because their specialist herbivores are absent in their new habitat. Numerous examples support this hypothesis. For example, Norway maple has higher levels of leaf herbivory in its native Europe

Box 9.2. Seed predation experiments.

A variety of types of seed predation experiments can be done. For example, a researcher could survey natural populations to assess levels of predispersal seed predation, or alternatively, the researcher could set up experiments to explicitly test a hypothesis. Fenner and Lee (2001) conducted a survey of predispersal seed predators in 13 species native to Britain but invasive in New Zealand. They collected flowering heads from 1000 individuals of each species at three locations in each country, dissected them, noted whether predatory larvae were present and calculated the percentage infestation rate of each species (Table 9.3). Their results showed that the infestation rate was higher in the native country (Britain) and that seed predators were almost absent in the invaded habitat (New Zealand). This is a survey approach and provides general information about the presence of predispersal predators. Note that predators were not identified and their abundance per inflorescence was not counted.

Swanton *et al.* (1999) used an experimental approach to examine whether farming practices affected the rate of predispersal seed predation of two pigweed species (red-root pigweed, *A. retroflexus* and green pigweed, *Amaranthus powellii*) when grown under a maize (*Z. mays*) crop. They created different environments by varying the maize row width (37.5 cm and 75 cm) and maize density (75,000 and 100,000 plants/ha). They found that predispersal seed predation was higher when maize was planted at low density but that row width had no effect.

Table 9.3 Percentage infestation of inflorescences of 13 species in the Asteraceae family at three locations (1–3) in Britain (native habitat) and in New Zealand (invaded habitat) (adapted from data in Fenner and Lee, 2001).

Species	Britain (infestation %)				New Zealand (infestation %)			
	1	2	3	Mean	1	2	3	Mean
Yarrow (*Achillea millefolium*)	0	0	0	0	0	0	0	0
English daisy (*Bellis perennis*)	0	5.0	1.5	2.2	0	0	0	0
Canada thistle (*Cirsium arvense*)	1.5	1.0	0	0.8	0	0	0	0
Bull thistle (*Cirsium vulgare*)	2.0	28.0	4.0	11.3	0	0	0	0
Smooth hawk's-beard (*Crepis tectorum*)	6.0	0	1.5	2.5	0	0	0	0
Mouse-ear hawkweed (*Hieracium pilosella*)	0	0.5	0	0.2	0	0	0	0
Nipplewort (*Lapsana communis*)	0	0	0	0	0	0	0	0
Ox-eye daisy (*Leucanthemum vulgare*)	35.5	30.5	12.0	26.0	0	0	0	0
Tansy ragwort (*Senecio jacobaea*)	0	1.5	2.5	1.3	0	0	0	0
Common groundsel (*Senecio vulgaris*)	0	0	0	0	0	0	0.5	0.2
Dandelion (*Taraxacum officinale*)	0	3.0	3.0	2.0	0	0	0	0
Scentless mayweed (*Tripleurospermum inodorum*)	23.5	48.0	27.5	33.0	0	0	0	0

Continued

Box 9.2. Continued.

Cromar *et al.* (1999) looked at postdispersal seed predation in a similar agricultural situation. They conducted two experiments which examined the effect of (i) tillage (mouldboard or chisel ploughing, or no till) and (ii) crop residues (maize; soybean, *Glycine max*; wheat, *Triticum aestivum*) on postdispersal seed predation of common lambsquarters (*C. album*) and barnyard grass (*E. crus-galli*). In both experiments, seeds were placed in Petri dishes which were buried flush with the soil surface, and soil residue then placed over the top. These dishes had small mesh cages placed over them to exclude various seed predators. Mesh of 1.5 mm excluded all organisms and was used as a control to calculate losses due to effects such as wind. A mesh size of 7 mm excluded mammals and birds, but allowed insects into enter. By comparing the retention rates of seeds under the two mesh sizes, seed loss due to predation by insects was estimated. Cromar *et al.* (1999) found that seed loss was lowest under chisel plough systems, and lowest under wheat and soybean residues (Table 9.4).

Table 9.4. Percentage seed lost to postdispersal seed predation of common lambsquarters (*C. album*) and barnyard grass (*E. crus-galli*) in tillage and crop residue cover (maize, *Z. mays*; soybean, *G. max*; wheat, *T. aestivum*) experiments (Cromar *et al.*, 1999). Percentage predation rates (SE, standard error) are based on spring and autumn sampling periods averaged over 3 years. Within experiments, treatments followed by the same letter (a, b) are not significantly different according to the Tukey test.

	Tillage experiment			Crop residue cover experiment		
	No till	Chisel plough	Mouldboard plough	Maize	Soybean	Wheat
Residue biomass (g dry wt/m²)	572 (15)	225 (37)	64 (21)	5.3 (41)	328 (40)	510 (66)
Predation (%) (SE)	32 (2) **a**	24 (2) **b**	32 (2) **a**	31 (2) **a**	24 (2) **b**	21 (2) **b**

than in North America (Adams *et al.*, 2009). However, enemy release does not occur in all invasive species; for example, invasive north American vines experience high rates of herbivore damage in their non-native habitat (Ashton and Lerdau, 2008).

As a result of enemy release, the population size of the plant species concerned may increase. Eventually, new genotypes may evolve that divert more energy to growth and reproduction, and less to herbivore defence (Ashton and Lerdau, 2008). Invasive populations of tansy ragwort (*S. jacobaea*) allocated more resources to growth and reproduction; however, there was greater protection against generalist herbivores and less against specialist herbivores (Joshi and Vrieling, 2005). This trade-off, however, did not occur in the broad-leaved paperbark (*Melaleuca quinquenervia*) or common St John's wort (*H. perforatum*) (Maron *et al.*, 2004; Franks *et al.*, 2008).

Biological control

As a consequence of herbivore release, one approach to managing invasive species is introducing specialist herbivores (often insects) as biological control agents. These herbivores may themselves be exotics introduced from the same area of origin as the plant. While biological control does not usually eradicate the plant species, even if the biological control agent consumes only a small proportion of biomass, it may be enough to alter the competitive interactions between it and otherwise less-competitive native species.

The biological control of prickly pear cacti (*Opuntia* spp.), introduced from Mexico and the southern USA into Australia, exemplifies both successful biological control and the risks involved. The cactus moth borer (*Cactoblastis cactorum*) was introduced from Argentina as a biological control agent to help control prickly pear cacti in 1926. As a caterpillar, the cactus moth borer consumes the tissue, and this introduces bacterial soft rot and other pathogens into the cactus. In 1925, the cactus covered 24 million ha of Australia, but by 1930 it was under control. The cactus still survives in small populations that last only until they are detected by the moths. Recently, however, the cactus moth borer was accidentally introduced into eastern North America from Argentina. Unlike Australia, prickly pear cacti are native to eastern North America and are not considered invasive. Eventually, the moth may spread to Mexico and south-western North America where it could find many more species of native cacti and cause serious damage (Cory and Myers, 2000). The cactus moth borer is a native part of the ecosystem in Argentina, a saviour in Australia, and a serious pest in North America. This is why using herbivores as biological control agents must be tested carefully.

Seed predators (especially predispersal) are a common method of biological control of invasive plants. For example, a weevil (*Erytenna consputa*) and moth (*Arposina autologa*) have reduced the spread of the highly invasive silky hakea (*Hakea sericea*) in South Africa (Le Maitre *et al.*, 2008). However, to be successful, high levels of seed predation may have to occur (Le Maitre *et al.*, 2008). High levels of seed loss are reduced when the life cycles of the plant and predator are not synchronized (Elzinga *et al.*, 2007). Total control of an invasive plant through seed predators is unlikely. Thomas and Reid (2007) review the effectiveness of biological control of invasive species.

9.3 Parasitism

A parasite is an organism that depends on another organism (its host) for nutrition, support or shelter; parasitic plants do this by physically infecting and/or climbing on their hosts (Table 9.5). Plants that are entirely dependent on their host (holoparasites) are usually white because they lack chlorophyll and cannot photosynthesize. Dodder (*Cuscuta*) and broomrape (*Orobanche*) are holoparasites. Plants that rely on their host for only some resources (hemiparasites) form either obligate or facultative relationships. Many mistletoes (order Santalales) are obligate hemiparasites whereas yellow rattle (*Rhinanthus minor*) is a facultative hemiparasite. Some parasitic species are dependent on their host only for physical support (epiphytes). Orchids, ferns, bromeliads, lichens and many mistletoes are epiphytic. Epiphytes live upon other plants and may or may not have a negative effect on their host.

Invasive species can be parasitic (Table 9.6). Parasitic agricultural weeds appear to be more important in developing countries (Zimdahl, 1999). For example, witchweeds (*Striga* spp.) infest approximately 44 million ha in Africa and infestations can reduce crop yield by more than 20%. North Americans are more concerned with parasitic forestry weeds such as dwarf mistletoes (Viscaceae), which are parasites of many conifer species; their impact on the North American forest industry is on the same scale as that of witchweeds in African agriculture (Baker and Knowles, 1992). The mistletoes preferentially

Table 9.5. Definitions of the various types of parasitism.

Type of parasite	Explanation
Holoparasite	Entirely dependent on host for carbon, water and nutrients
Hemiparasite	Rely on host for some resources, but are self-sufficient in others
Obligate hemiparasite	Can survive only when associated with the appropriate host
Facultative hemiparasite	Can survive without host, but are usually associated with the host
Epiphyte	Rely on host for physical support

Table 9.6. Examples of important agricultural parasitic weeds.

Common name	Genus	Host species	Comments
Dodder	*Cuscuta*	Non-specific	Twines around stems – the seedling makes contact with and inserts haustoria into the stem
Mistletoe	*Arceuthobium* *Loranthus* *Viscum*	Coniferous trees	Shoot hemiparasite
Broomrape	*Aeginatia* *Orobanche*	Carrots, broadbeans, tomatoes, sunflowers, red clover	Root holoparasite
Witchweed	*Alectra* *Striga*	Species specific to sorghum, millet, maize, cowpeas, groundnuts and other crops	Root hemiparasite. Called witchweed because it harms the crop before the parasite has emerged from the soil

parasitize young trees because the bark is thinner and easier to penetrate (Parker and Riches, 1993). Forestry practices have increased infestation rates because cutting and replanting trees means there are more younger trees. In addition, forestry practices now suppress fires which had once reduced dwarf mistletoe infestation. Managing parasitic weeds is difficult and often requires herbicides. Sometimes trap cropping is used to stimulate the germination and growth of parasitic weeds so that they can be managed in a crop that is not economically important before the weeds infest more important crops (Chittapur *et al.*, 2001). In these cases, the 'trap crop' is not usually a weed but a crop species such as flax (*Linum* spp.) that is not meant to be harvested for human use.

9.4   Mutualisms

Not all interactions between individuals produce negative effects. Mutualisms are interactions where both individuals benefit. Generally, a mutualist provides a service to its partner that the partner cannot provide for itself, and in return obtains a reward (Bronstein, 1994). The types of benefits that mutualists gain include:

- nutrition (each organism supplies different essential nutrients to the other)
- protection (one individual protects the other)
- transport (one mutualist gains mobility through the actions of the other).

The two organisms do not necessarily gain the same benefit. For example, one mutualist may provide nutrition while the other provides transport.

Types of mutualisms

Obligate mutualisms

In an obligate mutualism, both partners of the association require each other in order to survive. The most extreme examples of an obligate mutualism (often called a 'symbiosis') are lichens, which are associations between fungi and algae. The fungus forms the main body structure of the lichen and the alga provides carbohydrates. Because they no longer can live independently, neither the fungus nor the alga is considered to be a distinct species. Plants and their insect pollinators may be obligate mutualists, especially in tropical forests. For example, the over 900 species of fig trees (*Ficus* spp.) are pollinated by a separate species of agaonid wasp

(Janzen, 1979), although there are some exceptions to this (Cook and Rasplus, 2003). Similar obligate pollinator associations exist for yucca (*Yucca* spp.) (Pellmyr, 2003) and some orchids (Jersáková *et al.*, 2006). Invasive species generally do not form obligate mutualisms with pollinators because they are not co-adapted with the available fauna.

Obligate mutualisms are beneficial to both mutualists but they can reduce a species' ability to survive, i.e. the extinction of one partner will almost certainly lead to the extinction of the other. Species that are part of an obligate mutualism are not usually invasive unless both species invade at the same time and can withstand the environmental conditions of the new habitat. For example, of the 60 species of figs introduced into Florida, the three species that have become invasive only did so after their pollinator wasps were unintentionally introduced (Nadel *et al.*, 1992; Richardson *et al.*, 2000). Similarly, two species of figs introduced into New Zealand did not set seed until their specific pollinating wasps arrived, apparently through long-distance dispersal, from Australia (Gardner and Early, 1996). This is not to claim that plants that are obligate mutualists can never become invasive because it is possible for the plants to adapt to local fauna or vice versa; hence, the relationship between figs and their pollinator wasps may not be obligate in all cases (Richardson *et al.*, 2000).

Facultative mutualisms

In a facultative mutualism, both species can survive independently, but both benefit when they are found together. Facultative mutualisms are common. For example, mutualistic animals provide plants with 'services' like pollination (Chapter 4), and seed dispersal (Chapter 6). Pollen and seed dispersal is facilitated by the presence of native and non-native animal dispersers. In North America, species with simple flowers, such as ox-eye daisy (*L. vulgare*) and orange hawkweed (*Hieracium aurantiacum*), are visited by a wide range of insects that include European honeybees (*A. mellifera*) and native bumblebees (*Bombus* spp.) (Murphy and Aarssen, 1995). This is true of plants with more complex flowers, such as cow vetch (*Vicia cracca*), where native bumblebees have had sufficient 'experience' with similar native flowers that pollinating cow vetch is not a difficult task to master (Murphy and Aarssen, 1995). Yellow star-thistle and European honeybees act as invasive mutualists where the honey bees prefer yellow star-thistle, and the star-thistles are assured of pollination (Barthell *et al.*, 2001).

Mutualistic fungi and bacteria provide access to more or different sources of nutrients (Richardson *et al.*, 2000). Mycorrhizal associations are formed when a fungus infects plant roots. The fungus increases the effective 'root' surface area and therefore increases the supply of nutrients and water. In turn, the plant provides carbohydrates from photosynthesis to the fungus. Ectomycorrhizal fungi penetrate the intercellular spaces of roots whereas endomycorrhizal fungi penetrate the cells themselves. Mycorrhizae are so prevalent and important that most native flora and some exotics will grow only if certain fungal species are present in sufficient densities (Schroth, 1998; van der Heijden *et al.*, 1998; Jordan *et al.*, 2000; Dahlberg, 2001).

Mycorrhizal invasive plant species may be specialists or generalists. When Monterey pines were introduced into New Zealand plantations, their spread was limited until spores from the right species of mycorrhizal fungi from plantation soil had accumulated (Richardson and Higgins, 1998). In contrast, other invasive species, such as Russian thistle (*Salsola kali*), are generalists; because they can use most mycorrhizal fungal species, they can invade most habitats. While mycorrhizal invasive plant species often have negative effects on ecosystems, they actually can have beneficial effects if the net result of having additional mycorrhizal species is to improve soil quality or attract organisms that attack invasive species (Jordan *et al.*, 2000).

Many invasive species are not mycorrhizal (Goodwin, 1992). This is why they are found in highly disturbed habitats, i.e. where the mycorrhizal system has been disrupted by human activities. When mycorrhizal fungi are absent from soil, non-mycorrhizal species, such as most mustards (Brassicaceae) and lambsquarters, are favoured. However, a non-mycorrhizal invasive species may use an existing mycorrhizal network in a relatively undisturbed habitat to that species' advantage. For example in North American prairie, weedy spotted knapweed (*Centaurea maculosa*) is able to outcompete the native Idaho fescue (*Festuca idahoensis*), but only if mycorrhizal fungi are present (Marler *et al.*, 1999). Normally, neither species benefits individually from mycorrhizae as both are infected with the fungi but don't provide carbohydrates to them (hence the fungi are merely a passive infection). If, however, an extensive mycorrhizal network is formed among prairie native species, spotted knapweed can 'steal' nutrients from Idaho fescue via their common fungal connections.

Plants often form associations with mutualistic bacteria that use (or fix) atmospheric nitrogen (N_2). Nitrogen in this form is normally unavailable to plants because they lack the enzymes needed to capture it. The plant gets another source of nitrogen, while the bacteria receive carbohydrates and protection. These nitrogen-fixing bacteria are generally ubiquitous (Richardson *et al.*, 2000) and invasive nitrogen-fixing plants usually have no difficulty finding the required bacteria. e.g. gorse (*Ulex europaeus*) and Scotch broom (*C. scoparius*) (Peterson and Prasad, 1998; Clements *et al.*, 2001).

Nitrogen-fixing species may have a strong influence on their new habitat because they may change the nutrient dynamics. This is especially true in places like Hawaii, where nitrogen fixers are not a part of the native flora. For example, the fire tree (*M. faya*) enriches soil nitrogen and provides shade that improves germination and seedling growth of the native 'ōhià lehua (*Metrosideros polymorpha*). However, the increased amount of nitrogen in the soil also permitted the invasion of species formerly constrained by limited nitrogen, e.g. strawberry guava (*Psidium cattleianum*) (Wall and Moore, 1999). The same phenomenon can occur when nitrogen-fixing crops such as clover (*Trifolium* spp.) or lucerne (*Medicago* spp.) are used; the increased soil nitrogen is of benefit to non-nitrogen-fixing crops (like maize) that will be planted the next year in rotation, but it also benefits weeds. Further, if the nitrogen-fixing crops colonize non-agricultural areas like forests, then the invasive species also will be able to colonize.

Mutualistic relationships, such as pollination and seed dispersal, between native species can be disrupted by invasive species (Traveset and Richardson, 2006). Invasive plants may be preferred by pollinators if they are larger and showier, or if they offer larger nectar rewards than native species. Visits to non-native species may result in lower seed set or in hybridization if species are closely related. Similarly, invasive species with more attractive fruits may disrupt seed dispersal patterns. Mycorrhizal associations of native plants and fungi can be disrupted by invasive plants. Garlic mustard, for example, suppressed the growth of a number of native species in North America that rely on mycorrhizal fungi (Callaway *et al.*, 2008).

9.5 Complexity in the Real World: Interactions Between Ecological Processes

The net effect of interacting ecological processes is not straightforward. A good example is the interaction of competition and herbivory between the invasive exotic Japanese honeysuckle (*Lonicera japonica*) and the native trumpet honeysuckle (*Lonicera sempervirens*) (Schierenbeck *et al.*, 1994). In the absence of herbivory, Japanese honeysuckle will be outcompeted by trumpet honeysuckle because it has a higher growth rate and accumulates more biomass (Fig. 9.2). However, being a native, trumpet honeysuckle is more vulnerable to herbivores, and the resultant damage makes it less competitive than Japanese honeysuckle. Why doesn't trumpet honeysuckle adapt by developing strong defences against herbivores so it can be both impervious and a good competitor? The main reason is because there is a trade-off of limited resources allocated to competitive traits like growth versus those allocated to defences against herbivores (Herms and Mattson, 1992).

A more complicated example of interactions between ecological processes is that of invasive tree of heaven (*A. altissima*) (Facelli, 1994). White oak (*Quercus alba*) leaf litter provides a habitat for herbivores that attack tree of heaven saplings and adults, and delays the seedling emergence of tree of heaven. Separate experiments also show that competition from species such as giant foxtail (*Setaria faberii*) reduces seedling biomass of the tree of

heaven. We might expect that a combination of white oaks and species such as giant foxtail would harm tree of heaven even further. However, when white oak litter and competition from species such as giant foxtail occur simultaneously, the leaf litter has a greater effect on giant foxtail than it does on tree of heaven. Contrary to expectations from the separate examination of leaf litter and competition, a combination of both actually increases the ability of tree of heaven to survive.

9.6 Summary

In this chapter we discussed the processes of herbivory, parasitism and mutualism. Herbivory is usually non-fatal because plants have adapted avoidance, tolerance and inducible defence mechanisms, but it can alter the population dynamics of invasive species. The absence of herbivores may give invasive species an advantage. The introduction of herbivorous biological control agents attempts to remove this advantage. Parasitic plant species, such as invasive mistletoes and witchweeds, are especially troublesome in agricultural and forestry systems. Established mutualisms in native species can be disrupted by invasive species. Most invasive species species do not form obligate mutualisms. While studying species interactions separately is useful, the real world is a place where interactions may occur unpredictably. In the next chapter, we study communities as a whole and look at community structure and diversity.

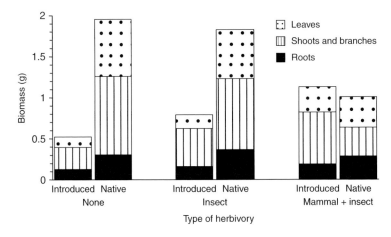

Fig. 9.2. Distribution of biomass to the roots, shoots and branches, and leaves of the introduced Japanese honeysuckle (*Lonicera japonica*) and the native trumpet honeysuckle (*Lonicera sempervirens*) when subject to no herbivory, insect herbivory, and mammal and insect herbivory (adapted and redrawn from data in Schierenbeck *et al.*, 1994).

9.7 Questions

1. Herbivores have three immediate effects on plants: (i) the loss of photosynthetic tissue; (ii) the removal of meristems; and (iii) the loss of stored nutrients (Noy Meir, 1993). Explain how *each* of these can influence the growth and survival of a plant.

2. Design an experiment to test the enemy release hypothesis.

3. Research the relationship between an invasive plant and its biological control agents. Was the agent successful at controlling the species? Explain why or why not.

4. Explain how parasitic weeds might be controlled by the use of trap crops.

5. How would an invasive nitrogen-fixing plant affect the nitrogen cycle in an ecosystem?

Further Reading

Kareiva, P.M. and Bertness, M.D. (1997) Re-examining the role of positive interactions in communities. *Ecology* 78, 1945–1951.

Kolb, A., Ehrlen, J. and Eriksson, O. (2007) Ecological and evolutionary consequences of spatial and temporal variation in pre-dispersal seed predation. *Perspectives in Plant Ecology, Evolution and Systematics* 9, 79–100.

Pringle, A., Bever, J.D., Gardes, M., Parrent, J.L., Rillig, M.C. and Klironomos, J.N. (2009) Mycorrhizal symbioses and plant invasions. *Annual Review of Ecology, Evolution, and Systematics* 40, 699–715.

Richardson, D.M., Allsopp, N., D'Antonio, C.M., Milton, S.J. and Rejmánek, M. (2000) Plant invasions – the role of mutualisms. *Biological Reviews* 75, 65–93.

Schmitz, O.J. (2008) Herbivory from individuals to ecosystems. *Annual Review of Ecology, Evolution and Systematics* 39, 133–152.

References

Aarssen, L.W. (1995) Hypotheses for the evolution of apical dominance in plants: implications for the interpretation of overcompensation. *Oikos* 74, 149–156.

Adams, J.M., Fang, W., Callaway, R.M., Cipollini, D., Newell, E. and Transatlantic *Acer platanoides* Invasion Network (TRAIN) (2009) A cross-continental test of the enemy release hypothesis: leaf herbivory on *Acer platanoides* is three times lower in North America than in its native Europe. *Biological Invasions* 11, 1005–1016.

Agrawal, A.A. (2000) Overcompensation of plants in response to herbivory and the by-product benefits of mutualism. *Trends in Plant Science* 5, 309–313.

Andrews, T.S. (1995) Dispersal of seeds of giant *Sporobolus* spp. after ingestion by grazing cattle. *Australian Journal of Experimental Agriculture* 35, 353–356.

Ashton, I.W. and Lerdau, M.T. (2008) Tolerance to herbivory, and not resistance, may explain differential success of invasive, naturalized, and native North American temperate vines. *Diversity and Distribution* 14, 169–178.

Baker, F.A. and Knowles, K. (1992) Impact of dwarf mistletoes on jack pine forests in Manitoba. *Plant Disease* 76, 1256–1259.

Barthell, J.F., Randall, J.M., Thorp, R.W. and Wenner, A.M. (2001) Promotion of seed set in yellow star-thistle by honey bees: evidence of an invasive mutualism. *Ecological Applications* 11, 1870–1883.

Becerra, P.I. and Bustamante, R.O. (2008) The effect of herbivory on seedling survival of the invasive exotic species *Pinus radiata* and *Eucalyptus globulus* in a Mediterranean ecosystem of Central Chile. *Forest Ecology and Management* 256, 1573–1578.

Belsky, A.J., Carson, W.P., Jensen, C.L. and Fox, G.A. (1993) Overcompensation by plants: herbivore optimization or red herring. *Evolutionary Ecology* 7, 109–121.

Best, R.J. and Arcese, P. (2009) Exotic herbivores directly facilitate the exotic grasses they graze: mechanisms for an unexpected positive feedback between invaders. *Oecologia* 159, 139–150.

Blaney, C.S. and Kotanen, P.M. (2001) Post-dispersal losses to seed predators: an experimental comparison of native and exotic old field plants. *Canadian Journal of Botany* 79, 284–292.

Boege, K. and Marquis, R.J. (2005) Facing herbivory as you grow up: the ontogeny of resistance in plants. *Trends in Ecology and Evolution* 20, 441–448.

Bronstein, J.L. (1994) Our current understanding of mutualism. *Quarterly Review of Biology* 69, 31–51.

Cahill, J.F., Castelli, J.P. and Caspers, B.B. (2000) The herbivory uncertainty principle: visiting plants can alter herbivory. *Ecology* 82, 307–312.

Callaway, R.M., Kim, J. and Mahall, B.E. (2006) Defoliation of *Centaurea solstitialis* stimulates compensatory growth and intensifies negative effects on neighbors. *Biological Invasions* 8, 1389–1397.

Callaway, R.M., Cipollini, D., Barto, K., Thelen, G.C., Hallett, S.G., Prati, D., Stinson, K. and Klironomos, J. (2008) Novel weapons: invasive plant suppresses fungal mutualists in America but not in its native Europe. *Ecology* 89, 1043–1055.

Cates, R.G. (1975) The interface between slugs and wild ginger: some evolutionary aspects. *Ecology* 56, 391–400.

Chittapur, B.M., Hunshal, C.S. and Shenoy, H. (2001) Allelopathy in parasitic weed management: role of catch and trap crops. *Allelopathy Journal* 8, 147–159.

Clark, C.J., Poulsen, J.R., Levey, D.J. and Osenberg, C.W. (2007) Are plant populations seed limited? A critique and meta-analysis of seed addition experiments. *American Naturalist* 170, 128–142.

Clear Hill, B.H. and Silvertown, J. (1997) Higher order interaction between molluscs and sheep affecting seedling numbers in grassland. *Acta Oecologia* 18, 587–597.

Clements, D.R., Peterson, D.J. and Prasad, R. (2001) The biology of Canadian weeds. 112. *Ulex europaeus* L. *Canadian Journal of Plant Science* 81, 325–337.

Cook, J.M. and Rasplus, J.-Y. (2003) Mutualists with attitude: coevolving fig wasps and figs. *Trends in Ecology and Evolution* 18, 241–248.

Cory, J.S. and Myers, J.H. (2000) Direct and indirect ecological effects of biological control. *Trends in Ecology and Evolution* 15, 137–139.

Crawley, M.J. (1989) The relative importance of vertebrate and invertebrate herbivores in plant population dynamics. In: Bernays, E.A. (ed.) *Insect–Plant Interactions*, Vol. 1. CRC Press, Boca Raton, Florida, pp. 45–71.

Crawley, M.J. (1992) Seed predators and plant population dynamics. In: Fenner, M. (ed.) *Seeds: the Ecology of Regeneration in Plant Communities*. CAB International, Wallingford, UK, pp. 157–191.

Cromar, H.E., Murphy, S.D. and Swanton, C.J. (1999) Influence of tillage and crop residue on postdispersal predation of weed seeds. *Weed Science* 47, 184–194.

Cyr, H. and Pace, M.L. (1993) Magnitude and patterns of herbivory in aquatic and terrestrial ecosystems. *Nature* 361, 148–150.

Dahlberg, A. (2001) Community ecology of ectomycorrhizal fungi: an advancing interdisciplinary field. *New Phytologist* 150, 555–562.

Elliot, S.L., Sabelis, M.W., Janssen, A., van der Geest, L.P.S., Beerling, E.A.M. and Fransen, J. (2000) Can plants use entomopathogens as bodyguards? *Ecology Letters* 3, 228–235.

Elzinga, J.A., Atlan, A., Biere, A., Gigord, L., Weis, A.E. and Bernasconi, G. (2007) Time after time: flowering phenology and biotic interactions. *Trends in Ecology and Evolution* 22, 432–439.

Eschtruth, A.K. and Battles, J.J. (2008) Acceleration of exotic plant invasion in a forested ecosystem by a generalist herbivore. *Conservation Biology* 23, 388–399.

Facelli, J.M. (1994) Multiple indirect effects of plant litter affect the establishment of woody seedlings in old fields. *Ecology* 75, 1727–1735.

Fenner, M. and Lee, W.G. (2001) Lack of pre-dispersal seed predators in introduced Asteraceae in New Zealand. *New Zealand Journal of Ecology* 25, 95–99.

Foroni, J. and Nunez-Farfan, J. (2000) Evolutionary ecology of *Datura stramonium*: genetic variation and costs for tolerance to defoliation. *Evolution* 54, 789–797.

Franks, S.J., Pratt, P.D., Dray, F.A. and Simms, E.L. (2008) No evolution of increased competitive ability or decreased allocation to defense in *Melaleuca quinquenervia* since release from natural enemies. *Biological Invasions* 10, 455–466.

Gardner, R.O. and Early, J.W. (1996) The naturalisation of banyan figs (*Ficus* spp., Moraceae) and their pollinating wasps (Hymenoptera: Agaonidae) in New Zealand. *New Zealand Journal of Botany* 34, 103–110.

Gonzales, E.K. and Arcese, P. (2008) Herbivory more limiting than competition on early and established native plants in an invaded meadow. *Ecology* 89, 3282–3289.

Goodwin, J. (1992) The role of mycorrhizal fungi in competitive interactions among native bunchgrasses and alien weeds: a review and synthesis. *Northwest Science* 66, 251–260.

Herms, D.A. and Mattson, W.J. (1992) The dilemma of plants: to grow or defend. *Quarterly Review of Biology* 67, 283–335.

Janzen, D.H. (1979) How to be a fig. *Annual Review of Ecology and Systematics* 10, 13–51.

Jennings, J.C., Apel-Birkhold, P.C., Bailey, B.A. and Anderson, J.D. (2000) Induction of ethylene biosynthesis and necrosis in weed leaves by a *Fusarium oxysporum* protein. *Weed Science* 48, 7–14.

Jersáková, J., Johnson, S.D. and Kindlmann, P. (2006) Mechanisms and evolution of deceptive pollination in orchids. *Biological Reviews* 81, 219–235.

Jordan, N.R., Zhang, J. and Huerd, S. (2000) Arbuscular-mycorrhizal fungi: potential roles in weed management. *Weed Research* 40, 397–410.

Joshi, J. and Vrieling, K. (2005) The enemy release and EICA hypothesis revisited: incorporating the fundamental difference between specialist and generalist herbivores. *Ecology Letters* 8, 704–714.

Keane, R.M. and Crawley, M.J. (2002) Exotic plant invasions and the enemy release hypothesis. *Trends in Ecology and Evolution* 17, 164–170.

Kolb, A., Ehrlen, J. and Eriksson, O. (2007) Ecological and evolutionary consequences of spatial and temporal variation in pre-dispersal seed predation. *Perspectives in Plant Ecology, Evolution and Systematics* 9, 79–100.

Le Maitre, D.C., Krug, R.M., Hoffmann, J.H., Gordon, A.J. and Mgidi, T.N. (2008) *Hakea sericea*: development of a model of impacts of biological control on population dynamics and rates of spread of an invasive species. *Ecological Modelling* 212, 342–358.

Lewis, O.T. and Gripenberg, S. (2008) Insect seed predators and environmental change. *Journal of Applied Ecology* 45, 1593–1599.

Marler, M.J., Zabinski, C.A. and Callaway, R.M. (1999) Mycorrhizae indirectly enhance competitive effects of an invasive forb on a native bunchgrass. *Ecology* 80, 1180–1186.

Maron, J.L. and Crone, E. (2006) Herbivory: effects of plant abundance, distribution and population growth. *Proceedings of the Royal Society B* 273, 2575–2584.

Maron, J.L., Vilà, M. and Arnason, J. (2004) Loss of enemy resistance among introduced populations of St. John's Wort, *Hypericum perforatum*. *Ecology* 85, 3243–3253.

Mauchline, A.L., Watson, S.J., Brown, V.K. and Froud-Williams, R.J. (2005) Post-dispersal seed predation of non-target weeds in arable crops. *Weed Research* 45, 157–164.

Murphy, S.D. and Aarssen, L.W. (1995) *In vitro* allelopathic effects of pollen from three *Hieracium* species (Asteraceae) and pollen transfer to sympatric Fabaceae. *American Journal of Botany* 82, 37–45.

Musser, R.O., Hum-Musser, S.M., Eichenseer, H., Peiffer, M., Ervin, G., Murphy, B. and Felton, G.W. (2002) Herbivory: caterpillar saliva beats plant defences. *Nature* 416, 599–600.

Nadel, H., Frank, J.H. and Knight, R.J. (1992) Escapees and accomplices: the naturalization of exotic *Ficus* and their associated faunas in Florida. *Florida Entomologist* 75, 29–38.

Noy Meir, I. (1993) Compensating growth of grazed plants and its relevance to the use of rangelands. *Ecological Applications* 3, 32–34.

Olson, D. and Wallander, R. (1999) Carbon allocation in *Euphorbia esula* and neighbours after defoliation. *Canadian Journal of Botany* 77, 1641–1647.

Parker, C. and Riches, C.R. (1993) *Parasitic Weeds of the World: Biology and Control*. CAB International, Wallingford, UK.

Pellmyr, O. (2003) Yuccas, yucca moths, and coevolution. *Annals of the Missouri Botanical Garden* 90, 35–55.

Peterson, D.J. and Prasad, R. (1998). The biology of Canadian weeds. 109. *Cytisus scoparius* (L.) Link. *Canadian Journal of Plant Science* 78, 497–504.

Price, P.W., Bouton, C.E., Gross, P., McPheron, B.A., Thompson, J.N. and Weis, E.A. (1980) Interactions among three trophic levels: influence of plants on interactions between insect herbivores and natural enemies. *Annual Review of Ecology and Systematics* 11, 41–65.

Richardson, D.M. and Higgins, S.I. (1998) Pines as invaders in the southern hemisphere. In: Richardson, D.M. (ed.) *Ecology and Biogeography of Pinus*. Cambridge University Press, Cambridge, UK, pp. 450–473.

Richardson, D.M., Allsopp, N., D'Antonio, C.M., Milton, S.J. and Rejmánek, M. (2000) Plant invasions – the role of mutualisms. *Biological Reviews* 75, 65–93.

Sabelis, M.W., van Baalen, M., Bakker, F.M., Bruin, J., Drukker, B., Egas, M., Janssen, A.R.M., Lesna, I.K., Pels, B., van Rijn, P.C.J. and Scutareanu, P. (1999) The evolution of direct and indirect plant defence against herbivorous arthropods. In: Olff, H., Brown, V.K. and Drent, R.H. (eds) *Herbivores: Between Plants and Predators*. Blackwell Scientific, Oxford, UK, pp. 109–166.

Sandras, V.O. (1996) Cotton compensatory growth after loss of reproductive organs as affected by availability of resources and duration of recovery. *Oecologia* 106, 432–439.

Schierenbeck, K.A., Mack, R.N. and Sharitz, R.R. (1994) Effects of herbivory on growth and biomass allocation in native and introduced species of *Lonicera*. *Ecology* 75, 1661–1672.

Schroth, G. (1998) A review of belowground interactions in agroforestry, focussing on mechanisms and management options. *Agroforestry Systems* 43, 5–34.

Silvertown, J. (1980) The evolutionary ecology of mast seedling in trees. *Biological Journal of the Linnaean Society* 14, 235–250.

Strauss, S.Y. and Agrawal, A.A. (1999) The ecology and evolution of plant tolerance to herbivory. *Trends in Ecology and Evolution* 14, 179–185.

Swanton, C.J., Griffiths, J.T., Cromar, H.E. and Booth, B.D. (1999) Pre- and post-dispersal weed seed predation and its implications to agriculture. In: *Proceedings of the 1999 Brighton Conference – Weeds*. Brighton, UK, pp. 829–834.

Thomas, M.B. and Reid, A.M. (2007) Are exotic natural enemies an effective way of controlling invasive plants? *Trends in Ecology and Evolution* 22, 447–453.

Tiffin, P. and Inouye, B.D. (2000) Measuring tolerance to herbivory: accuracy and precision of estimates made using natural versus imposed damage. *Evolution* 54, 1024–1029.

Traveset, A. and Richardson, D.M. (2006) Biological invasions as disruptors of plant reproductive mutualisms. *Trends in Ecology and Evolution* 21, 208–216.

van der Heijden, M.G.A., Klironomos, J.N., Ursic, M., Moutoglis, P., Streitwolf-Engel, R., Boller, T., Wiemken, A. and Sanders, I.R. (1998) Mycorrhizal fungal diversity determines plant biodiversity, ecosystem variability, and productivity. *Nature* 396, 69–72.

Venecz, J.I. and Aarssen, L.W. (1998) Effects of shoot apex removal in *Lythrum salicaria* (Lythraceae): assessing the costs of reproduction and apical dominance. *Annales Botanici Fennici* 35, 101–111.

Wall, D.H. and Moore, J.C. (1999) Interactions underground. Soil biodiversity, mutualism, and ecosystem processes. *BioScience* 49, 109–117.

Zimdahl, R.C. (1999) *Fundamentals of Weed Science*, 2nd edn. Academic Press, San Diego, California.

10 Basic Community Concepts and Diversity

Concepts

- A community consists of groups of species found together in the same space and time. Communities exist at many spatial scales.
- Species diversity is a measure of the number of species present (richness) and their relative abundances (evenness).
- Diversity can be calculated using a variety of indices; the method chosen depends on the ecological information needed.
- Diversity is generally higher at lower latitudes.
- On a local scale, the level of disturbance influences diversity (Intermediate Disturbance Hypothesis).
- Scientists have long debated whether increased species diversity leads to a more stable ecosystem.

10.1 Introduction

A community is a group of populations of different species that occur in the same space and time (Begon *et al.*, 1990). Definitions of communities are generally vague on where the community boundaries are. We can define boundaries based on the needs of our study. A further difficulty with defining a community is deciding what organisms to include. Do we look at just plants, animals, fungi or all three? Clearly, we *should* include all organisms within the boundaries of our community because any one may have an important function. However, because of the practical limitations placed on researchers, this is rarely done. Really, a community is a human construct: a group of species lumped together for our convenience, and not necessarily reflective of an ecological reality. That does not necessarily mean that communities are not an ecological reality. It is just not a precondition.

We can describe communities in terms of their structure and function. Community structure refers to the external appearance of the community. Species composition (species lists, diversity), species traits (lifespan, morphology), and strata characteristics (canopy, shrubs, vines, herbs) are used to describe community structure. Function refers to how the community works. Nutrient allocation and cycling, biomass production and allocation, and plant productivity are ways to describe community function. Community ecologists ask questions such as:

- How does community structure change over time?
- Can we predict community changes over time?
- Why are there so many (or so few) species in this community?
- How does the addition (or loss) of one species effect the distribution or abundance of other species?

This chapter addresses aspects of community structure and of diversity (the number and relative abundances of species present in a community). We discuss how to define and describe communities, and then discuss patterns, causes and consequences of diversity. Chapter 11 then addresses community dynamics; how communities change over time, while Chapter 12 addresses how and why species invade communities and their effects on community structure and dynamics. As you will see, we are beginning to integrate the information you have learned on populations in earlier chapters as we begin to examine communities as a whole.

10.2 Describing Communities

Much of the ecological literature is taken up with discussions on whether communities exist at all and, if they do, how will we recognize them (Clements, 1916, 1936; Gleason, 1917, 1926; Drake, 1990; Wilson, 1991, 1994; Dale, 1994). While this appears to be a somewhat semantic argument, it does highlight the importance of considering what physical entity is being studied and what criteria are being used to define it. We can identify communities in a number of ways (Morin, 1999). Here we will present three ways: physical, taxonomic and statistical.

Defining communities based on natural physical boundaries is simple for a community in a pond or on a cliff because these have distinct boundaries. Managed ecosystems, such as agricultural fields or woodlots, also tend to have distinct boundaries but this is only because edges are imposed and maintained by human activities. In most communities,

however, there are no distinct boundaries; one community blends into another. Furthermore, there is movement of plants and animals across community boundaries. In ecological studies, community boundaries are usually set based upon our perception of the community structure rather than upon how the community actually functions. Thus, we view a forest, field or bog as a community whether or not we know how they function (Booth and Swanton, 2002). We must make reasonable decisions about community boundaries, but be cognizant that they are not 'real' entities and that these decisions may affect the interpretation of data.

One way to observe whether there are relatively distinct communities is to graph species' abundances across an environmental gradient. If communities are not tight associations of interacting species, then species' distributions will overlap and there will be no discrete boundaries between them (Fig. 10.1a). If species do occur in close association, then their distributions along a gradient will be similar and species' boundaries will coincide (Fig. 10.1b). The area of transition between communities is called an ecotone. Ecotones usually have many species

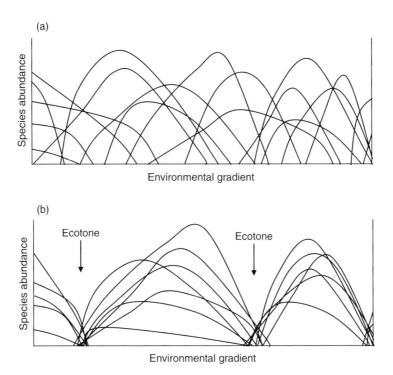

Fig. 10.1. Species' distributions may be (a) individualistic, without tight associations, or (b) in close associations (based on Whittaker, 1975).

because members of both communities will be present, albeit in low abundance. Figure 10.1b shows three communities with two ecotones.

A second way to define a community is based on taxonomic structure. We do this when we talk about a field of maize, a tall-grass prairie or a maple/beech forest. We may not know the exact species composition and abundances, but we will know which species are likely to be present. We have an instinctive knowledge of how these communities differ from each other, and could probably list their dominant plant and animal species and their important ecological processes.

A third method to define communities is based on statistically detected associations among species. Methods used to do this involve examining a large data set of species abundances taken from multiple sites (Jensen, 1998; Shrestha *et al.*, 2002; Rew *et al.*, 2005). Several types of statistical analysis sort these data into sites that have similar species composition. When data points are separated into distinct groups, then we can say that two (or more) community types are present. The benefit of this method is that it removes any personal biases that we may have about what a community looks like.

Community composition

Most community studies only consider part of the community. We talk about a plant community, a bird community or weed/crop community. This reflects the taxonomic bias of individual researchers, but is also done for practical reasons. Considering only a subset of an entire community, of course, has limitations because the results may not be relevant to complex natural communities (Carpenter, 1996). Communities, however, are often intractable when we attempt to examine them as a whole because we cannot control all the variables (Drake *et al.*, 1996). With the exception of microcosm and mesocosm experiments, no studies that we know of examine the dynamics of an entire community. This is likely to remain out of necessity; however, we must remember that community dynamics may be caused by species or factors not included in the study.

When researchers ignore groups of organisms, ecological patterns might be missed. Alternatively, observed patterns may not be explainable if they arise through interactions with organisms excluded from the study (Booth and Swanton, 2002). For example, in a situation where soil-borne organisms

control the community structure of plants (Jordán *et al.*, 2000), interactions between the plant species may be incorrectly used to explain a pattern if soil-borne organisms are omitted from the study. Mycorrhizal fungi, for example, can influence the competitive outcome in a tall-grass prairie (Smith *et al.*, 1999), and their interaction with vegetation should be considered as part of community dynamics (Chapter 9).

Often the importance of a species will not be obvious from its size or abundance. A keystone species has a disproportionate effect on community function relative to its biomass (Paine, 1966, 1969; Power *et al.*, 1996; Jordán, 2009). For example, kangaroo rats (*Dipodomys* spp.) are keystone species in the Sonoran and Chihuahuan deserts (Brown and Munger, 1985; Brown and Heske, 1990; Waser and Ayers, 2003). The rats preferentially feed on large-seeded plant species, giving small-seeded species an advantage they would not otherwise have. Their burrowing influences soil structure and provides habitats to other organisms. Parasitic mistletoe can be a keystone species because it provides nesting and roosting habitat for about 50 species and is a high-quality food source for about 100 mammal and bird species. (Watson, 2001). An invasive species could become a keystone if it alters nutrient cycles or soil properties, or provides food for invasive animals. For example, when the fire tree (*M. faya*) invaded Hawaii, it changed the nitrogen dynamics, which in turn influenced which other species could survive (Walker and Vitousek, 1991).

Matters of scale

Community ecologists recognize the importance of scale when designing and interpreting ecological experiments (Levin, 1986; Allen and Hoekstra 1990, 1991; Menge and Olsen, 1990; Hoekstra *et al.*, 1991). The effect of scale means that a community may be responding to changes that may be local (e.g. succession), regional (e.g. landscape processes or mesoclimatic shifts) or global (e.g. anthropogenic changes to global climate or plate tectonics) (Delcourt *et al.*, 1983; Davis, 1987). Conceptually and pragmatically, it helps if one recognizes that these scales are holarchical – they are nested one within another. Thus, impacts of anthropogenic climate change or large-scale deforestation will influence nutrient cycling, which, in turn, will influence species' interactions (and succession) and the selection pressures on phenotypes.

This is not a one-way process though; once different phenotypes succeed or fail, they influence population dynamics, species' interactions, succession, nutrient cycling and landscape processes. The spatial patterns we observe in a community are the result of phenotypes (and their genotypes) responding and influencing these multiple scales (Menge and Olson, 1990; Woodward and Diament, 1991; Díaz *et al.*, 1998).

Ecologists are also aware that the scale one chooses to study at is important in that it must address a research question being asked, but it also means that other scales may be missed (Levin, 1986). This can be a problem because a process or a pattern may emerge in one scale but not in others. For example, the number of species in a community may remain constant over longer timescales, while the species composition changes (Brown *et al.*, 2001).

10.3 Community Attributes

In Chapter 2, we discussed attributes of populations that can be measured: distribution, abundance and demography. Communities, too, have specific attributes that are used to characterize and compare them. Community attributes include species composition, physiognomy and diversity (Barbour *et al.*, 1999).

The first and most basic way to describe a community is to list all of the species present. However, as already mentioned, it is not usually possible to list *all* species, therefore this option is often not possible. Instead, we may list the dominant species.

A second and more general approach is to describe the general appearance of a community (physiognomy). Physiognomy includes such variables as:

- vertical structure of the vegetation (e.g. canopy, shrub layer, understorey)
- spacing of individuals (e.g. random versus clumped, sparse versus dense)
- life forms of the dominant species (e.g. tree, shrub, herb).

We can go a long way towards understanding vegetation physiognomy by describing a few simple features. For example, we could describe the physiognomy of a tropical rainforest by dividing it into five stratified layers of vegetation: ground-level vegetation, the shrub and sapling layer, and three canopy layers (lower, mid-crown and emergent).

The third community attribute listed above is biological diversity (or biodiversity). Diversity is a measure of the variety of organisms in a community. The remainder of the chapter will discuss diversity in detail, because it is a widely used measure in community ecology. Diversity captures many of the attributes of a community, such as the number of species, and the number of individuals of each species and their relative dominance. It is a useful way to describe a community and it is easy to measure. More advanced methods do exist and readers are directed to the many references in this chapter for that purpose.

10.4 Diversity

Diversity can be examined at many spatial scales (Franklin, 1993; Angermeier, 1994). We usually think of diversity in terms of species, but we could also consider diversity at smaller scales, such as genetic diversity, or at larger scales, such as plant family diversity. In this text we focus on species diversity.

Basic components of diversity: richness and evenness

One component of species diversity is richness. This is the number of species present in an area or in a community. You could calculate species richness of your backyard by counting the number of species there. The second component of species diversity is evenness. Evenness compares the abundance of each species in a community and tells you whether there are many rare species and a few common ones, or whether most species are equally common. Evenness is more informative than species richness because it indicates whether the community is dominated by one or a few species, or whether most species are represented by approximately equal numbers of individuals.

We can illustrate richness and evenness with the following example, which uses four sample fields (Fig. 10.2). The species richness is the same for all the fields (four species in each); however, evenness differs among the fields. Fields 1 and 3 have the same evenness as all species are equally represented. Field 2 is dominated by downy brome (*Bromus tectorum*) and Field 4 is dominated by viper's bugloss (*E. vulgare*), and therefore abundance is uneven. The fact that different species dominate in two fields gives us information that species richness alone would miss.

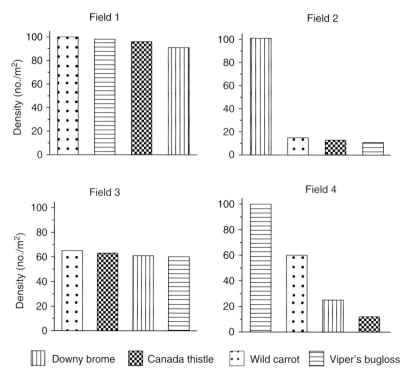

Fig. 10.2. Rank abundance curves of four hypothetical field communities, each with four species: downy brome (*B. tectorum*), Canada thistle (*C. arvense*), wild carrot (*D. carota*) and viper's bugloss (*E. vulgare*). In Fields 1 and 3 these are equally represented, but Field 2 is dominated by downy brome and Field 4 by viper's bugloss.

Downy brome prefers field margins or abandoned farm fields, whereas viper's bugloss is more typical of gravelly areas.

10.5 Measurement of Diversity

There are dozens of methods of measuring diversity, all with varying advantages and disadvantages, and popularity (Table 10.1; see Magurran, 1988; Cousins, 1991; Schlesinger *et al.*, 1994; Conroy and Noon, 1996; Yorks and Dabydeen, 1998; Stiling, 1999; Wilson *et al.*, 1999). We discuss both within community diversity (alpha-diversity) and between community diversity (beta-diversity).

Within community diversity

One common measure of within-community diversity is Margalef's Diversity Index (D_{Mg}) This index is a rough and quick estimate of the species diversity based on richness. Margalef's Diversity Index

is sensitive to the total area sampled, the number of individuals sampled, and whether all or most of the species that exist were represented in the sample. This is because a researcher who samples a small area in a forest may obtain a lower estimate of richness than one who samples extensively because you are likely to find more species (especially the rare ones) if you sample more area.

A second commonly used diversity index is the Shannon–Wiener Diversity Index (H') (note that this is sometimes incorrectly called the Shannon–*Weaver* Index) (Magurran, 1988; Magnussen and Boyle, 1995). Unlike Margalef's index, this index incorporates both species richness and evenness.

A third common index is Simpson's Dominance Index. Dominance measurements preferentially increase the importance of the abundances of the most common species. They do not provide a good assessment of species richness, but are useful when examining evenness. Simpson's Dominance Index is based on the probability that any two individuals sampled will turn out to be the same species. The more a community is dominated by any one species, the more likely that you will be able to keep sampling the same species over and over. The advantage of Simpson's Dominance Index is that it is simple to calculate. It requires only that you sample and estimate the number of individuals of any given species and the total number of individuals in a site.

Table 10.1. Comparing the scales, sophistication and biases/limitations of diversity measurements discussed in this chapter (modified from Magurran, 1988).

Name	Scale	Level of sophistication	Bias and limitations
Margalef's Index	Within community diversity	Measures only gross species richness	Very sensitive to number and thoroughness of sample
Shannon–Wiener Diversity Index and Evenness Measure	Within community diversity	Measures species richness and evenness, and compares diversity between two different sampling areas within a given habitat	Moderately sensitive to sample size but sample meets criteria of randomness and completeness (all possible species represented)
Simpson's Dominance Index	Within community diversity	Measures dominance of one species versus others	Low sensitivity to sample size but says little about species richness
Sørensen Coefficient Index or Coefficient of Similarity	Between community diversity	Measures the similarity of species composition between two communities and the number of individuals in each species; based on presence/absence data	Moderately sensitive to sample completeness
Steinhaus Coefficient Index or Coefficient of Similarity	Between community diversity	Measures the similarity of species composition between two communities; accounts for differences in numbers of individuals of different species and relative success of species common between communities; includes abundance data	Moderately sensitive to sample completeness

---

Box 10.1. Calculating within community diversity.

Here we show how to calculate three within community (alpha-diversity) indices: Margalef's Diversity Index (D_{Mg}), the Shannon–Weiner Diversity Index (H') and Simpson's Dominance Index (D) for two communities – a sandy meadow and a wet meadow – dominated by invasive species. We use the following symbols, although we caution that different symbols may be used elsewhere:

- n = population density or number
- n_i = density or number of the ith species (i.e. any particular species you choose)
- N = total number of individuals of all species in the community
- S = species richness (total number of species)
- 'Σ' means 'sum of all the factors that follow'
- p_i = proportional abundance of a given species (the ith species). $p_i = n/N$

We will use the same basic data set for all the calculations. This is given in Table 10.2.

Continued

Box 10.1. Continued.

Table 10.2. Population size (density or number of the ith species, n_i) of the weed species found in two portions of the same habitat: the sandy and wet areas of a meadow. Data were collected using a stratified random sample.

Weed taxa	Sandy area n_i	Wet area n_i
Garlic mustard (*A. petiolata*)	32	43
Chicory (*Cichorium intybus*)	41	11
Canada thistle (*C. arvense*)	58	15
Deptford pink (*Dianthus armeria*)	48	14
Purple loosestrife (*L. salicaria*)	0	44
Common reed (*Phragmites australis*)	0	36
Lady's thumb (*P. persicaria*)	0	26
Common purslane (*P. oleracea*)	59	17
Common chickweed (*S. media*)	78	24
N^a	316	230
S^b	6	9

N^a = total no. of all species in the community; S^b = species richness (total no. of species).

Margalef's Diversity Index: $D_{Mg} = S - 1/\ln N$

$$D_{Mg\,(sandy\,meadow)} = S - 1/\ln N$$
$$= 6 - 1/\ln 316$$
$$= 5/\ln 316$$
$$= 5/5.7557$$
$$= 0.869$$

$$D_{Mg\,(wet\,meadow)} = S - 1/\ln N$$
$$= 9 - 1/\ln 230$$
$$= 8/5.4381$$
$$= 8/5.4381$$

As one would expect from inspecting the data, Margalef's Index indicates that the wet meadow is more diverse than the sandy meadow.

The Shannon–Weiner Diversity Index: $H' = -[p_i\,(\ln p_i)]$

Using a table makes it easy to do the calculations (Table 10.3). For example, to calculate $-p_i \ln p_i$ for garlic mustard (*A. petiolata*) in the sandy meadow:

$n_i = 32$
$N = 316$
$p_i = n_i/N = 32/316 = 0.101$
$\ln p_i = \ln (0.101) = -2.291$
$-p_i(\ln p_i) = -(0.101)(-2.291) = -(-0.232) = 0.232$

The values of H' are calculated by summing all the $-p_i\,(\ln p_i)$ values for each meadow community, as in Table 10.3, which gives H'(sandy meadow) = 1.753, while H' (wet meadow) = 2.089.

We also can calculate evenness using the numbers from Table 10.3 with the calculated values of H':

Evenness = $E = H'/\ln S$

$$E_{(wet\,meadow)} = H'_{(wet\,meadow)}/\ln S_{(wet\,meadow)}$$
$$= 2.089/\ln 9 = 2.089/2.197 = 0.951$$

$$E_{(sandy\,meadow)} = H'_{(sandy\,meadow)}/\ln S_{(sandy\,meadow)}$$
$$= 1.753/\ln 6 = 1.753/1.792 = 0.978$$

Overall, these calculations indicate that $H'_{(sandy\,meadow)} = 1.753$, while $H'_{(wet\,meadow)} = 2.089$, and that $E_{(sandy\,meadow)} = 0.978$, while $E_{(wet\,meadow)} = 0.951$.

Higher diversity values indicate a more diverse community; however, this is an arbitrary scale. There is no pre-determined value of H' that indicates a community is diverse or not. However, as we will see below, there are ways to compare the diversity indices of two communities. For evenness, values of zero indicate that the habitat is

Continued

Box 10.1. Continued.

Table 10.3. An example of how to calculate the Shannon–Weiner Diversity Index (H'). This example compares the weed species found in two portions of the same habitat: the sandy and wet areas of a meadow. Data were collected using a stratified random sample.

Weed taxa	Sandy area				Wet area			
	n_i	p_i	$\ln p_i$	$-p_i(\ln p_i)$	n_i	p_i	$\ln p_i$	$-p_i(\ln p_i)$
Garlic mustard (*A. petiolata*)	32	0.101	−2.291	0.232	43	0.187	−1.677	0.314
Chicory (*C. intybus*)	41	0.130	−2.042	0.265	11	0.048	−3.040	0.145
Canada thistle (*C. arvense*)	58	0.184	−1.695	0.311	15	0.065	−2.730	0.178
Deptford pink (*D. armeria*)	48	0.152	−1.885	0.286	14	0.061	−2.799	0.170
Purple loosestrife (*L. salicaria*)	0	0.000			44	0.191	−1.654	0.316
Common reed (*P. australis*)	0	0.000			36	0.157	−1.855	0.290
Lady's thumb (*P. persicaria*)	0	0.000			26	0.113	−2.180	0.246
Common purslane (*P. oleracea*)	59	0.187	−1.678	0.313	17	0.074	−2.605	0.193
Common chickweed (*S. media*)	78	0.247	−1.399	0.345	24	0.104	−2.260	0.236
Σ (sum of the columns)	316	1.000	$H' = 1.753$		230	1.000	$H' = 2.089$	

Key: n_i = density or number of the *i*th species; p_i = proportional abundance of the *i*th species.

extremely uneven (dominated by one species), whereas values approaching one indicate that the habitat is extremely even (maximum species diversity exists, no one species dominates).

The results in Table 10.3 indicate that the wet areas of the meadow have greater species richness but that the sandy meadow has greater species evenness. Looking at the data set, the wet areas of the meadow do indeed have more species ($S = 9$), but are dominated by a few species, mainly garlic mustard, purple loosestrife and common reed. In this example, we can simply look at the data and verify that these calculations make sense; in reality, most data sets are too complex to allow the luxury of visual inspection, hence diversity indices are necessary.

Simpson's Dominance Index: $D = \Sigma\{[n_i(n_i - 1)]/[N(N - 1)]\}$

Table 10.4. Calculation of Simpson's Dominance Index (D^{-1}) using weed species data from the sandy area of a meadow habitat. Data were collected using a stratified random sample.

Weed taxa	Sandy area of meadow habitat						
	n_i	$n_i - 1$	$n_i(n_i - 1)$	N	$N - 1$	$N(N - 1)$	$n_i(n_i - 1)/N(N - 1)$
Garlic mustard (*A. petiolata*)	32	31	992	316	315	99540	0.010
Chicory (*C. intybus*)	41	40	1640	316	315	99540	0.016
Canada thistle (*C. arvense*)	58	57	3306	316	315	99540	0.033
Deptford pink (*D. armeria*)	48	47	2256	316	315	99540	0.023
Common purslane (*P. oleracea*)	59	58	3422	316	315	99540	0.034
Common chickweed (*S. media*)	78	77	6006	316	315	99540	0.060
Σ (sum of the columns)	316						$D = 0.177$

Key: n_i = density or number of the *i*th species; N = total number of individuals of all species.

Continued

Box 10.1. Continued.

From Table 10.4, Simpson's Dominance Index (D) for the sandy meadow ($S = 6$) is:

$$D_{(sandy\ meadow)} = \Sigma\{[n_i\ (n_i - 1)]/[N(N-1)]\}$$
$$= 0.010 + 0.016 + 0.033 + 0.023 + 0.034 + 0.060$$
$$= 0.177$$

By convention, Simpson's Dominance Index is usually written as the reciprocal value (D^{-1}), so the value becomes $D^{-1}_{(sandy\ meadow)} = 1/0.177 = 5.65$. Following similar calculations, $D_{(wet\ meadow)} = 0.132$, and $D^{-1}_{(wet\ meadow)} = 7.58$. By using this convention, the higher the index value, the more diversity there is (though, more accurately, this actually means that there is more species evenness). In this example, no one species dominates in either meadow, hence the values of D^{-1} are relatively high (i.e. it is a relatively even community). For our data set, this concurs with the results obtained using the Shannon–Weiner Diversity Index and Evenness calculation.

Between community diversity

Between community diversity (beta-diversity) is a measure of whether there are separate communities within an area. The advantage of measuring between community diversity, rather than within community diversity is that it is explicitly meant to compare different areas.

One common way to calculate between community diversity is to use indices of similarity such as the Sørenson and Steinhaus indices (Table 10.1, Box 10.2). The Sørensen Coefficient Index (Sørensen, 1948)

emphasizes the number of individuals of species common to both communities, i.e. it is a similarity index (and hence in Box 10.2 is designated S_S). The Steinhaus Coefficient Index is based on abundance data (and hence in Box 10.2 is designated S_A); it takes the smallest abundance for each species as a proportion of the average community abundance (Motyka, 1947). Before proceeding, please note that there are now multiple approaches and formulas used for these indices; we present one version of each here.

Box 10.2. Calculating between community diversity.

The Sørensen Coefficient Index (S_S) is represented by the equation:

$$S_S = \{2j/(a+b)\} \times 100$$

where j = the number of species common to each community, and $a + b$ = the sum of the total numbers of species in each community.
The Steinhaus Coefficient Index (S_A) is represented by the equation:

$$S_A = W/\{(A+B)/2\} = 2W/(A+B)$$

where W = the sum of the lower of the two abundances of each species in the community, and $A + B$ = the sum of abundances all species in each community.

We will use the relevant data from our example of the Shannon–Weiner calculation (see Box 10.1, Tables 10.2 and 10.3), which compared the communities in sandy and wet areas of a meadow, to illustrate how the Sørensen and Steinhaus indices are calculated (Table 10.5). For both indices, we interpret their values on a scale from 0 (complete dissimilarity) to 1 (complete similarity).

Sørenson Index
$$S_S = \{2j/(a + b)\} \times 100$$
$$= \{(2 \times 6) / (6 + 9)\} \times 100$$
$$= \{12 / 15\} \times 100$$
$$= 0.8 \times 100$$
$$= 80\%$$

Continued

Box 10.2. Continued.

Table 10.5. Indices used to calculate similarity between communities: the Sørensen Coefficient Index (S_S) and the Steinhaus Coefficient Index (S_A). This example compares the weed species found in two portions of the same habitat: the sandy and wet areas of a meadow. Data were collected using a stratified random sample.

Species	Sandy area Abundance	Wet area Abundance	Minimum abundance value
Garlic mustard (*A. petiolata*)	32	43	32
Chicory (*C. intybus*)	41	11	11
Canada thistle (*C. arvense*)	58	15	15
Deptford pink (*D. armeria*)	48	14	14
Purple loosestrife (*L. salicaria*)	0	44	0
Common reed (*P. australis*)	0	36	0
Lady's thumb (*P. persicaria*)	0	26	0
Common purslane (*P. oleracea*)	59	17	17
Common chickweed (*S. media*)	78	24	24
Total no. of individuals per area	316	230	
Total no. of species (*N*)	6	9	
No. of common species (*j*)		*j* = 6	
Sum of the lower of the two abundances (*W*)			*W* =113

Steinhaus Index

$S_A = 2WI(A + B)$

$= 2(113) / (316 + 230)$

$= 226 / (316 + 230)$

$= 226 / 546 = 0.414$

The Sørensen Index suggests that the communities are quite (80%) similar (a value of 100% would indicate completely similar). However, we have the luxury of being able to inspect the data with relative ease here. If you look at the data, the communities do differ – in the sense that three species are found only in the wet meadow and the numbers of individuals of each species are rather dissimilar. This is why we use the Steinhaus Index – it accounts for differences in abundance and hence is more accurate than Sørensen's Index. One can use other approaches (such as Mountford's Index of Similarity) that also resolve this issue. We also note that one other reason that Sørenson's Index can be a bit misleading is that it is sensitive to small sample sizes – and one can argue that our example is a small sample.

10.6   Patterns of Species Diversity

With the increasing attention paid to introduced species, it is important to ask what effect invasive species have on diversity, and whether this effect is good, bad or neither. For example, invasive species may reduce biodiversity if their presence leads to the extinction of other species. Alternatively, introduced species may increase biodiversity, but this may or may not have an impact on ecosystem function.

If we use species richness as a measure of diversity and compare this among communities around the world, there are observable global patterns. In general, diversity is much higher in the tropical regions than in temperate or polar regions. For example, the species richness of woody plants decreases from the equator to the poles (Currie and Paquin, 1987) (Fig. 10.3). This pattern is not exact, however. For example, at the same latitude, there are more tree species in eastern North America than in the west. A number of mechanisms have been proposed to explain these patterns (Fraser and Currie, 1996; O'Brian, 1998; Whittaker *et al.*, 2001). At large scales, climatic and historical factors are important determinants

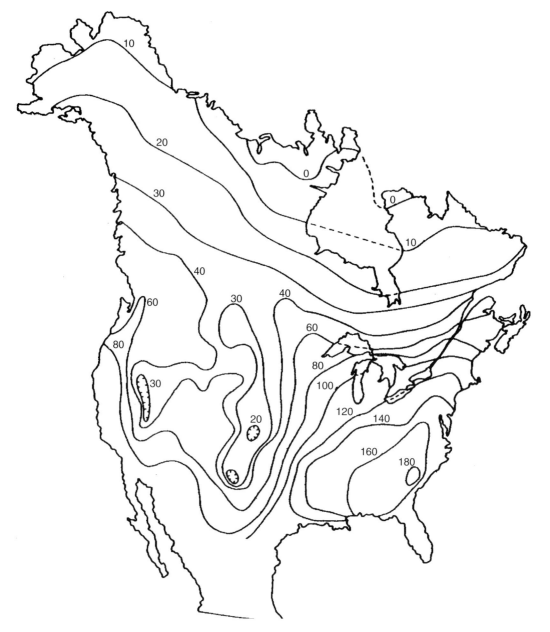

Fig. 10.3. Species richness patterns of trees in North America (as numbers of species) (after Currie and Paquin, 1987; in Stiling, 1999).

of diversity. Historical factors include the time since glaciation, speciation rate and dispersal effects. Communities closer to the equator have had much more time since glaciation and therefore have had more time for evolution to occur and for new species to arise.

On smaller scales, local factors moderate general richness patterns. For example, topography, species' interactions and disturbance will affect local richness. Disturbance is any event that disrupts the existing structure or function of the environment. Disturbance can vary in temporal scale, i.e. it can be periodic (e.g. a tree is blown over, a forest fire erupts or a hurricane occurs) or continual (e.g. grazing, burial, yearly high tides, ice scouring). Disturbances vary in intensity. For example, a forest fire that burns at high temperatures and destroys all vegetation and mineralizes almost all the nutrients is an intense disturbance. Disturbances also vary in spatial scale (e.g. a mole digging a hole is small scale and a hurricane is large scale). A habitat's structure and function result from many frequencies and intensities of disturbance. For example, grasslands may have areas of high- and low-intensity grazing as well as patches of high nutrients caused by piles of animal faeces. The various frequencies and intensities of disturbances help to open new gaps that allow new individuals to colonize.

The Intermediate Disturbance Hypothesis

The Intermediate Disturbance Hypothesis relates diversity to the frequency and intensity of disturbance in an ecosystem (Connell, 1978; Petratis *et al.*, 1989; Wilkinson, 1999; Buckling *et al.*, 2000; Whittaker *et al.*, 2001). This hypothesis assumes that there is a trade-off between a plant's ability to compete and its ability to withstand disturbance. At one extreme, when disturbances are frequent and/or intense, the number of species that survive will be small. Longer-lived, late-successional species will be eliminated by excessive disturbance. As the frequency and/or intensity of disturbance declines, diversity will increase because:

- there is sufficient disturbance to open a niche for new species, and
- there is enough time between major disturbances to allow more species to colonize successfully.

The Intermediate Disturbance Hypothesis predicts that diversity will be highest at intermediate levels of disturbance (Fig. 10.4). When the frequency of disturbance is very low, a few well-adapted species can dominate and competitively exclude most would-be colonizers. As a result, few new species establish themselves and diversity is low.

Bongers *et al.* (2009) tested this hypothesis to determine if disturbance could explain the diversity of trees in wet, moist and dry tropical forests. They found that variation in diversity (species richness) was explained by the Intermediate Disturbance Hypothesis in dry tropical forests but not wet tropical forests. In dry forests, species adapted to disturbance increased while shade-tolerant species

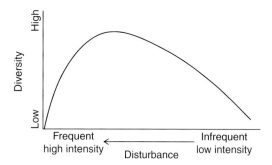

Fig. 10.4. The Intermediate Disturbance Hypothesis (based on Connell, 1978).

decreased as disturbance level increased. Similarly, Sasaki *et al.* (2009) reported that intermediate disturbance, in the form of grazing intensity, increased diversity in landscapes where the environmental conditions were relatively stable. In harsh environments, however, intermediate disturbance could not account for the observed patterns of species richness.

A main problem with the Intermediate Disturbance Hypothesis is related to defining what exactly high, intermediate and low frequencies of disturbance represent. For example, does high disturbance mean a 1000-ha forest fire that burns at 300°C and occurs every year? If it does, is a 10-ha forest fire that burns at 100°C and occurs every 10 years also a high-intensity fire? What if there was a mix of these conditions, e.g. a 1000-ha forest fire that burns at 100°C every 50 years – is this high, intermediate or low disturbance? There are no standard criteria to define frequencies and intensities, hence there is disagreement over how, when and if the Intermediate Disturbance Hypothesis explains diversity patterns. Perhaps the best approach is to clearly define the spatial and temporal scales of interest (Roxburgh *et al.*, 2004). One approach is to establish the historical pattern of disturbance (if possible) and then relate the current disturbance frequency and intensity to this. It is important to understand the context of the study rather than try to apply too general a concept such as intermediate disturbance. The basic theory is sound; its application requires careful analysis within the particular situation that is being studied.

10.7   Is Species Diversity Important to Ecosystem Function?

In the last 50 years, people have debated whether high diversity makes ecosystems less vulnerable to destruction. That is, are diverse ecosystems more stable? The diversity–stability hypothesis originally focused on the idea that complex trophic structures resulted in more stable communities (MacArthur, 1955). Elton (1958) refined this further when he hypothesized that human-disturbed communities were prone to pest outbreaks and unpredictable fluctuations in populations. May (1974) asked what is, in hindsight, an obvious question: does diversity confer ecosystem stability or does ecosystem stability allow for increased diversity? He found that stability really depends on a relatively constant environment. A constant environment was the limiting factor and therefore stability causes diversity.

This reversal of the diversity–stability hypothesis caused scientists to test the hypothesis further and offer alternatives or refinements. There are a number of hypotheses used to describe the relationship between diversity and stability (Table 10.6).

The main series of questions concerning this relationship are:

- Does diversity create stability?
- Does stability create diversity?
- Is there a constant feedback so that diversity and stability cannot be separated?

Alternatively, there may be no relationship. Most likely, the two concepts are actually subject to the

Table 10.6. Hypothetical relationships between (bio) diversity (*x*-axis) and stability/function (*y*-axis).

Relationship	Description	
Linear Relationship/Singular Hypothesis (MacArthur, 1955; Elton, 1958)	Direct linear correlation between diversity and ecosystem function – all species equally important	
Species Redundancy (Walker, 1992)	Species are interchangeable Conservation of most species is not necessary in terms of ecosystem function	
Keystone Species (Walker, 1992)	Loss of a keystone species will cause drastic reduction in ecosystem function	
Rivet–Popper Hypothesis (Ehrlich and Ehrlich, 1981)	There is redundancy built into an ecosystem (like an aeroplane wing) Loss of one or several species (rivets) may not cause any change At some unpredictable point, the loss of a species will cause a change because of the cumulative effect Ecosystem function is compromised as species loss increases, but the number required to cause problems is unknown	

Table 10.7. Terms associated with descriptions of community stability.

Term	Definition	Source
Persistence	'the ability of a community to remain relatively unchanged over time'	Barbour *et al.* (1999)
Resistance	'the ability of a community to remain unchanged during a period of stress'	Barbour *et al.* (1999)
Resilience	'the ability of a community to return to its original state following stress or disturbance'	Barbour *et al.* (1999)
Elasticity	'the speed at which the system returns to its former state following a perturbation'	Putman (1994)

same influences and therefore any tests to separate diversity and stability will be context specific and therefore hard to generalize. The real question is what factors (including diversity) affect how ecosystems function, considering that stability is only one feature (Ives and Carpenter, 2007).

One further problem with separating the relationship between diversity and stability is semantic. Both terms can mean many things (Ives and Carpenter, 2007). We have already identified the many ways of measuring diversity (richness, evenness). The term *stability* is often used to describe how communities resist change in response to disturbance or stress, but it is a vague term and has been defined in many ways (Lepš *et al.*, 1982; Pimm, 1991). Community stability is often broken into three components: persistence, resistance and resilience (Table 10.7).

- Persistence describes how long a community remains the same.
- Resistance describes how well a community resists stress or disturbance.
- Resilience is the ability of the community to return to its original state following a disturbance.

There are still other ways of examining stability (Ives and Carpenter, 2007). Because stability has many definitions, there are many diversity–stability relationships (Ives and Carpenter, 2007).

Many papers about the role of diversity role in ecosystem function have been written (Naeem, 1998; Loreau *et al.*, 2002; Hooper *et al.*, 2005; Srivastava and Vellend, 2005; Ives and Carpenter, 2007; Jiang *et al.*, 2009; Loreau, 2010). The general conclusion of these papers has been that diversity and stability may be related but that there are enough problems with experimental designs and semantics that conclusive evidence about the exact nature of the relationship between diversity and stability is still lacking.

10.8 Summary

A community can be the organisms living on a log or the expanse of the Canadian boreal forest. Either way, we can look at community structure and diversity in similar ways. We can consider communities as physical, taxonomic or statistical entities. Often we look at only a subgroup of all the species that coexist. Diversity is a measure of species richness and evenness. There are many indices that measure within and between species diversity. On a broad scale, diversity is higher towards the equator than at the poles; however, this is only a general pattern because many local factors influence this general trend. The Intermediate Disturbance Hypothesis describes how communities with intermediate levels of disturbance tend to have the highest levels of diversity; however, disturbance is difficult to measure. Scientists have long discussed whether diversity creates stability, or stability creates diversity. In reality, while diversity and stability are related, the relationship is fairly complex and we do not yet understand it. In this chapter we considered how communities are described at one point of time. In the next chapter we look at how and why communities change over time.

Box 10.3. Invasive species case study: basic community concepts and diversity.

- Describe the communities where your selected invasive species occurs.
- Is your invasive species a keystone species?
- Has your species influenced the diversity and/or ecosystem function of any community in which it is found? If so, is there an explanation for how it exerts its influence?

10.9 Questions

1. Do plant communities really exist?
2. Why is it so difficult to determine the role of diversity in ecosystems?
3. If invasive species were to increase species diversity in a habitat, would this be interpreted as a benefit to the ecosystem? How would your interpretation differ using the various hypotheses about the role of diversity in ecosystems?
4. Examine the raw data (Table 10.2) from Box 10.1 and describe the diversity of the wet and dry meadow areas. Do the diversity indices support your conclusions? Explain why or why not.

Further Reading

Duffy, J.E. (2009) Why biodiversity is important to the functioning of real-world ecosystems. *Frontiers in Ecology and the Environment* 7, 437–444.

Ives, A.R. and Carpenter, S.R. (2007) Stability and diversity of ecosystems. *Science* 317, 58–62.

Loreau, M. (2010) Linking biodiversity and ecosystems: towards a unifying ecological theory. *Philosophical Transactions of the Royal Society B* 365, 49–60.

Whittaker, R.J., Willis, K.J. and Field, R. (2001) Scale and species richness: towards a general, hierarchical theory of species diversity. *Journal of Biogeography* 28, 453–470.

References

Allen, T.F.H. and Hoekstra, T.W. (1990) The confusion between scale-defined levels and conventional levels of organization in ecology. *Journal of Vegetation Science* 1, 5–12.

Allen, T.F.H. and Hoekstra, T.W. (1991) Role of heterogeneity in scaling of ecological systems under analysis. In: Kolasa, J. and Pickett, S.T.A. (eds) *Ecological Heterogeneity*. Springer-Verlag, New York, pp. 47–68.

Angermeier, P.L. (1994) Does biodiversity include artificial diversity? *Conservation Biology* 8, 600–602.

Barbour, M.G., Burk, J.H., Pitts, W.D., Gilliam, F.S. and Schwartz, M.W. (1999) *Terrestrial Plant Ecology*, 3rd edn. Benjamin/Cummings, Menlo Park, California.

Begon, M., Harper, J.L. and Townsend, C.R. (1990) *Ecology: Individuals, Populations and Communities*. Blackwell Scientific, Oxford.

Bongers, F., Poorter, L., Hawthorne, W.D. and Sheil, D. (2009) The intermediate disturbance hypothesis applies to tropical forests, but disturbance contributes little to tree diversity. *Ecology Letters* 12, 798–805.

Booth, B.D. and Swanton, C.J. (2002) Assembly theory applied to weed communities. *Weed Science* 50, 2–13.

Brown, J.H. and Heske, E.J. (1990) Control of a desert–grassland transition by a keystone rodent guild. *Science* 250, 1705–1707.

Brown, J.H. and Munger, J.C. (1985) Experimental manipulation of a desert rodent community: food addition and species removal. *Ecology* 66, 1545–1563.

Brown, J.H., Ernest, S.K.M., Parody, J.M. and Haskell, J.P. (2001) Regulation of diversity: maintenance of species richness in changing environments. *Oecologia* 126, 321–332.

Buckling A., Kassen, R., Bell, G. and Rainey, P.B. (2000) Disturbance and diversity in experimental microcosms. *Nature* 408, 961–964.

Carpenter, S.R. (1996) Microcosm experiments have little relevance for community and ecosystem ecology. *Ecology* 77, 677–680.

Clements, F.E. (1916) *Plant Succession: an analysis of the development of vegetation*. Publication Number 242, The Carnegie Institution, Washington, DC.

Clements, F.E. (1936) The nature and structure of the climax. *Journal of Ecology* 22, 9–68.

Connell, J.H. (1978) Diversity in tropical rainforests and coral reefs. *Science* 199, 1302–1310.

Conroy, M.J. and Noon, B.R. (1996) Mapping of species richness for conservation of biological diversity: conceptual and methodological issues. *Ecological Applications* 6, 763–773.

Cousins, S.H. (1991) Species diversity measurement: choosing the right index. *Trends in Ecology and Evolution* 6, 190–192.

Currie, D.J. and Paquin, V. (1987) Large-scale biogeographical patterns of species richness of trees. *Nature* 329, 326–327.

Dale, M.B. (1994) Do ecological communities exist? *Journal of Vegetation Science* 5, 285–286.

Davis, M.B. (1987) Invasions of forest communities during the Holocene: beech and hemlock in the Great Lakes region. In: Gray, A.J., Crawley, M.J. and Edwards, P.J. (eds) *Colonization, Succession and Stability*. Blackwell Scientific, Oxford, pp. 373–393.

Delcourt, H.R., Delcourt, P.A. and Webb, T. III (1983) Dynamic plant ecology: the spectrum of vegetation change in space and time. *Quaternary Science Reviews* 1, 153–175.

Díaz, S., Cabido, M. and Casanoves, F. (1998) Plant functional traits and environmental filters at a regional scale. *Journal of Vegetation Science* 9, 113–122.

Drake, J.A. (1990) Communities as assembled structures: do rules govern pattern? *Trends in Ecology and Evolution* 5, 159–164.

Drake, J.A., Hewitt, C.L., Huxel, G.R. and Kolasa, J. (1996) Diversity and higher levels of organization. In: Gaston, K. (ed.) *Biodiversity: a Biology of Numbers and Difference*. Blackwell Science, Cambridge, Massachusetts, pp. 149–166.

Ehrlich. P.R. and Ehrlich, A.H. (1981) *Extinction: the Causes and Consequences of the Disappearance of Species*. Random House, New York.

Elton, C.S. (1958) *The Ecology of Invasions by Animals and Plants*. Chapman and Hall, London.

Franklin, J.F. (1993) Preserving biodiversity: species, ecosystems, or landscapes? *Ecological Applications* 3, 202–205.

Fraser, R.H. and Currie, D.J. (1996) The species richness–energy hypothesis in a system where historical factors are thought to prevail: coral reefs. *American Naturalist* 148, 138–159.

Gleason, H.A. (1917) The structure and development of the plant association. *Bulletin of the Torrey Botany Club* 44, 463–481.

Gleason, H.A. (1926) The individualistic concept of the plant association. *Bulletin of the Torrey Botany Club* 53, 7–26.

Hoekstra, T.W., Allen, T.F.H and Flather, C.H. (1991) Implicit scaling in ecological research. *BioScience* 41, 148–154.

Hooper, D.U., Chapin, F.S. III, Ewel, J.J., Hector, A., Inchausti, P., Lavorel, S., Lawton, J.H., Lodge, D.M., Loreau, M., Naeem, S., Schmid, B., Setälä, H., Symstad, A.J., Vandermeer, J. and Wardle, D.A. (2005) Effects of biodiversity on ecosystem functioning: a consensus of current knowledge. *Ecological Monographs* 75, 3–35.

Ives, A.R. and Carpenter, S.R. (2007) Stability and diversity of ecosystems. *Science* 317, 58–62.

Jensen, K. (1998) Species composition of soil seed bank and seed rain of abandoned wet meadows and their relation to aboveground vegetation. *Flora* 193, 345–359.

Jiang, L., Wan, S. and Li, L. (2009) Species diversity and productivity: why do results of diversity-manipulation experiments differ from natural patterns? *Journal of Ecology* 97, 603–608.

Jordán, F. (2009) Keystone species and food webs. *Philosophical Transactions of the Royal Society B* 364, 1733–1741.

Jordan, N.R., Zhang, J. and Huerd, S. (2000) Arbuscular-mycorrhizal fungi: potential roles in weed management. *Weed Research* 40, 397–410.

Lepš, J., Osbornová-Kosinová, J. and Rejmánek, M. (1982) Community stability, complexity and species life history strategies. *Vegetatio* (now *Plant Ecology*) 50, 53–63.

Levin, S.A. (1986) Pattern, scale, and variability: an ecological perspective. In: Hastings, A. (ed.) *Community Ecology*. Springer-Verlag, Berlin, pp. 1–13.

Loreau, M., Naeem, S. and Inchausti, P. (eds) (2002) *Biodiversity and Ecosystem Functioning: Synthesis and Perspectives*. Oxford University Press, Oxford, UK.

Loreau, M. (2010) Linking biodiversity and ecosystems: towards a unifying ecological theory. *Philosophical Transactions of the Royal Society B* 365, 49–60.

MacArthur, R. (1955) Fluctuations of animal populations and a measure of community stability. *Ecology* 36, 533–536.

Magnussen, S. and Boyle, T.J.B. (1995) Estimating sample size for inference about the Shannon–Weaver and the Simpson indices of species diversity. *Forest Ecology and Management* 78, 71–84.

Magurran, A.E. (1988) *Ecological Diversity and its Measurement*. Croom Helm, London.

May, R.M. (1974) *Stability and Complexity in Model Ecosystems*, 2nd edn. Princeton University Press, Princeton, New Jersey.

Menge, B.A. and Olson, A.M. (1990) Role of scale and environmental factors in regulation of community structure. *Trends in Ecology and Evolution* 5, 52–57.

Morin, P.J. (1999) *Community Ecology*. Blackwell Science, Malden, Massachusetts.

Motyka, J. (1947) O zadaniach i metodach badan geobotanicznych. Sur les buts et les méthodes des recherches géobotaniques. *Annales Universitatis Mariae-Curie Sklodowska, Sectio C, Supplementum I*.

Naeem, S. (1998) Species redundancy and ecosystem reliability. *Conservation Biology* 12, 39–45.

O'Brian, E.M. (1998) Water–energy dynamics, climate, and prediction of woody plant species richness: an interim general model. *Journal of Biogeography* 25, 379–398.

Paine, R.T. (1966) Food web complexity and species diversity. *American Naturalist* 100, 65–75.

Paine R.T. (1969) A note on trophic complexity and community stability. *American Naturalist* 103, 91–93.

Petratis, P.S., Latham, R.E. and Niesenbaum, R.A. (1989) The maintenance of species diversity by disturbance. *Quarterly Review of Biology* 64, 393–418.

Pimm, S.L. (1991) *The Balance of Nature? Ecological Issues in the Conservation of Species and Communities*. University of Chicago Press, Chicago, Illinois.

Power, M.E., Tilman, D., Estes, J.A., Menge, B.A., Bond, W.J., Mills, L.S., Daily, G., Castilla, J.C., Lubchenco, J. and Paine, R.T. (1996) Challenges in the quest for keystones. *BioScience* 46, 609–620.

Putman, R.J. (1994) *Community Ecology*. Chapman and Hall, New York.

Rew, L.J., Medd, R.W., Van de Ven, R., Gavin, J.J., Robinson, G.R., Tuitee, M., Barnes, J. and Walker, S. (2005) Weed species richness, density and relative abundance on farms in the subtropical grain region of Australia. *Australian Journal of Experimental Ecology* 45, 711–723.

Roxburgh, S.H., Shea, K. and Wilson J.B. (2004) The intermediate disturbance hypothesis: patch dynamics and mechanisms of species coexistence. *Ecology* 85, 359–371.

Sasaki, T., Okubo, S., Okayasu, T., Jamsran, U., Ohkuro, T. and Takeuchi, K. (2009) Management applicability

---

of the intermediate disturbance hypothesis across Mongolian rangeland ecosystems. *Ecological Applications* 19, 423–432.

Schlesinger, R.C., Funk, D.T., Roth, P.L. and Myers, C.C. (1994) Assessing changes in biological diversity over time. *Natural Areas Journal* 14, 235–240.

Shrestha, A., Knezevic, S.Z., Roy, R.C., Ball-Coelho, B.R. and Swanton, C.J. (2002) Effect of tillage, cover crop and crop rotation on the composition of weed flora in a sandy soil. *Weed Research* 42, 76–87.

Smith, M.D., Hartnett, D.C. and Wilson, G.W.T. (1999) Interacting influence of mycorrhizal symbiosis and competition on plant diversity in tallgrass prairie. *Oecologia* 121, 574–582.

Sørensen, T. (1948) A method of establishing groups of equal amplitude in plant sociology based on similarity of species content and its application to analysis of the vegetation on Danish commons. *Biologiske Skrifter Kongelige Danske Videnskabernes Selskab* 5, 1–34.

Srivastava, D.S. and Velland, M. (2005) Biodiversity-ecosystem function research: is it relevant to conservation? *Annual Review of Ecology, Evolution, and Systematics* 36, 267–294.

Stiling, P.D. (1999) *Ecology: Theories and Applications*, 3rd edn. Prentice Hall, Upper Saddle River, New Jersey.

Walker, B.H. (1992) Biodiversity and ecological redundancy. *Conservation Biology* 6, 18–23.

Walker, L.R. and Vitousek, P.M. (1991) An invader alters germination and growth of a native dominant tree in Hawai'i. *Ecology* 72, 1449–1455.

Waser, P.M. and Ayers, J.M. (2003) Microhabitat use and population decline in banner-tailed kangaroo rats. *Journal of Mammalogy* 84, 1031–1043.

Watson, D.M. (2001) Mistletoe – a keystone resource in forests and woodlands worldwide. *Annual Review of Ecology and Systematics* 32, 219–249.

Whittaker, R.J. (1975) *Communities and Ecosystems*, 2nd edn. MacMillian Publishing, New York.

Whittaker, R.J., Willis, K.J. and Field, R. (2001) Scale and species richness: towards a general, hierarchical theory of species diversity. *Journal of Biogeography* 28, 453–470.

Wilkinson, D.M. (1999) The disturbing history of intermediate disturbance. *Oikos* 84, 145–147.

Wilson, J.B. (1991) Does vegetation science exist? *Journal of Vegetation Science* 2, 189–190.

Wilson, J.B. (1994) Who makes the assembly rules? *Journal of Vegetation Science* 5, 275–278.

Wilson, J.B., Steel, J.B., King, W.M. and Gitay, H. (1999) The effect of spatial scale on evenness. *Journal of Vegetation Science* 10, 463–468.

Woodward, F.I. and Diament, A.D. (1991) Functional approaches to predicting the ecological effects of global change. *Functional Ecology* 5, 202–212.

Yorks, T.E. and Dabydeen, S. (1998) Modification of the Whittaker sampling technique to assess plant diversity in forested areas. *Natural Areas Journal* 18, 185–189.

11 Community Dynamics: Succession and Assembly

Concepts

- Communities are dynamic; their composition is constantly changing.
- Primary succession occurs on newly created land, whereas secondary succession occurs after a disturbance removes only part of the plant biomass and substrate.
- Processes that influence succession are facilitation, tolerance and inhibition. Succession is directed by site availability, species availability and species performance.
- Communities are assembled over time. Membership of a community will be determined by the interaction of species traits and biotic and abiotic filters.
- Species present in a community must be members of the total, habitat, geographical and ecological species pools, all of which have different filters.
- A species must pass through all filters to become a member of the community.

11.1 Introduction

In Chapter 10, we focused on the basic definitions, structure (mainly diversity) and issues around how structure and function are related in ecological communities. The composition of any community, however, is rarely static and will fluctuate seasonally and change over decades and centuries in response to environmental factors (Fig. 11.1). Even when we try to maintain a stable community through management, composition is dynamic. Succession is the directional change in community composition following a disturbance; therefore, it is different from seasonal or random fluctuations in vegetation. All plant communities face periodic disturbances. As discussed in Chapter 10, disturbances vary in their intensity, frequency and extent, ranging from a tree being blown over in a wind storm, to a hurricane, to a widespread and intense forest fire. Following a disturbance, the ecological community will respond, changing in response to the changes in the biotic and abiotic environment.

11.2 History and Development of Community Dynamics Theory

Early views of succession

The term succession was originally used by Thoreau in 1860 to describe changes in forest trees (McIntosh, 1999). It remained largely unused until Cowles (1899, 1901) studied primary succession on sand dunes of Lake Michigan near Chicago. He described how dunes developed through various community types into relatively stable forests. Clements (1916, 1936), a contemporary of Cowles, was a more forceful individual and therefore his rather dogmatic writings on succession overshadowed those of Cowles (McIntosh, 1999). Clements believed that a community was greater than the sum of the individual species and would have emergent properties unforeseen based on species alone. He described succession as a directional, progressive, orderly change in vegetation that would ultimately converge to a stable, predictable climax community. Clements believed that

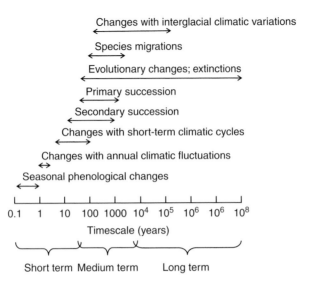

Fig. 11.1. Vegetation change over different timescales (redrawn from Miles, 1987).

early-establishing species facilitated the establishment of later species. Thus, Clements' view of succession proposed that autogenic (internal) processes controlled the development of the community climax. Two criticisms arose about Clements' ideas. First, Clements invoked climate as the sole determinant of community composition and neglected other factors. As we have seen in earlier chapters, other biotic and abiotic factors can be important community determinants. Second, Clements did not recognize the possibility of multiple successional pathways. Nevertheless, his work is important for his observations of community dynamics and his recognition of the importance of disturbance to the process of succession (Walker, 1999).

In response to Clements' work, Gleason (1917, 1926) noted that successional events were *not* predictable and that succession proceeded *independently* following a disturbance. Gleason saw communities as collections of co-occurring species with similar environmental tolerances. One criticism of Gleason's work was that he ignored the importance of species interactions in determining community composition, and instead focused almost entirely on abiotic processes (Tansely, 1935). Gleason's views were greeted sceptically at first, and were not taken seriously until decades later, in the 1950s. By then, ecologists had begun to recognize the work of Watt (1947) and others who said that communities were a mosaic of patches at different successional stages. While overall community structure might remain constant, individual patches were dynamic.

In reality, both Clements' and Gleason's views contribute to our understanding of ecological communities. Gleason's model is closer to current ideas, and most plant communities seem to follow his individualistic model. Certainly, most species will have unique sets of environmental tolerances and will therefore have a unique distribution. However, species interactions can change how and where a plant will live (Chapters 8 and 9), and therefore environmental tolerances alone do not determine distribution. As Clements suggested, some species are interdependent; we have seen this in our discussion of mutualisms (Chapter 9).

Equilibrium and non-equilibrium models

As we have seen, natural communities were once thought to exist in a state of equilibrium in which they develop over time into a specific stable community type (often called a climax community). Equilibrium communities were thought to be controlled primarily by competition, and species coexistence was thought to be dependent on niche differentiation and resource partitioning. The theory of communities reaching equilibrium has now been replaced by non-equilibrium concepts that focus on community-level processes and changes over time rather than on any single climax community state (DeAngelis and Waterhouse, 1987; Pickett *et al.*, 2009).

Non-equilibrium concepts recognize that while some communities may be at equilibrium at some scales, this is not necessarily the normal situation. Pickett *et al.* (1992) suggested that we consider nature to be in flux rather than in balance.

Implicit in the equilibrium model was the idea that following a disturbance, a community would return to its original state (Perrings and Walker, 1995). This idea seems to persist amongst laypersons, even though it has been rejected by most empirical tests for the last 80 years (Pickett *et al.*, 1992; Holling *et al.*, 1995; Odion *et al.*, 2010). For example, we now know that following a disturbance, different types of communities can develop and therefore there is no such thing as a single climax community (Walker, 1981; McCune and Allen, 1985; Roscher *et al.*, 2009; Warman and Moles, 2009; Odion *et al.*, 2010). In a classical example, Abrams *et al.* (1985), for example, showed that clear-cut jack pine (*Pinus banksiana*) stands can develop into either sedge meadows, hardwood and shrub communities, or return to a jack pine community, depending on the season of cutting and whether the site was burned by natural or controlled burns (Fig. 11.2).

Once we accept that communities are not necessarily at equilibrium, we can begin to describe how they change over time. In Chapter 10, we discussed how community stability can be described in terms of persistence, resistance and resilience. A community very strong in one component is not necessarily strong in the others. For example, a community may be persistent but not resistant or resilient, and therefore will be susceptible to disturbance or invasive species. A community may be resilient, but once a threshold is passed (such as when a new species invades), a transition will occur and a different community will form. Such changes in community structure and function can be irreversible and quite abrupt (Perrings and Walker, 1995). Murphy (personal communication), for example, studied the threshold under which northern deciduous forest communities can resist invasive buckthorn (*Rhamnus* and other species). In the deciduous forest, there is a relatively narrow range of biotic and abiotic factors; however, these tend to change once forests are cut down, edges made abrupt and the patches isolated. The patches resist invasion by species such as buckthorn for decades, but once a threshold is passed, there is a rapid (10-year) shift in which the patch becomes nearly totally dominated by buckthorn. Not all buckthorn species react similarly during succession. At least one species of buckthorn (glossy buckthorn, *Frangula alnus*) may be gradually suppressed by light and nutrient changes during succession, which gradually reduces its dominance (Cunard and Lee, 2009).

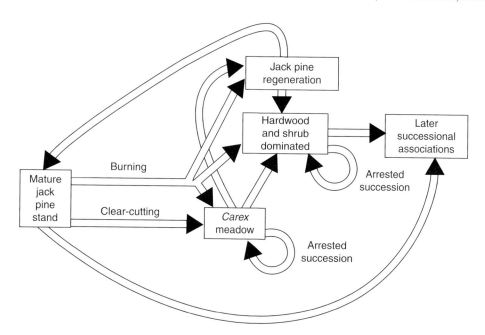

Fig. 11.2. Clear-cut jack pine (*Pinus banksiana*) forests have multiple successional pathways (Abrams *et al.*, 1985).

11.3 Primary and Secondary Succession

There are two main types of succession, primary and secondary. Primary succession occurs on newly established land (such as following a volcanic eruption) where no plants have grown previously or where there is no effective seed bank. Secondary succession occurs after a disturbance removes only part of the plant biomass and substrate (such as following a forest fire). Secondary succession proceeds faster than primary succession because there are more resources available and there are remaining plant propagules and fragments (Pickett *et al.*, 2009).

Primary succession

Habitats undergoing primary succession are usually environmentally harsh because there is no vegetation to ameliorate the abiotic environment, and there tend to be few nutrients and little water. The 1980 eruption of Mount St Helens in Washington state has provided an ideal venue to study primary succession (Dale *et al.*, 2005; del Moral and Lacher, 2005; del Moral, 2007; del Moral *et al.*, 2009). In general, researchers found succession proceeded slowly. While species richness increased fairly rapidly, even after 10 years, the percentage cover was low (Fig. 11.3) (del Moral

and Bliss, 1993). During succession, species composition differed among the habitat types, with the characteristics of early-colonizing species influencing how succession proceeded. Once an early-colonizing species established, its own seed production often outnumbered propagules of other species. Furthermore, early colonizers tended to be located in the more favourable microhabitats, thus improving their own survival. Colonizing species often influenced subsequent species by altering the biotic and abiotic environment. For example, nitrogen-fixing species, such as lupins (*Lupinus* spp.) and red alder (*Alnus rubra*), changed the soil fertility and influenced what species subsequently colonized. There were also areas with remnant populations remaining, leading to secondary succession.

Secondary succession

Secondary succession is initiated by a natural or human-caused disturbance such as a fire, hurricane or flooding. In all cases, there are seed banks or existing vegetation available to aid regeneration. The rate at which secondary succession proceeds is dependent on the type of soil substrate remaining and on whether established vegetation and other organisms are nearby and can provide propagules

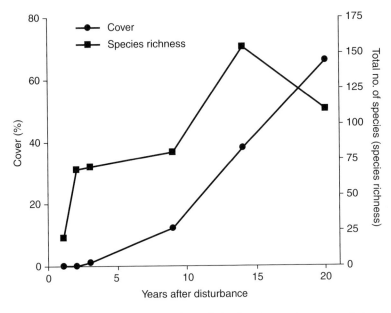

Fig. 11.3. Change in species richness (species number/250 m² plots) and percentage cover during primary succession at five sites on Mount St Helens (redrawn from del Moral and Bliss, 1993).

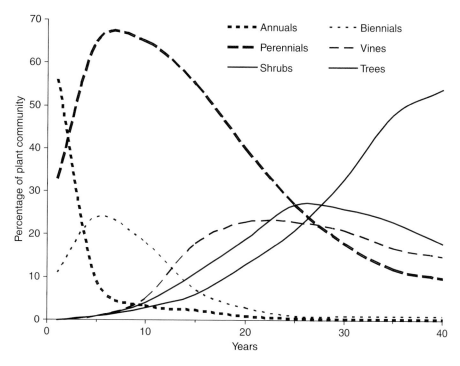

Fig. 11.4. Changes in percentages of annuals, biennials, perennials, vines, shrubs and trees during old-field secondary succession (redrawn from Meiners *et al.*, 2008).

for regeneration. The rate of succession is also dependent on the type of disturbance initiating succession. For example, a hurricane, clear-cut or small brush fire can initiate secondary succession, but the vegetation will develop at different rates.

Old-field succession is probably the most studied type of secondary succession. It follows the cessation of farming activities such as ploughing and herbicide usage. A classic example of secondary succession is the Buell-Small Succession study which started in 1958 in New Jersey (Meiners *et al.*, 2008) (Fig. 11.4). Annual species dominated the first year but quickly decreased in subsequent years, while the biennial and perennial herbs were the main species for the next 10 years. Trees and shrubs gradually increased over time. Non-native species made up over 50% of the species during the first 10 years but decreased to less than 30% after 30 years.

11.4 Patterns and Processes of Succession

We have looked at specific examples of primary and secondary succession, but what generalizations can

we make about communities undergoing succession, and what ecological processes are important over the course of a successional pathway?

Patterns of succession

As communities develop, the types of plants growing there will change, often in a broadly predictable manner. During the early stages of succession, plant cover, biomass and species richness tend to increase. During the later stages, these may level off or decrease. In addition, the canopy height increases from small annual and herbaceous perennials to shrubs and trees during many successional pathways.

Penman *et al.* (2008) present a model of how communities change over time by using a dissimilarity index that shows how much a community changes from its original state (Fig. 11.5). An unchanging community will be a flat line (A), whereas a constantly changing ecosystem (for example, in response to climate change) would increase as the community became progressively more different from the initial state (B). Following a classical succession model, the community would

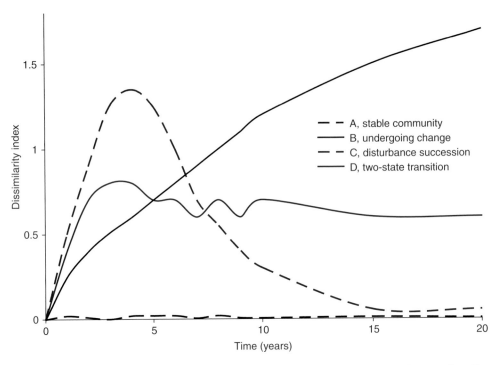

Fig. 11.5. Models of how communities may change over time, based on a dissimilarity index (redrawn from Penman *et al.*, 2008).

rapidly change and then gradually return to the original state (C). Alternatively, a disturbance can affect a community so that it never returns to its original state (D).

Plant traits associated with early and late successional species differ (Table 11.1). During the early stages of succession there is a high rate of species change and replacement, but as the community ages, species turnover rate declines. This is primarily caused by the rapid growth rate and short lifespan of early-successional species compared with late-successional species. Grime (1977) illustrated patterns of succession in his C-S-R model (Fig. 11.6; see also Chapter 3). In early secondary succession, species are ruderals (R) because there is abundant light and nutrients. The successional pathway favours competitive species (C), and then long-lived stress tolerators (S), as nutrients and light become limiting. The direction of the pathway will vary with the level and consistency of potential productivity. Higher potential productivity will lead to species that are competitive in the intermediate stages. Potential productivity rarely stays constant throughout a

pathway. Nutrients can also be added or removed through human actions.

Using this model, successional pathways of specific habitats can be mapped. Caccianiga *et al.* (2006) found support for this pathway from ruderals to stress tolerators in primary succession following a glacial retreat, as did Navas *et al.* (2010) in a study of secondary succession in Mediterranean France. This way of illustrating succession, however, is unlikely to fit all situations. Ecke and Rydin (2000) examined the C-S-R model and found that primary succession on some sea-coast meadows did not follow the expected trend. Here, they found that ruderals were not able to colonize the disturbed sites, but that species tolerant of stress were dominant in early succession. Thus, we can gain general insight using the C-S-R model, but it may not be applicable in all situations.

Processes of succession: facilitation, inhibition and tolerance

There are many models of succession and they are nuanced in the sense that their differences are

Table 11.1. Comparison of physiological and life history traits and population dynamics of plants from early and late stages of succession (from Pianka, 1970; Bazzaz, 1979; Huston and Smith, 1987).

Trait	Early succession	Late succession
Seed size and number	Many, small seeds	Few, large seeds
Seed dispersal distance	Long	Short
Seed dispersal mechanisms	Wind, birds, bats	Gravity, mammals
Seed viability	Long	Short
Size at maturity	Small	Large
Maximum lifespan	Short (often <1 yr)	Long (usually > 1 yr)
Timing and frequency of reproductive events	Early, often monocarpic	Late, usually polycarpic
Growth rate	Fast	Slow
Structural strength	Low	High
Survivorship	Often Deevy Type III	Often Deevy Types I and II
Population size	Often variable over time	Fairly constant over time
Resource acquisition rate	High	Often low
Recovery from nutrient stress	Fast	Slow
Root-to-shoot ratio	Low	High
Photosynthetic rate	High	Low
Photosynthetic rate at low light	Low	High
Respiration rate	High	Low

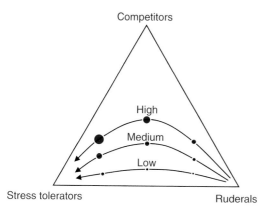

Fig. 11.6. Succession represented on Grime's C-S-R (competitor–stress tolerator–ruderal) triangle, showing general successional patterns under conditions of high, medium and low potential productivity. Circle size indicates the amount of plant biomass present at that stage of succession (redrawn from Grime, 1977, 1987).

subtle and represent refinements rather than radical changes (Fig. 11.7). Egler (1954) proposed two models to explain succession. The 'initial floristics model' described how most species were present at the initial stages of succession. Later successional species became more prominent over time as a result of longer lifespan, slower growth rate and larger size at maturity. The second of Egler's models, the 'relay floristics model', proposed that species prepare the environment for later-appearing species by making it inhospitable to themselves. For example, early species might provide shade for germinating trees, which then grow tall and shade out early-successional species. Later, Drury and Nisbet (1973) expanded on Egler's initial floristics model and tested whether physical stress and competition might also be important processes in succession.

Connell and Slatyer (1977) expanded on some of these ideas and presented three models to explain the mechanisms of succession: facilitation, inhibition and tolerance. The facilitation model develops Clements' and Egler's original ideas that early species aid the invasion of later ones. Inhibition occurs when existing plants prevent or inhibit the establishment of subsequent species. This is caused by a combination of physical, chemical or biotic mechanisms. Usually, the early species become established in a site and pre-empt biological space. In the tolerance model, all species are able to establish following disturbance but the species composition will change owing to the different resource-use strategies. Eventually, late-successional species become dominant because they are more tolerant of competition.

Walker and Chapin (1987) and Connell *et al.* (1987) both clarified (and debated) how many models of succession there might be and began a discussion of the successional mechanisms. On the one hand there was the desire to simply classify successional models with as much parsimony as

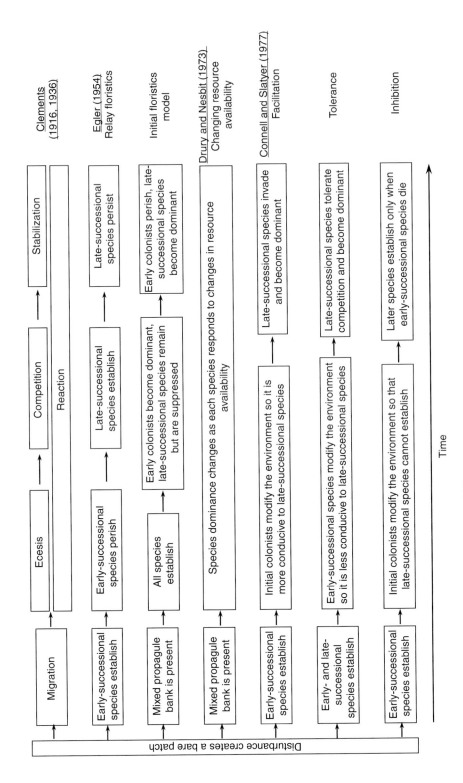

Fig. 11.7. Models of succession (redrawn from Luken, 1990).

possible, i.e. the facilitation, inhibition and tolerance models. On the other hand, Connell *et al.* (1987) clarified that this is at odds with the reality that there are many drivers of succession. Facilitation, inhibition and tolerance are still used as a basis to describe the processes underlying succession. These processes interact and the development of most pathways is a combination of all three (Pickett *et al.*, 2009). Furthermore, the relative importance of the three processes will change over time.

The contemporary view of succession is that while classifications are handy, they belie the complexity of how succession proceeds along trajectories that depend on nutrient and species pools in the context of land use and its history (Pickett *et al.*, 2009; Xiao *et al.*, 2009). Effectively, the outcome of succession in one case may be more dependent on classical facilitation as colonizing species build soil and nutrients and are replaced by new ones, but in other cases it may be that species persist as they tolerate abiotic or biotically influenced environmental change (Xiao *et al.*, 2009).

Interacting successional processes

The types of processes that are important in determining species change will vary with the stage of succession (early, mid and late), the type of succession (primary versus secondary) and the level of available resources (water and nutrients) (Walker and Chapin, 1987). For example, seed arrival is highly important in the early stages of primary succession because seeds will be limited (Fig. 11.8a). The first species establishing in a newly exposed environment can determine which subsequent species will be successful. Conversely, buried seeds and propagules will be more important in secondary succession because these will determine early species composition. Facilitation is more important in nutrient-poor environments, but its importance decreases over the course of succession because vegetation will moderate the environment (Fig. 11.8b).

There are many examples of interacting processes in natural communities. For example, after Mount St Helens erupted, nitrogen-fixing lupins (Pacific lupin, *Lupinus lepidus*) were among the first species to colonize. When they died, they facilitated the establishment of other species that could take advantage of the released nitrogen (del Moral and Bliss, 1993; Dale *et al.*, 2005). The same lupins, however, also inhibited the establishment of

other individuals while they were alive. It was only after their death that lupins facilitated the establishment of other species.

Hierarchy of successional processes

When Connell and Slatyer (1977) developed their three models, they did not look at hierarchy among processes. Therefore, Pickett *et al.* (1987a,b; 2009) proposed a hierarchy of successional processes; at the largest scale, they said that succession is determined by three factors: site availability, species availability and species performance, which are influenced by contributing factors and their modifiers (Fig. 11.9).

Site availability

The process of succession is initiated when a disturbance creates or alters a site. Disturbances can be characterized by their extent (area affected), frequency and magnitude (Walker and Willig, 1999). Magnitude is a combination of intensity (physical force) and severity (impact). The extent of a disturbance determines the environmental conditions and the heterogeneity of the habitat, whereas the magnitude will determine space and light availability and the numbers and types of propagules available for regeneration (Pickett *et al.*, 1987a,b).

Species availability

Succession following an intense and extensive disturbance can be more dependent on long-distance dispersal, whereas the seed bank and remnant vegetation will be the source of propagules following less intense disturbance (Walker and Chapin, 1987). For example, del Moral and Bliss (1993) found that proximity to vegetation influenced succession on volcanoes. Lupins and other large-seeded species were early colonists because there were remnant populations serving as seed sources. Sites further from remnant vegetation were composed mostly of small-seeded, wind-dispersed species.

This mechanism of recolonization from local species pools is a key process in succession and leads to some interesting considerations with respect to invasive non-native species. The first is semantic: technically, all new species colonizing during primary or secondary succession are invasive. Some may not be local and might be somewhat

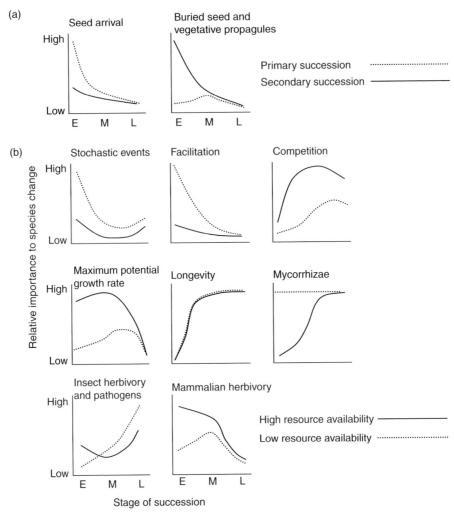

Fig. 11.8. Influence of (a) type of succession, and (b) level of resource availability on the successional process over the course of succession (E = early; M = mid; L = late) (redrawn from Walker and Chapin, 1987).

non-native, though still native to a larger watershed region – they may be transported or have survived an environmental change and can now dominate. Recolonization after a disturbance would usually proceed with similar species that gradually disperse from the unaffected or less disturbed surrounding communities. Unless the scale of disturbance is over thousands of square kilometers and near total in its destructive capacity, the surrounding communities are normally similar to the ones disturbed or destroyed, and recolonization would proceed with similar species. However, if there are large numbers of non-native invasive species in these surrounding communities, it is likely that these will colonize. It is possible that these colonizing non-native species may fulfil a similar ecosystem function to the former native colonizers, and that succession proceeds on a similar trajectory, but the cumulative evidence indicates that successional trajectories are altered dramatically (Kassi N'Dja and Decocq, 2008; Ainsworth and Kauffmann, 2009; Cunard and Lee, 2009; Fowler and Simmons, 2009; Laungani and Knops, 2009; Motta *et al.*, 2009; Peltzer *et al.*, 2009; Reid *et al.*, 2009; Tallent-Halsell and Watt, 2009; Tassin *et al.*, 2009).

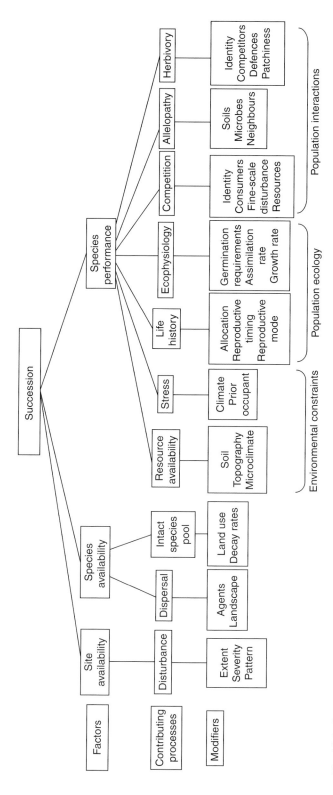

Fig. 11.9. Causes of succession and their contributing processes and modifying factors (redrawn from Pickett *et al.*, 1987a).

Species performance

A species' performance depends on its population ecology, on how it interacts with other populations, and on environmental factors such as resource level and abiotic stress. Performance can be influenced by idiosyncratic events that occur during succession. For example, the timing of the Mount St Helens eruption in May 1980 may have changed which species were successful (del Moral and Bliss, 1993). When the eruption occurred, much of the ground was still covered in thick snow, and this allowed some plants to persist despite ash covering much of the surface. Also, ice blocks left crevices when the ice melted, and this created moist habitats suitable for seedling establishment. Furthermore, the favourable conditions during the three summers following the eruption are likely to have increased seedling establishment by allowing new recruits to get a foothold. If conditions had been dry during these first summers, succession might have been delayed or altered.

Using succession to manage invasive species and restore ecosystems

Understanding succession can help to determine how, when and if invasive species should be managed.

Some colonizing species that are invasive may disappear on their own during succession and therefore no management is needed (Pyšek et al., 2004). For example, the proportion of non-native species decreased over time during succession of old fields in New Jersey (Meiners et al., 2008). In New Zealand, however, gorse (U. europaeus) interrupts the usual course of succession and must be removed to allow the early stages of succession to occur (Sullivan et al., 2007). In Hawaiian lowland wet forests, only the later stages of succession (after 200 to 300 years) were susceptible to invasive species (Zimmerman et al., 2008). During the early stages, the harsh, low nutrient conditions in Hawaii limit invasive species. Hence we should be cognizant that management may be necessary in the later stages of succession.

Invasive species may be used to change the direction of the successional pathway. For example, non-native species were sown on Mount St Helens to reduce soil erosion (Dale et al., 2005). After 15 years, plots with non-native species had higher native species richness and higher cover because the non-native species facilitated the establishment of the native species. Care must be taken when using non-native species for management, however, as this practice may have unintentional results.

Box 11.1. Detecting change in vegetation over time.

There are two general ways to examine succession. Long-term studies are used to follow how one community changes over time, whereas chronosequence studies are used to compare communities of the same type but of different ages.

Long-term studies

The most logical way to study how communities change over time is to simply watch and measure changes as they occur. You (the researcher) could regularly visit a community (say every year) and measure biotic and abiotic factors. This was done by Meiners et al. (2002, 2008), who studied agricultural fields that were sequentially abandoned. To conduct this study, the researchers used 48 permanent plots in each of ten agricultural fields that were sequentially abandoned starting in 1958 (see also Small et al., 1971). Plots were sampled annually for 11 years, and then every second year until year 40. On each sampling date, percentage cover of each species was recorded in permanent plots. From these data, the researchers could calculate the proportional species richness and cover of native and non-native species (Fig. 11.4).

Long-term studies are an excellent way to follow precise changes in community structure and function over time. However, this type of study requires time, and this is not always available. It also requires that a patch of land be preserved for the use of the researcher and protected from development (unless this is what is being studied).

Chronosequences

To avoid the problems of long-term studies, some researchers use chronosequences to study changes in vegetation over time. To do this, plots of different successional stages are compared and the researcher recreates a chronological sequence (chronosequence) of the successional pathway. Pickett et al. (1987a) called this 'space for time substitution'.

Continued

Box 11.1. Continued.

Table 11.2. Plant functional groups used by Csecserits and Rédei (2001).

Type	Description
Weeds	Ruderal species
Sand generalists	Disturbance-resistant, pioneer species of open sand steppe
Sand specialists	Species of open sand steppe that are less resistant to disturbance
Steppe generalists	Disturbance-resistant species of closed sand steppe
Steppe specialists	Species of closed sand steppe that are less resistant to disturbance

Csecserits and Rédei (2001) used the chronosequence approach to study whether natural secondary succession was adequate to restore plant communities following field abandonment, or whether active restoration efforts were required. Selected fields were divided into four age classes according to years since abandonment. Species were divided into four life history strategies (annual, biennial, perennial, woody), as well as into five functional groups (Table 11.2). The authors found that the relative abundance of annuals tended to decrease over time, while the abundance of perennials and woody plants increased. In the functional groups, weed species richness decreased in the first 10 years, whereas sand and steppe specialist groups increased during this interval. The patterns of abundance of the functional groups were slightly different. From the first to third age group, weed abundance decreased, while the abundance of sand specialists increased (Fig. 11.10). Csecserits and Rédei (2001) concluded that there was no need to have active restoration efforts because weed abundance decreased over time and late-successional species (sand and steppe specialists) had appeared after 10 years of abandonment. Therefore, the authors were able to determine that weeds would not cause persistent problems and that the process of natural succession was enough to return abandoned fields to semi-natural communities.

There are limitations to the use of chronosequences (Bakker *et al.*, 1996; Pickett *et al.*, 1987a; Foster and Tilman, 2000). First, this approach assumes that conclusions drawn from spatial relationships are the same as conclusions drawn from temporal relationships. That is, if you sample sites of different ages, you will observe the same patterns as you would have if you had observed one site over time. In addition, because you are averaging the effects observed at different sites, you can only obtain a general level of detail to explain observations. Finally, since site-specific factors can confound results, correlations between species abundances and community attributes may not be directly related to the successional processes. Nevertheless, chronosequences are a useful and commonly used technique.

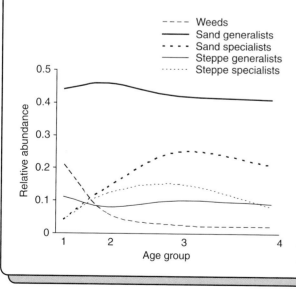

Fig. 11.10. Change in the relative abundance of weeds, sand generalists and specialists, and steppe generalists and specialists, following the abandonment of agricultural fields in central Hungary. Fields were divided into four age classes: 1: 1–5 yrs, 2: 6–10 yrs, 3: 11–23 yrs, 4: 24–33 years (redrawn from data in Csecserits and Rédei, 2001).

11.5 Community Assembly

Communities are complex entities and, as a result, the thought of looking at whole communities is daunting because there are so many interacting parts. Some scientists deal with complexity by trying to understand all of the community parts individually, and then trying to put the parts together. Another way to deal with complexity is to look at the community as a whole entity and to ignore the details. As the large-scale processes are understood, then more detail can be added. Community assembly theory allows us to do this (Booth and Swanton, 2002). It is a way of looking at community development as a process of filtering or sorting of species as they change over time (Keddy, 1992; Booth and Swanton, 2002; Shipley *et al.*, 2006).

Communities follow trajectories through time

To help us think about communities as dynamic entities we can think of a community as following a trajectory through time (Drake *et al.*, 1999). A trajectory is a path through a series of community states. In a traditional Clementsian view of succession, community development follows a single trajectory: that is, given a species pool, only one community type develops along a deterministic trajectory (Fig. 11.11a). Alternatively, many types of communities can result from one species pool if trajectories are divergent (i.e. they are indeterministic) (Fig. 11.11b). Sometimes divergent trajectories may converge, producing identical community states (Fig. 11.11c).

It may be difficult to distinguish between a deterministic and an indeterministic trajectory because, over time, trajectories can repeatedly diverge and converge (Fig. 11.12a), and this may or may not be predictable (Rodriguez, 1994; Samuels and Drake, 1997). In field experiments, for example, Inouye and Tilman (1988, 1995) found that old-field communities converged after 4 years of nitrogen addition, but that after 11 years, these communities had diverged. Sometimes a community reaches a state of predictable cycling and, if viewed over too short a timespan, the trajectory will appear indeterministic. Furthermore, two communities following the same trajectory, but offset in time, will appear different, even though their dynamics are the same (Figure 11.12b). Such communities must be observed over a long enough timescale to distinguish them from random trajectories.

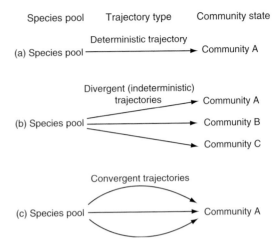

Fig. 11.11. Examples of how different community configurations may be produced from one species pool. (a) Assembly trajectories are deterministic when a species pool consistently produces the same extant community. These deterministic trajectories are relatively immune to historical influences such as invasion sequence. This is the classical view of succession. (b) Different communities may be produced from one species pool when trajectories diverge. These indeterministic trajectories are more sensitive to historical influences such as invasion sequence and changes in the assembly environment. (c) One community type may be produced when assembly trajectories converge. (adapted from Drake, 1990).

While many old fields develop along a predictable trajectory to re-form the pre-disturbance community type, some old-field trajectories are not predictable (Cramer *et al.*, 2008). Repeatable trajectories are more likely to occur after traditional

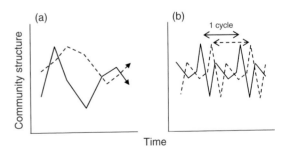

Fig. 11.12. Examples of the divergence and convergence of community trajectories over time. In (a) communities are random or chaotic and trajectories will continue to converge and diverge. In (b) one trajectory is reached, but the two communities are cycling out of synchronization (adapted from Samuels and Drake, 1997).

agricultural practices have been followed and where there is intact vegetation nearby. Trajectories are more likely to remain in a degraded state when establishment is limited by a lack of seeds or when there is competition from non-native grasses. The early stages of community development are often the most important in determining the type of trajectory. For example, MacDougall *et al.* (2008) found that degraded semi-arid grasslands returned to their native community type only if dominant exotic species were removed and if sufficient rain occurred during seedling establishment.

Species pools

Only some members of the available species pool will become part of a community. Other species will be removed or filtered out by biotic and abiotic processes at various life stages. Every community is composed of a subset of the total species pool – that is, the group of species that are available to colonize (Fig. 11.13). Belyea and Lancaster (1999) differentiated among five types of species pools.

- total species pool – a large-scale species pool determined by landscape-scale ecological and evolutionary processes
- habitat species pool – the subset of the total species pool that could establish and survive in the habitat
- geographical species pool – species able to disperse into the habitat

- ecological species pool – the overlap of species present in both the geographical and habitat species pool
- actual species pool – species present in the above-ground community.

The only time when all types of species pools would be the same is in a closed and stable community where there is no dispersal into a community.

Ecological filters

The processes that remove species from a community are commonly called filters. Thus, biotic and abiotic filters limit membership to each species pool, and different types of filters will operate under different conditions. Dispersal filters determine the geographical species pool, environmental filters determine the habitat species pool, and biotic filters determine the actual species pool. Some consider filters to work at two levels (Keddy, 1992; Cingolani *et al.*, 2007). First, filters determine which species are more likely to pass through filters and be present in the community. Second, they influence which of these species will be likely to be dominant. Different traits will be advantageous in the two types of filters.

Dispersal filters – arriving at the party

Dispersal filters determine what species arrive at a site. Communities do not have an unlimited and continuous supply of propagules because propagules

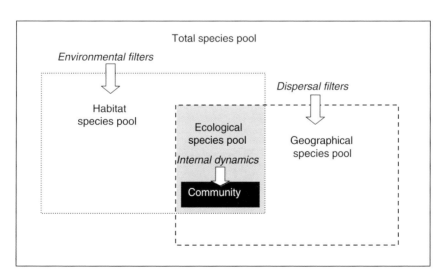

Fig. 11.13. Types of species pools (redrawn from Belyea and Lancaster, 1999).

are not produced at a constant rate, nor do they disperse evenly over space (Belyea and Lancaster, 1999). Earlier in this chapter, we discussed the importance of proximity to seed sources. In addition to this, seed characteristics, plant phenology and abiotic conditions determine when and whether a propagule can arrive at a site. There has to be the right combination of seed type, dispersal agent and environmental conditions for dispersal to occur.

The timing, sequence, frequency (number of times a species' invasion is repeated) and rate (how quickly invasions are repeated) of species' introductions into a community can alter trajectories. The effect of invasion sequence is the most understood of these. Numerous studies using natural communities (Abrams et al., 1985; McCune and Allen, 1985) and microcosms (Robinson and Dickerson, 1987; Drake, 1991; Drake et al., 1993) have shown that the order that species are introduced can influence the community trajectory. Early invaders may have the advantage simply because they occupy biological space, inhibiting the invasion of late species, but this is not always the case. Later invaders may drive early ones to extinction by direct or indirect means. They may directly outcompete early invaders, or they may change the abiotic environment, making it inhospitable to the earlier species.

Less studied are the effects of invasion rate and frequency on the trajectory. In experimental studies, species introductions are done singly and at a constant rate, but this is not how invasions occur in natural situations. Generally, increasing the invasion rate and/or frequency increases a community's richness and decreases the likelihood of there being a single stable trajectory because different species will be favoured over time (Hraber and Milne, 1997; Lockwood et al., 1997). A high invasion rate and frequency minimize the influence of historical events (Lockwood et al., 1997). Communities will be more persistent when the invasion rate and frequency is low because the assembly process is not disrupted.

Environmental filters – crashing the party

After a species is dispersed into a community, it must be able to survive in the physical environment. We can get some indication of a plant's suitability to an environment, for example, by looking at a plant's growth rate versus average temperature and rainfall (Chapter 2), but the environment can also have subtle persistent effects on a community (Chesson, 1986). When we consider only average or typical environ-

mental conditions, we neglect occasional environmental extremes which could have long-term persistent effects on a community. For example, the distribution of the saguaro cactus (*Carnegiea gigantea*) in Arizona is limited by periodic frosts that kill seedlings, rather than by the cactus' physiological response to average temperature (Hastings and Turner, 1965). Periods of stress, or environmental fluctuation or extremes, may have a greater impact on the long-term community dynamics than average, relatively predictable environmental conditions. An extreme event may cause some species to go extinct, or may severely reduce their abundance, allowing other species to gain an edge. Environmental variation will alter a community's susceptibility to invasion. Species not usually able to establish may gain an advantage during a period of unusual environmental conditions.

Internal filters – being the life of the party

Seeds or propagules can arrive at a site and may be able to survive the abiotic conditions, but not all species will become part of the extant community. Internal filters act on the ecological species pool. Population interactions (Chapters 8 and 9) drive internal dynamics. We cannot predict the outcome of all interactions between all species under all dispersal and environmental constraints, nor would this have any predictive value. To make this approach possible and useful, some researchers have used plant characteristic or traits to examine assembly dynamics (Keddy, 1992; Díaz et al., 1999a,b; McIntyre et al., 1999a,b; Weiher et al., 1999).

Plant traits and functional groups

We have been taught to classify plants according to their phylogeny – that is, we place them in their family, genus and species, but it might also be useful to classify plants by their traits. Traits are the physical and physiological characteristics that determine a species' ecological function. Examples of traits are leaf or seed size, water-use efficiency, rooting depth and pollination syndrome. Using traits rather than species allows researchers to make broad-scale comparisons because they are not tied to species identities (Aubin at al., 2009).

Environmental filters act to remove species that lack specific traits (Fig. 11.14), and thus *traits*, rather than *species* are filtered (Weiher and Keddy, 1999; Booth and Swanton, 2002; Funk et al., 2008). Species without the suite of requisite traits will not

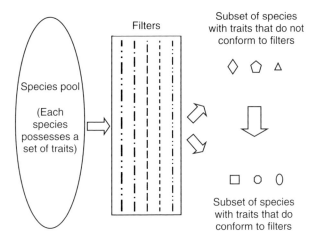

Fig. 11.14. A conceptual model of the trait-based approach to community assembly. A large pool of species is available but the species must pass through a series of biotic and abiotic filters which remove species that do not possess specific traits (adapted from Weiher and Keddy, 1999).

be able to pass through the series of environmental filters. Plants respond at scales from climate to disturbance to biotic interactions (Woodward and Diament, 1991). Funk *et al.* (2008) categorized the filters used in restoration from regional to local scale as dispersal filters (seeding) to environmental filters (burning and mowing) and biotic interactions (weeding to biotic control). However you classify them, each type of filter selects against a unique set of traits; therefore, the ability of a species to pass through one filter will not necessarily affect its ability to pass through another. Different traits may be required for each. When we work with traits, we first have to decide which traits are important.

Selecting plant traits and forming functional groups

How do we determine which traits are biologically relevant? We need to select traits that address the multiple scales of environmental filters. Traits associated with both growth and reproduction should be included (Díaz *et al.*, 1999a). The first influences resource acquisition and storage, and the second recolonization and regeneration. Furthermore, both physiological and morphological traits are important. One trait may be important to several processes. For example, seed size affects dispersal, germination, risk of predation and seedling competitive ability. The selection of traits will depend on the habitat type, on the

regional flora and on the goals of the study. Furthermore, the traits that are advantageous at one stage of invasion may not be advantageous at another stage (Pyšek and Richardson, 2007). Table 11.3 presents a list of potential traits to consider.

We have reduced the complexity of our community from a list of species to a list of traits. We can simplify it still further by constructing functional groups from the traits. A functional group contains species with a similar set of traits. They serve similar ecological functions in a community and are therefore filtered from species pools in a similar manner. But how do we divide our traits into functional groups? There are a number of ways to do this (Lavorel *et al.*, 1997; Smith *et al.*, 1997; Aubin *et al.*, 2007, 2009). Functional groups can be formed based on a researcher's ecological understanding of traits important to the community type (Nobel and Slatyer, 1980). Alternatively, there are a variety of statistical clustering techniques that group species with similar traits (Kleyer, 1999; McIntyre *et al.*, 1999b; Deckers *et al.*, 2004; Aubin *et al.*, 2007). Then a matrix analysis can be used to look at the relative occurrence of the emergent groups in different habitats; for example, in an invaded and an uninvaded forest.

Aubin *et al.* (2009) compared two trait-based approaches to characterize vegetation changes in an old-field to deciduous forest succession. In the first approach, when 13 biological traits were considered separately, traits were clearly associated with specific successional stages. In the second

Table 11.3. List of traits that could be used as a starting point for a trait-based approach to community assembly and possible ways to quantify them. Some or all of the traits could be selected to record, and of these only some traits would be ecologically important. Based on lists in Díaz *et al.* (1999b), Díaz Barradas *et al.* (1999), Kleyer (1999) and McIntyre *et al.* (1999b).

Trait	Classification of trait
Vegetative traits	
Plant height	<10 cm; 11–20 cm; 21–50 cm; 51–100 cm; >100 cm
Plant height (h):width (w) ratio	h:w>1; h:w = 1; h:w<1
Specific leaf area	Aphyllous; <1 cm^2; 1.1–2 cm^2; 2.1–3 cm^2; 3.1–5 cm^2; >5 cm^2
Life cycle	Summer annual, winter annual, biennial, perennial monocarpic, perennial polycarpic
Lifespan	<1 yr; 2–5 yr; 5–10 yr; 11–20 yr; >20 yr
General form	Prostrate, rosette; erect; tussock; vine; shrub
Leaf form	Aphyllous; evergreen; deciduous
Leaf angle	<90°; >90°
Leaf size	Aphyllous; <1 cm^2; 2–5cm^2; 6–10 cm^2; 11–25 cm^2; 26–50 cm^2; >50 cm^2
Leaf shape (length (l):width (w))	l:w>1; l:w = 1; l:w<1
Photosynthetic metabolism	CAM; C$_3$; C$_4$
Potential relative growth rate	Low; medium; high
Drought avoidance	None; succulent stem, tap root or other storage organ
Palatability	Unpalatable; low or just at juvenile stage; moderate; high
Leaf texture	Smooth; hairy; spines
Root morphology	Tap root; mostly horizontal; mostly vertical
Maximum rooting depth	<10 cm; 10–25 cm; 26–50 cm; 51–100 cm; >100 cm
Extent of clonal expansion	None; some (dm); high degree (m)
Resprouting ability	None; moderate (daughter plants remain attached to parent plant for some period of time); high (daughter plants rapidly become independent)
Mycorrhizal associations	None; ectomycorrhizal; vesicular–arbuscular
Storage organs	None; tubers, bulbs, rhizomes present
Reproductive traits	
Seed size (max. length)	<1 mm; 1–2 mm; 3–5 mm; 6–10 mm; >10 mm
Seed shape (variance of seed length, width and depth)	<0.15; 0.15–<1; 1–5; >5
Seed number (per plant)	<100; 100–999; 1000–5000; >5000
Weight of dispersal unit (fruit or seed)	<0.2 mg; 0.3–0.5 mg; 0.6–1 mg; 1–2 mg; >2 mg;
Seed dispersal vector	No mechanism; wind; highly mobile animals (birds, bats); low-mobility animals (ants, rodents)
Fruit type	Dry indehiscent; dry dehiscent; fleshy
Season of germination	Variable; early spring; late spring; summer; autumn
Age of first reproduction	<3 months; 3 months–1 yr; 1–3 yr; >3 yr
Peak period of flower and fruit production	None; autumn to early spring; spring; late spring to late summer; late summer to autumn
Pollination mode	Wind; specialized animals; non-specialized animals
Position of dormant buds (physiognomic types)	Therophyte; geophyte; hemiphyte; cryptophyte; chamaephyte
Agriculture-specific traits	
Herbicide tolerance	–
Weed size relative to crop	Smaller; same; larger

Table 11.4. Results from emergent functional group analysis (Aubin *et al.*, 2009).

Code	Functional group	No. of species	Proportional changes over chronosequence
W1	Tall, mammal-dispersed woody species	11	Increase with stand age
W2	Wind-dispersed woody species	22	Increase with stand age
W3	Short, bird-dispersed, multi-stemmed woody species	22	Decrease with stand age
W4	Summer-flowering, bird-dispersed, woody species	8	Peak at tall shrub/young tree stage
H1a	Short, non-native species dispersed by mammals	21	Decrease with stand age
H1b	Tall, wind-dispersed, non-native species	20	Decrease with stand age
H2	Annuals	16	Decrease with stand age
H3	Summer-flowering perennials	27	No change
H4	Late-flowering, wind-dispersed perennials	28	Peak at shrub stage
H5	Ferns and allies	16	Increase with stand age
H6	Spring-flowering, mammal-dispersed perennials	29	Decrease with stand age
H7a	Wind- or gravity-dispersed herbaceous species	18	No change
H7b	Ant-dispersed herbaceous species	15	No change

approach, when functional traits were analysed together, Aubin *et al.* (2009) identified 11 emergent functional groups (Table 11.4). With the exception of groups H3 and H7 (a and b), all groups varied with the stage of succession. Aubin *et al.* (2009) concluded that the second analysis is an easy-to-use approach that gives a synthetic view of the community. However, this method is reliant on the traits selected, and whether they are correlated. Aubin *et al.* (2009) found that emergent functional groups were too general and that the results obtained from the individual trait analysis were more detailed. For example, the importance of seed size, light requirements, plant form, and height were lost in the functional group analysis whereas they were significant traits in the trait-based analysis.

11.6   Summary

Plant communities are in a state of non-equilibrium, changing over time. The change in community composition following a disturbance is called succession. Secondary succession occurs faster than primary succession because there are more resources and propagules available. Facilitation, inhibition and tolerance are three underlying processes that act during succession. Their relative importance will change over time. Habitat availability and the availability and characteristics of a species will influence the course of succession. Community assembly theory allows us to consider how species composition changes over time as traits, and therefore species, are filtered out of the species pool. As we improve our understanding of assembly, we may be able to predict changes in community structure and function.

Box 11.2. Invasive species case study: community dynamics – succession and assembly.

- What role does your species play in community dynamics?
- Is it an early-, mid- or late-successional species?

11.7   Questions

1. Why are the terms 'stability' and 'equilibrium' misleading when describing communities?
2. Refer to Fig. 11.4 and explain the observed successional changes in terms of the plant traits that each of the six groups will have.

3. Explain why more competitors (C) occur in Grime's model (Fig. 11.6) when there are higher levels of potential productivity.
4. Refer to Fig. 11.8 from Walker and Chapin (1987) and explain each of the ten patterns displayed.

Further Reading

Booth, B.J. and Swanton, C.J. (2002) Assembly theory applied to weed communities. *Weed Science* 50, 2–13.

Funk, J.L., Cleland, E.E., Suding, K.N. and Zavaleta, E.S. (2008) Restoration through reassembly: plant traits and invasion resistance. *Trends in Ecology and Evolution* 23, 695–703.

Pickett, S.T.A., Cadenasso, M.L. and Meiners, S.J. (2008) Ever since Clements: from succession to vegetation dynamics and understanding to intervention. *Applied Vegetation Science* 12, 9–21.

Walker, L.R. and del Moral, R. (2003) *Primary Succession and Ecosystem Rehabilitation.* Cambridge University Press, Cambridge, UK.

References

Abrams, M.C., Sprugel, D.G. and Dickmann, D.I. (1985) Multiple successional pathways on recently disturbed jack pine sites in Michigan. *Forest Ecology and Management* 10, 31–48.

Ainsworth, A. and Kauffman, J.B. (2009) Response of native Hawaiian woody species to lava-ignited wildfires in tropical forests and shrublands. *Plant Ecology* 201, 197–209.

Aubin, I., Gachet, S., Messier, C. and Bouchard, A. (2007) How resilient are northern hardwood forests to human disturbance? An evaluation using a plant functional group approach. *Ecoscience* 14, 259–271.

Aubin, I., Ouellette, M.-H., Legendre, P., Messier, C. and Bourchard, A. (2009) Comparison of two plant functional approaches to evaluate natural restoration along an old-field-deciduous forest chronosequence. *Journal of Vegetation Science* 20, 185–198.

Bakker, J.P., Olff, H.J., Willems, J.H. and Zobel, M. (1996) Why do we need permanent plots in the study of long-term vegetation dynamics? *Journal of Vegetation Science* 7, 147–156.

Bazzaz, F.A. (1979) The physiological ecology of plant succession. *Annual Review of Ecology and Systematics* 10, 351–371.

Belyea, L.R. and Lancaster, J. (1999) Assembly rules within a contingent ecology. *Oikos* 86, 402–416.

Booth, B.J. and Swanton, C.J. (2002) Assembly theory applied to weed communities. *Weed Science* 50, 2–13.

Caccianiga, M., Luzzaro, A., Pierce, S., Ceriani, R.M. and Cerabolini, B. (2006) The functional basis of a primary succession resolved by CSR classification. *Oikos* 112, 10–20.

Chesson, P.L. (1986) Environmental variation and the coexistence of species. In: Diamond, J.M. and Case, T.J. (eds) *Community Ecology.* Harper and Row, New York, pp. 240–256.

Cingolani, A.M., Cabido, M., Gurvich, D.E., Renison, D. and Díaz, S. (2007) Filtering processes in the assembly of plant communities: are species presence and abundance driven by the same traits? *Journal of Vegetation Science* 18, 911–920.

Clements, F.E. (1916) *Plant Succession: an analysis of the development of vegetation.* Publication Number 242, The Carnegie Institution, Washington, D.C.

Clements, F.E. (1936) The nature and structure of the climax. *Journal of Ecology* 22, 9–68.

Connell, J.H. and Slatyer, R.O. (1977) Mechanisms of succession in natural communities and their role in community stability and organization. *American Naturalist* 111, 1119–1144.

Connell, J.H., Noble, I.R. and Slatyer, R.O. (1987) On the mechanisms producing successional change. *Oikos* 50, 136–137.

Cowles, H.C. (1899) The ecological relationships of vegetation on sand dunes of Lake Michigan: Parts I–IV. *Botanical Gazette* 27, 95–117, 167–202, 281–308, 361–391.

Cowles, H.C. (1901) The physiographic ecology of Chicago and vicinity. *Botanical Gazette* 31, 73–108.

Cramer, V., Hobbs, R.J. and Standish, R.J. (2008) What's new about old fields? Land abandonment and ecosystem assembly. *Trends in Ecology and Evolution* 23, 104–112.

Csecserits, A and Rédei, T. (2001) Secondary succession on sandy old-fields in Hungary. *Applied Vegetation Science* 4, 63–74.

Cunard, C. and Lee, T.D. (2009) Is patience a virtue? Succession, light, and the death of invasive glossy buckthorn (*Frangula alnus*). *Biological Invasions* 11, 577–586.

Dale, V.H., Campbell, D.R., Adams, W.M., Crisafulli, C.M., Dains, V.I., Frenzen, P.M. and Holland, R.F. (2005) Plant succession on the Mount St Helens debris-avalanche deposit. In: Dale, V.H., Swanson, F.J. and Crisafulli, C.M (eds) *Ecological Responses to the 1980 Eruption of Mount St. Helens.* Springer, New York, pp. 59–73.

DeAngelis, D.L. and Waterhouse, J.C. (1987) Equilibrium and nonequilibrium concepts in ecological models. *Ecological Monographs* 57, 1–21.

Deckers, B., Hermy, M. and Muys, B. (2004) Factors affecting plant species composition of hedgerows: relative importance and hierarchy. *Acta Oecologia* 26, 23–37.

del Moral, R. (2007) Limits to convergence of vegetation during early primary succession. *Journal of Vegetation Science* 18, 479–488.

del Moral, R. and Bliss, L.C. (1993) Mechanisms of primary succession: insights resulting from the eruption of Mount St Helens. *Advances in Ecological Research* 24, 1–66.

del Moral, R. and Lacher, I.L. (2005) Vegetation patterns 25 years after the eruption of Mount St. Helens, Washington, USA. *American Journal of Botany* 92, 1948–1956.

del Moral, R., Sandler, J.E. and Muerdter, C.P. (2009) Spatial factors affecting primary succession on the muddy River Lahar, Mount St. Helens, Washington. *Plant Ecology* 202, 177–190.

Díaz Barradas, M.C., Zunzunegui, M., Tirada, R., Ain-Lhout, F. and García Novo, F. (1999) Plant functional types and ecosystem function in Mediterranean shrubland. *Journal of Vegetation Science* 10, 709–716.

Díaz, S., Cabido, M. and Casanoves, F. (1999a). Functional implications of trait-environment linkages in plant communities. In: Weir, E. and Keddy, P.A. (eds) *Ecological Assembly Rules: Perspectives, Advances, Retreats.* Cambridge University Press, Cambridge, UK, pp. 338–362.

Díaz, S., Cabido, M., Zak, M., Martínez Carretero, E. and Araníbar, J. (1999b) Plant functional traits, ecosystem structure and land-use history along a climatic gradient in central-western Argentina. *Journal of Vegetation Science* 10, 651–660.

Drake, J.A. (1990) Communities as assembled structures: do rules govern pattern? *Trends in Ecology and Evolution* 5, 159–164.

Drake, J.A. (1991) Community-assembly mechanics and the structure of an experimental species ensemble. *American Naturalist* 137, 1–26.

Drake, J.A., Flum, T.E., Witteman, G.J., Voskuil, T., Hoylman, A.M., Creson, C., Kenny, D.A., Huxel, G.A., Larue, C.S. and Duncan, J.R. (1993) The construction and assembly of an ecological landscape. *Journal of Animal Ecology* 62, 117–130.

Drake, J.A., Zimmerman, C.R., Purucker, T. and Rojo, C. (1999) On the nature of the assembly trajectory. In: Weiher, E. and Keddy, P.A. (eds) *Ecological Assembly Rules: Perspectives, Advances, Retreats.* Cambridge University Press, Cambridge, UK, pp. 233–250.

Drury, W.H. and Nisbet, I.C.T. (1973) Succession. *Journal of the Arnold Arboretum* 54, 331–368. Reprinted In: Golley, F.B. (ed.) (1977) *Ecological Succession. Benchmark Papers in Ecology*, Vol. 5. Dowden, Hutchison and Ross, Stroudsburg, Pennsylvania, pp. 287–324.

Ecke, F. and Rydin, H. (2000) Succession on a land uplift coast in relation to plant strategy theory. *Annales Botanical Fennici* 37, 163–171.

Egler, F.E. (1954) Vegetation science concepts. I. Initial floristic composition, a factor in old field vegetation development. *Vegetatio* 4, 412–417.

Foster, B.L. and Tilman, D. (2000) Dynamic and static views of succession: testing the descriptive power of the chronosequence approach. *Plant Ecology* 146, 1–10.

Fowler, N.L. and Simmons, M.T. (2009) Savanna dynamics in central Texas: just succession? *Applied Vegetation Science* 12, 23–31.

Funk, J.L., Cleland, E.E., Suding, K.N. and Zavaleta, E.S. (2008) Restoration through reassembly: plant traits and invasion resistance. *Trends in Ecology and Evolution* 23, 695–703.

Gleason, H.A. (1917) The structure and development of the plant association. *Bulletin of the Torrey Botany Club* 44, 463–481.

Gleason, H.A. (1926) The individualistic concept of the plant association. *Bulletin of the Torrey Botany Club* 53, 7–26.

Grime, J.P. (1977) Evidence for the existence of three primary strategies in plants and its relevance to ecological and evolutionary theory. *American Naturalist* 111, 1169–1194.

Grime, J.P. (1987) Dominant and subordinate components of plant communities: implications for succession, stability and diversity. In: Gray, A.J., Crawley, M.J. and Edwards, P.J. (eds) *Colonization, Succession and Stability.* Blackwell Scientific, Oxford, UK, pp. 413–428.

Hastings, J.R. and Turner, R.M. (1965) *The Changing Mile.* University of Arizona Press, Tucson, Arizona.

Holling, C.B., Schindler, D.W., Walker, B.H. and Roughgarden, J. (1995) Biodiversity in the functioning of ecosystems: an ecological synthesis. In: Perrings, C.A., Mäler, K.-G., Folke, C., Holling, C.S. and Jansson, B.-O. (eds) *Biodiversity Loss: Economic and Ecological Issues.* Cambridge University Press, Cambridge, UK, pp. 44–48.

Hraber, P.T. and Milne, B.T. (1997) Community assembly in a model ecosystem. *Ecological Modelling* 104, 267–285.

Huston, M. and Smith, T. (1987) Plant succession: life history and competition. *American Naturalist* 130, 168–198.

Inouye, R.S. and Tilman D. (1988) Convergence and divergence of old-field plant communities along experimental nitrogen gradients. *Ecology*, 69, 995–1004.

Inouye, R.S. and Tilman, D. (1995) Convergence and divergence of old-field vegetation after 11 years of nitrogen addition. *Ecology* 76, 1872–1887.

Kassi N'Dja, N.J.K. and Decocq, G. (2008) Successional patterns of plant species and community diversity in a semi-deciduous tropical forest under shifting cultivation. *Journal of Vegetation Science* 19, 809–820.

Keddy, P. A. (1992) A pragmatic approach to functional ecology. *Functional Ecology* 6, 621–626.

Kleyer, M. (1999) Distribution of plant functional types along gradients of disturbance intensity and resource supply in an agricultural landscape. *Journal of Vegetation Science* 10, 697–708.

Laungani, R. and Knops, J.M.H. (2009) Species-driven changes in nitrogen cycling can provide a mechanism for plant invasions. *Proceedings of the National Academy of Sciences of the USA* 106, 12400–12405.

Lavorel, S., McIntyre, S., Landsberg, J. and Forbes, T.D.A. (1997) Plant functional classifications: from general groups to specific groups based on response to disturbance. *Trends in Ecology and Evolution* 12, 474–478.

Lockwood, J.L., Powell, R.D., Nott, M.P. and Pimm, S.L. (1997) Assembling ecological communities in time and space. *Oikos* 80, 549–553.

Luken, J.O. (1990) *Directing Succession.* Chapman and Hall, London.

MacDougall, A.S., Wilson, S.D. and Bakker, J.D. (2008) Climatic variability alters outcome of long-term community assembly. *Journal of Ecology* 96, 346–354.

McCune, B. and Allen, T.F.H. (1985) Will similar forests develop on similar sites? *Canadian Journal of Botany* 63, 367–376.

McIntosh, R.P. (1999) The succession of succession: a lexical chronology. *Bulletin of the Ecological Society of America* 80, 256–265.

McIntyre, S., Díaz, S., Lavorel, S. and Cramer, W. (1999a) Plant functional types and disturbance dynamics: introduction. *Journal of Vegetation Science* 10, 604–608.

McIntyre, S., Lavorel, S., Landsberg, J. and Forbes T.D.A. (1999b) Disturbance response in vegetation: towards a global perspective on functional traits. *Journal of Vegetation Science* 10, 621–630.

Meiners, S.J., Pickett, S.T.A. and Cadenasso, M.L. (2002) Exotic plant invasions over 40 years of old field successions: community patterns and associations. *Ecography* 25, 215–223.

Meiners, S.J., Rye, T.A. and Klass, J.R. (2008) On a level field: the utility of studying native and non-native species in successional systems. *Applied Vegetation Science* 12, 45–53.

Miles, J. (1987) Vegetation succession: past and present perceptions. In: Gray, A.J., Crawley, M.J. and Edwards, P.J. (eds) *Colonization, Succession and Stability.* Blackwell Scientific, Oxford, UK, pp. 1–29.

Motta, R., Nola, P. and Berretti, R. (2009) The rise and fall of the black locust (*Robinia pseudoacacia* L.) in the "Siro Negri" Forest Reserve (Lombardy, Italy): lessons learned and future uncertainties. *Annals of Forest Science* 66 (4), Art. No. 410.

Navas, M.-L., Roumet, C., Bellmann, A., Laurent, G. and Garnier, E. (2010) Suites of plant traits in species from different stages of a Mediterranean secondary succession. *Plant Biology* 12, 183–196.

Nobel, I.R. and Slatyer, R.O. (1980) The use of vital attributes to predict successional changes in plant communities subject to recurrent disturbances. *Vegetatio* 43, 5–21.

Odion, D.C., Moritz, M.A. and DellaSala, D.A. (2010) Alternative community states maintained by fire in the Klamath Mountains, USA. *Journal of Ecology* 98, 96–105.

Peltzer, D.A., Bellingham, P.J., Kurokawa, H., Walker, L.R., Wardle, D.A. and Yeates, G.W. (2009) Punching above their weight: low-biomass non-native plant species alter soil properties during primary succession. *Oikos* 118, 1001–1014.

Penman, T.D., Binns, D.L. and Kavanagh, R.P. (2008) Quantifying successional changes in response to forest disturbance. *Applied Vegetation Science* 11, 261–268.

Perrings, C.A. and Walker, B.H. (1995) Biodiversity loss and the economic of discontinuous change in semi-arid rangelands. In: Perrings, C.A., Mäler, K.-G., Folke, C., Holling, C.S. and Jansson, B.-O. (eds) *Biodiversity Loss: Economic and Ecological Issues.* Cambridge University Press, Cambridge, UK, pp. 190–210.

Pianka, E.R. (1970) On r and K selection. *American Naturalist* 104, 592–597.

Pickett, S.T.A., Collins, S.L. and Armesto, J.J. (1987a) A hierarchical consideration of causes and mechanisms of succession. *Vegetatio* 69,109–114.

Pickett, S.T.A., Collins, S.L. and Armesto, J.J. (1987b) Models, mechanisms and pathways of succession. *Botanical Review* 53, 335–371.

Pickett, S.T.A., Parker, V.T. and Fiedler, P. (1992) The new paradigm in ecology: implications for conservation biology above the species level. In: Fiedler P. and Jain S. (eds) *Conservation Biology: the Theory and Practice of Nature Conservation, Preservation and Management.* Chapman and Hall, New York, pp. 65–88.

Pickett, S.T.A., Cadenasso, M.L. and Meiners, S.J. (2009) Ever since Clements: from succession to vegetation dynamics and understanding to intervention. *Applied Vegetation Science* 12, 9–21.

Pyšek, P. and Richardson, D.M. (2007) Traits associated with invasiveness in alien plants: where do we stand? In: Nentwig, W. (ed.) *Biological Invasions*, Ecological Studies, Volume 139. Springer-Verlag, Berlin, pp. 97–125.

Pyšek, P., Davis, M.A., Daehler, C.C. and Thompson, K. (2004) Plant invasions and vegetation succession: closing the gap. *Bulletin of the Ecological Society of America* 85, 105–109.

Reid, A.M., Morin, L., Downey, P.O., French, K. and Virtue, J.G. (2009) Does invasive plant management aid the restoration of natural ecosystems? *Biological Conservation* 142, 2342–2349.

Robinson, J. V. and Dickerson, J. E. Jr (1987) Does invasion sequence affect community structure? *Ecology* 68, 587–595.

Rodriguez, M. A. (1994) Succession, environmental fluctuations, and stability in experimentally manipulated microalgal communities. *Oikos* 70,107–120.

Roscher, C., Temperton, V.M., Buchmann, N. and Schulze, E.-D. (2009) Community assembly and biomass production in regularly and never weeded grasslands. *Acta Oecologica* 35, 206–217.

Samuels, C.L. and Drake, J.A. (1997) Divergent properties on community convergence. *Trends in Ecology and Evolution* 12, 427–432.

Shipley, B., Vile, D. and Garnier, E. (2006) From plant traits to plant communities: a statistical mechanistic approach to biodiversity. *Science* 314, 812–814.

Small, J.A., Buell, M.E. and Siccama, T.G. (1971) Old-field succession on the New Jersey Piedmont – the first year. *William L. Hutcheson Memorial Forest Bulletin* 2, 26–30.

Smith, T.M., Shugart, H.H. and Woodward, F.I. (eds) (1997) *Plant Functional Types.* Cambridge University Press, Cambridge, UK.

Sullivan, J.J., Williams, P.A. and Timmins, S.M. (2007) Secondary forest succession differs through naturalised gorse and native kanuka near Wellington and Nelson. *New Zealand Journal of Ecology* 31, 22–38.

Tallent-Halsell, N.G. and Watt, M.S. (2009) The invasive *Buddleja davidii* (butterfly bush). *Botanical Review* 75, 292–325.

Tansely, A.G. (1935) The use and abuse of vegetational concepts and terms. *Ecology* 16, 284–307.

Tassin, J., Medoc, J.M., Kull, C.A., Riviere, J.N. and Balent, G. (2009) Can invasion patches of *Acacia mearnsii* serve as colonizing sites for native plant species on Réunion (Mascarene archipelago)? *African Journal of Ecology* 47, 422–432.

Walker, B.H. (1981) Is succession a viable concept in African savanna ecosystems? In: West, D.C., Shugart, H.H. and Botkin, D.B. (eds) *Forest Succession: Concepts and Applications*. Springer Verlag, New York, pp. 431–447.

Walker, L.R. (1999) Patterns and processes in primary succession. In: Walker, L.R. (ed.) *Ecosystems of Disturbed Ground*, Ecosystems of the World, Vol. 16. Elsevier, Amsterdam, pp. 585–610.

Walker, L.R. and Chapin, F.S. III (1987) Interactions among processes controlling successional change. *Oikos* 50, 131–135.

Walker, L.R. and Willig, M.R. (1999) An introduction to terrestrial disturbances. In: Walker, L.R. (ed.) *Ecosystems of Disturbed Ground*, Ecosystems of the World, Vol. 16. Elsevier, Amsterdam, pp. 1–16.

Warman, L. and Moles, A.T. (2009) Alternative stable states in Australian's wet tropics: a theoretical framework for the field data and a field-case for the theory. *Landscape Ecology* 24, 1–13.

Watt, A.S. (1947) Pattern and process in the plant community. *Journal of Ecology* 35, 1–22.

Weiher, E. and Keddy P.A. (1999) Assembly rules as general constraints on community composition. In: Weiher, E. and Keddy, P.A. (eds) *Ecological Assembly Rules: Perspectives, Advances, Retreats*. Cambridge University Press, Cambridge, UK, pp. 251–271.

Weiher, E., van der Werf, A., Thompson, K., Roderick, M., Garnier, E. and Eriksson, O. (1999) Challenging Theophrastus: a common core list of plant traits for functional ecology. *Journal of Vegetation Science* 10, 609–620.

Woodward, F.I. and Diament, A.D. (1991) Functional approaches to predicting the ecological effects of global change. *Functional Ecology* 5, 202–212.

Xiao, S., Michalet, R., Wang, G. and Chen, S.Y. (2009) The interplay between species' positive and negative interactions shapes the community biomass-species richness relationship. *Oikos* 118, 1343–1348.

Zimmerman, N., Hughes, R.F., Cordell, S., Hart, P., Chang, H.K., Perez, D., Like, R.K. and Ostertag, R. (2008) Patterns of primary succession of native and introduced plants in lowland wet forests in eastern Hawai'i. *Biotropica* 40, 277–284.

12 Landscape Scales and Invasive Species

> ### Concepts
>
> - Landscape ecology is the study of spatial patterns across landscapes.
> - Human activities fragment landscapes. This often benefits invasive species.
> - Landscape processes affect the growth and persistence of a population.
> - Habitat connectivity influences how species spread across landscapes.
> - Cross-scale modelling integrates the different scales at which ecological processes work.

12.1 Introduction

Landscape ecology is 'the study of spatial variation in landscapes at a variety of scales' (International Association of Landscape Ecology, 2010). This definition raises a subtle but key idea that despite the vernacular implications, landscape ecology does not always involve large areas of land or water. It involves spatial patterns and spatial movements of organisms between often isolated or fragmented habitats. These patterns and movements may occur over what humans subjectively would say are smaller scales, e.g. between wood-lots, wetland remnants or prairie fragments separated by only a few kilometres. None the less, because these patterns and movements cover relatively larger distances (from tens to thousands of kilometres), there is a tendency for the spatial scale of landscape ecology to be large, e.g. the size of watersheds.

In this chapter, we focus on how invasive species migrate across space. We examine spatially explicit processes, influenced by time, that facilitate or thwart the invasion process. We are often interested in long-distance transport or migration of invasive species (e.g. via air or ground vehicles), which occur much more rapidly than a historical rate of natural processes (e.g. postglacial migration) (Chapter 6).

12.2 Human Influences on Landscape Processes

Fragmentation is a natural process via fire, ice storms or hurricanes. However, humans fragment habitats more rapidly – for agriculture, mining, forestry, dams, trails, roads and all forms of rural and urban living space. While we have done this since agriculture and communal living began thousands of years ago, the industrial revolution facilitated fragmentation and transport of invasive species across larger swathes of land and water. In the last century, we have added the ability to move across thousands of kilometres in a day or two. In so doing, we alter and degrade massive areas of landscape. Many invasive species benefit from our actions and transport capabilities. This obviates the relatively slow pace of spatial processes that governed long-distance patch dynamics previously.

We can picture that the original human intent was economically driven with no ecological damage planned – but this reflects a historical lack of foresight. Thus, farmers cleared land for crop fields to maximize yield, and the wood-lot polygon left behind often had a large circumference and a low diameter. The same thing happened when housing lots were created and left isolated natural habitats in their wake with high edge (circumference) to interior (diameter) ratios. In such cases, most species that were adapted to deeper woods will decline and species adapted to abrupt transitions and open habitat will thrive. The latter are often invasive plants.

Thus, the ability of native species to form viable populations in suitable habitats is often low. Fragmented habitats tend to be places more amenable to colonization by exotic species transported on vehicles, pets or clothing. The process of fragmentation also may create narrow ecological corridors between degraded habitats that favour exotic species at the expense of native species, e.g. erosion-prone and open canopy trails that end at remnants of forests. Once an invasive species is able to establish in a degrading habitat, it may increase the degradation as the invasive species alters the biophysical microclimate that favours invasives over exotics. This cascading process means that habitats can change and lose ecological structure and function rapidly. This is why one key to mitigating the impacts of invasive species is to implement ecological restoration of the habitat structures and processes, especially by reintroducing larger numbers of native species.

12.3   Landscape Processes

So far, we have described the history of landscape fragmentation and its current state. But how do landscape-scale processes work? What theories and mechanisms define landscape ecology? These are difficult questions because they require an intensive conceptual and mathematical background that is not suited to the scope of our text. We will describe and explain theories and mechanisms of landscape ecology in terms of their fundamentals; the nuances will be left for readers to explore in more specialized texts or peer-reviewed articles (Hanski, 1998; Jongejans et al., 2008).

It begins with mapping

In studying landscape processes of invasive species, there is still much to be gained from basic mapping and the use of historical data such as aerial photographs, land-use histories or surveyors' notes (Brown and Carter, 1998; Andersen and Baker, 2006; Müllerová et al., 2005; Parks et al., 2005; Maheu-Giroux and de Blois, 2007). However, understanding landscape-scale invasion ecology requires a range of skill sets and at the least competence and confidence to attempt new approaches. These include grappling with an ability to use remote sensing to detect and spatially map invasives –

a tough job as it depends on the invasive species expressing a spectral signal that is distinguishable from those of other species (Andrew and Ustin, 2008; Walsh et al., 2008). New mapping techniques still require advanced knowledge and interest in applying these to better represent the realities of the distributions of invasive species (Barnett et al., 2007; Pande et al., 2007). Mapping is useful but can be atheoretical. Without explanations, maps of invasive species tell us little. Such explanations can begin with island biogeography.

Island biogeography

Early hypotheses on why and how species invaded owed their intellectual pedigree to the wider theory of island biogeography. One of the many uses of island biogeography is to help explain how the size and proximity of islands determine how populations and communities interact through immigration via dispersal. Island biogeography theory postulates that immigration is influenced most strongly by distance from a source population. The closer an island of habitat is to a population source of any species, the more likely that species will invade – as long as it is mobile. Immigration is also influenced by island size. The larger an island, the more likely a species will be dispersed into that island. In animals, behaviour may be involved so that the probability of invading a large island is augmented by some degree of choice. For plants, it is usually a matter of simple probability against random dispersal, although in animal-dispersed plants behaviour may indeed have some influence. The theory applies because isolated habitat fragments are quite similar to actual islands in an ocean – this is why island biogeography can be interpreted both literally and figuratively, but the ideas are the same: how do individuals of species manage to colonize habitats amid a larger aggregation of inhospitable locations?

Patch dynamics

Patch dynamics theory has an advantage over island biogeography. Patch dynamics explicitly accounts for the reality that fragments of habitat are not static – they change their shape, the populations and species within them change, and the ecosystem functions within and without change (Dillemuth et al., 2009). Dynamic patches mean that as individuals colonize, establish, compete, reproduce, migrate, decline and go extinct, the species composition (biological diversity) and ecosystem function of patches are always changing. Patch sizes and inter-patch distances within a landscape will influence the dynamics of populations of exotic invasive species. Exotic invasive species may be able to spread quickly and dominate small, isolated patches and then spread to larger ones if the species are favoured by the abiotic conditions. In addition, the invasive species may be favoured if it does not need obligate symbionts (such as mycorrhizae or animal pollinators) or if it is self-compatible (Chapters 4 and 9).

Deines *et al.* (2005) provide an example of how patch dynamics can be used in studying invasive ecology. They examined the effects of population growth, carrying capacity, competition between organisms, spatial arrangement of patches and migration between habitat patches. As expected, the most competitive species did best. There was a nuance in that the species that succeeded in the long term were not necessarily the ones that dominated immediately. It appeared that the long-term successful were actually poor colonizers, or even failed to invade. What happened was that small populations became established, and these small populations persisted for many years until the environmental conditions became more favourable, via more fragmentation, and then some species became (apparently) sudden and widespread dominants. Garlic mustard (*A. petiolata*) is one example of this type of species. It became dominant only after widespread fragmentation in rural and urban areas created a threshold of favourable forest edge habitat (Murphy, 2005).

Metapopulation dynamics

Metapopulations refer to a series of smaller, local and often isolated populations in a landscape that can interact via dispersal between isolated patches of populations (Valverde and Silvertown, 1997; Moilanen and Hanski, 1998). For plants, dispersal is via spores, pollen, seeds or vegetative propagules carried by wind or animals (Chapters 4 and 5). Each patch may have small populations of a given species, but with dispersal between patches the effective population may be larger than expected from surveys of smaller isolated patches (Lavorel *et al.*, 1999; Sebert-Cuvillier *et al.*, 2008). For example, with metapopulations, there may appear to be low genetic diversity in each isolated patch because of likely inbreeding depression (Chapter 4). However, across a landscape, there will be many small patches where there has been localized

genetic drift and selection. As long as there is some dispersal between patches, the realized genetic diversity may offer a large potential for selection in response to almost any environmental change and little risk of a single selection pressure eliminating all isolated populations. With invasive species, the existence of metapopulations may mean that one may underestimate the risk of spread and impacts because of this potential for greater-than-apparent genetic diversity (Radosevich *et al.*, 2003).

Source–sink dynamics simplify habitats by classifying them as either high or low quality (Diffendorfer, 1998). High-quality habitats can sustain native plant populations and will be a source of propagules to be dispersed into the landscape at a constant rate. Low-quality habitats are degraded and may become a sink for native species but still have sufficient resources for populations of invasive exotic species to establish. Once established, invasive exotic species can disperse into other habitats as they degrade into low quality. Source–sink dynamics hold dispersal between habitats as constant and generally disallow sink–source (reverse) migration. No one species is considered able to migrate to all possible habitats. Further, the density-dependent factors of disease, herbivores, competitors and resource limitation are not explicitly considered. Therefore, this is perhaps too simplistic an approach to apply to most real-world situations, but it can still provide insights into the potential for invasions at landscape scales.

In balanced dispersal dynamics, habitats are defined in a less amorphous manner: they are classified as having relatively high or low carrying capacities for native species. Importantly, dispersal rates between habitat patches are not constant, in contrast to source–sink dynamics. Balanced dispersal focuses on when there is an evolutionarily beneficial outcome to disperse from one habitat patch. This allows for the sink–source migration not allowed in the source–sink dynamics. Some patches will thus experience emigration and some will experience immigration, and eventually the numbers of individuals dispersing between all habitats will become equal or balanced (hence, the origin of the term). Like source–sink dynamics, balanced dispersal dynamics still does not consider migration to all possible habitats and ignores community interactions.

Despite their limitations, source–sink and balanced dispersal models can be applied in some real-world cases, for example, determining how barbed goatgrass (*Aegilops triuncialis*) will invade habitats (Thomson, 2007). That these approaches are of limited use is not surprising since neither was intended to be so specific – both were developed as a general theoretical framework for dispersal in landscapes. This is why ecologists have developed more advanced statistical models.

Statistical models of how invasives spread

Trying to analyse the history and future of invasive dispersal and population changes are landscape-scale problems because the processes depend on the spatial factors of habitat suitability and the rate of dispersal (Zhu *et al.*, 2007). There is a temptation to handle the implicit complexity by ignoring it or by making assumptions that are too simple. One of the more startling assumptions in some early landscape analyses of invasive species was that the invasion had stopped – when the whole point of the analysis was often to determine the future risk of invasion. Like the source–sink and balanced dispersal models, this assumption is not as naive as one might think

because early work explicitly recognized that the statistical approaches were too simple – the goal was gradually to develop more sophisticated approaches. Thus, early analyses used such statistical frameworks as regressions or covariates under a very simple model of invasions having stopped.

Concordant with advances in computing power, recent approaches no longer use this assumption (Collingham *et al.*, 2000; Fewster, 2003; Stephenson *et al.*, 2006; Verheyen *et al.*, 2007). For example, using georeferenced habitat data detailing the spread of invasive giant hogweed (*H. mantegazzianum*) across Britain in the 20th century, Cook *et al.* (2007) constructed a practical model that better predicts which habitats will be colonized by giant hogweed and how fast that will happen (Fig. 12.1). One outcome was that there was misplaced confidence and agreement that the habitats most at risk were urban or riparian/wet meadow areas. It turns out that the most vulnerable areas may be degraded woodlands – and there is empirical evidence to suggest that this is in fact what is happening in Europe (Pyšek and Pyšek, 1995; Pyšek, *et al.*, 1998; Cook *et al.*, 2007).

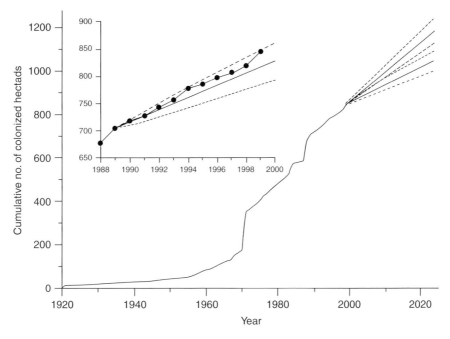

Fig. 12.1. A practical model was constructed using georeferenced habitat data detailing the spread of invasive giant hogweed (*H. mantegazzianum*) across Britain in the 20th century, which predicts the habitats that will be colonized by giant hogweed and how fast they will be colonized. The figures show the predicted spread by area colonized (hectads, 10 × 10 km), and the post-1989 spread into different habitats; the smaller figure is a more detailed representation of the larger one (Cook *et al.*, 2007).

12.4 Habitat Connectivity and Invasive Species Ecology

Thus far, the approaches presented tend to focus on the spatial aspects of dispersal as if there were no effect of the actual spatial locations of the habitat patches. To make landscape-scale analyses of invasive species more realistic, one must ask not only where patches are located but also how fragmented or how connected these habitats are. This can define the probability of whether increasing connections between habitat patches will promote ecological integrity via migration of native species or decrease it by providing avenues for invasive species (Donald and Evans, 2006). Calabrese and Fagan (2004) provided a primer of the various approaches to connectivity that will form the basis of our discussion in this section as it relates to invasive species.

Three approaches to connectivity

One approach is to use structural connectivity measures that examine the physical or effective connectivity via dispersal. A second approach is to use potential connectivity models that relate structural connectivity to predictive or hypothetical dispersal of organisms based on their traits, for example, how fast can they move and what dispersal modes are available. In plants, this might mean do insects carry pollen from habitat to habitat or can the wind penetrate the habitat and blow seeds to another location. A third class of analysis is to use measures of actual connectivity. This requires empirical data on both the structural connectivity and the dispersal of organisms. For example, in invasive species, this would mean measuring how many seeds of a species such as wild burdock (*Arctium minus*) are carried by fur or clothes and how far these are dispersed in eastern North American habitats.

Measuring structural and potential connectivity

Nearest-neighbour occupancy

With this method, we try and measure how far away we find habitats that are occupied by a species. This is a crude metric because it does not account for the sizes or shapes of habitat patches, the dispersal ability of organisms, or how more than a pair of patches is interacting. None the less,

if someone has been able to classify a series of habitats as being degraded and one of these has an invasive species that is easily dispersed, then nearest-neighbour occupancy can be a useful tool.

Scale-area slopes

In some cases, a researcher may have good historical data on the increasing presence of an invasive species. One can assign these historical records of species found across habitats to a series of cells within map grids. An analysis is then made of how the amount of space on the map that is occupied by a species changes (Kunin, 1998) (Figs 12.2 and 12.3). One of the uses of this method is that it yields a graph that calculates a scale-area slope. If the slope is shallow, species are found contiguously in a landscape; if it is steep, then the species are found in fragmented habitats. For invasive species, we can then tell if the invasion was and will be related to the extent of fragmentation.

Direct measures of connectivity

The availability of advanced software (geographical information systems, GIS) for spatially explicit data has meant that it has become possible to develop more accessible, yet sophisticated, landscape-level analyses for structural and potential connectivity. By measuring the shapes and sizes of habitat fragments, a researcher can calculate structural connectivity directly. The problem is that this does not necessarily allow a direct measure of what we really want – actual connectivity as it relates to the risk of spread of invasive species. Unless dispersal data are available, there is no means to measure actual connectivity.

An example of the benefits and challenges of using structural and potential connectivity in invasive ecology

Many studies have used the methods described so far to explain past invasions and predict new ones (Wangen *et al.*, 2006; Duguay *et al.*, 2007; Endress *et al.*, 2007; Burls and McClaugherty, 2008; Jodoin *et al.*, 2008; Kalwij *et al.*, 2008; Predick and Turner, 2008; Thiele and Otte, 2008; Thiel *et al.*, 2008; Pattison and Mack, 2009). We will use an example from one of the authors of this book to reflect on what connectivity may or may not tell us. Murphy's research group is measuring structural

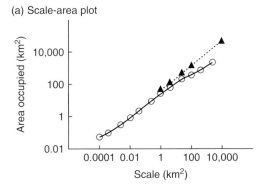

(a) Scale-area plot

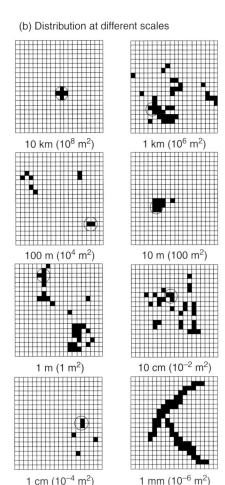

(b) Distribution at different scales

10 km (10^8 m^2)   1 km (10^6 m^2)

100 m (10^4 m^2)   10 m (100 m^2)

1 m (1 m^2)   10 cm (10^{-2} m^2)

1 cm (10^{-4} m^2)   1 mm (10^{-6} m^2)

Fig. 12.2. (a) Scale-area plots for two British plant species: wild gladiolus (*Gladiolus illyricus*) and beach pea (*Lathyrus japonicus*). The circles and solid line represent wild gladiolus; the triangles and dashed line represent beach pea. Beach pea data are courtesy of R. Quinn; wild gladiolus data at coarse resolutions (150 m) are courtesy of English Nature, and data at finer resolutions (10 m to 1 cm) are from S.D. Murphy's field surveys of three populations. (b) The distribution of British wild gladiolus populations at successively finer resolutions. Each grid displays a subset (circled) of the previous grid at 100-fold higher resolution (indicated below each grid). The finest grid (1 mm resolution) is hypothetical, to illustrate the potential use of pixel data to represent cover (from Kunin, 1998).

connectivity and edge-to-interior ratio in a series of urban natural areas in Ontario, Canada. Despite anticipated massive fragmentation, many habitats in each urban area were well connected and had much more interior habitat than expected. Empirical measures of invasive plants revealed that while there were patches of species such as garlic mustard on the edges, and some early invasion of dog-strangling vine (*V. rossicum*), none had spread into the interior woodlands – defined here as >5 m from the visual edge of a woodland. The habitats are not so fragmented that the interior habitat has become inadequate – it is still good enough to discourage edge-affinity invasive species. We can postulate that if urban woodland management continues to maintain connectivity and interior habitats, then invasives will neither spread rapidly nor dominate. However, the strength of this hypothesis comes with the caveat that dog-strangling vine may be better able to disperse and send out tendrils into interior areas. Connectivity may have prevented garlic mustard from rapid and wide invasions, but dog-strangling vine arrived less than a decade ago and may not be stopped in the same manner. With a relatively recent invasive, we cannot rely solely on structural connectivity and comparative scenarios with other invasives – even if the comparison is with a normally successful one (garlic mustard). We must measure and monitor how well dog-strangling vine (for example) is being dispersed within the existing structural connectivity of the landscape.

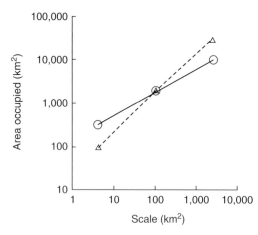

Fig. 12.3. Scale-area curves comparing the actual distribution and connectivity in Britain of two plant species: mat-grass fescue (*Vulpia unilateralis*) – triangles and dashed lines; and narrow-leaved lungwort (*Pulmonaria longifolia*) – circles and solid lines. The contrasting aggregation patterns of the two species are reflected in the slopes of their scale-area curves, with the more scattered distribution having a steeper curve. Even though the two species are equally common at the moderate resolution used here, they differ markedly at other scales. Also, extrapolating the scale-area curves suggests that lungwort may be much more common than fescue at still finer scales (Kunin, 1998).

Potential and actual connectivity measured using graph-theoretic approaches

More advanced methods of measuring potential and actual connectivity include graph-theoretic approaches. This type of approach relies on real-time observations of species dispersal and GIS data. It effectively marries the explanatory theories for connectivity and species dispersal with mathematical theories on measuring or graphing networks of connected or disconnected habitats in space. The reason for developing this more data-intensive method was to begin to grapple with potential and actual connectivity. The method works by considering how observed dispersal ability will or will not allow a species to migrate between all possible pairwise combinations of habitats. The researcher can analyse this by starting the dispersal at a random point in a habitat patch and letting a species migrate – again at random (this is called a random walk). The researcher also may use actual fixed points of observations where individuals of a species were actually located. Data on the extent to which individuals migrate imposed over a matrix

or collection of habitat fragments measures how connected those fragments are – at least in terms of how connected they are for the species being tested. Different species may need different degrees of connectivity, and this measure approaches connectivity in a way that might be described as being backward. We say this because connectivity is being defined by how well one species migrates, and not by how the relative connections and locations of habitats affect a species' migration.

The approach is complex but useful. For example, Tiébré *et al.* (2008) showed that Japanese knotweed (*Fallopia japonica*) tended to spread in certain habitats and was most especially abetted by linear networks such as roads. Further, a pair of studies on invasive American black cherry (*P. serotina*) in France demonstrated the gradual logic of graph-theoretic approaches. Sebert-Cuvillier *et al.* (2008) used a model that addressed connectivity in the sense of how spatial heterogeneity affects the probability of spread of black cherry. These researchers mapped invisibility spaces and showed the rate and extent of invisibility. The interesting result was that spatial heterogeneity (lower connectivity) could speed invasions because of edge effects, but lowered the final extent of area invaded because of distances between patches (Fig. 12.4). Concurrently, Deckers *et al.* (2008) showed that a landscape structure that provided perching sites encouraged birds carrying black cherries to drop seeds and increase invasion. The solution might be to focus on cutting large and reproductive black cherry trees as well as ensuring better forest practices that optimize spatial heterogeneity.

Connectivity does not always make the best indicator of risk of invasives at landscape scales

We caution that just because methods of measuring connectivity are becoming more sophisticated, they should not be the sole approach used for invasive species at landscape scales. For example, one might use a graph-theoretic method to test connectivity based on a native species that requires connected habitats. If the answer is that for this species there is sufficient connectivity, then the cluster of habitats also might be better able to resist invasion by exotics. However, this assumes that all invasive exotics will be unable to colonize fragments simply because they appear to be connected. To actually test the

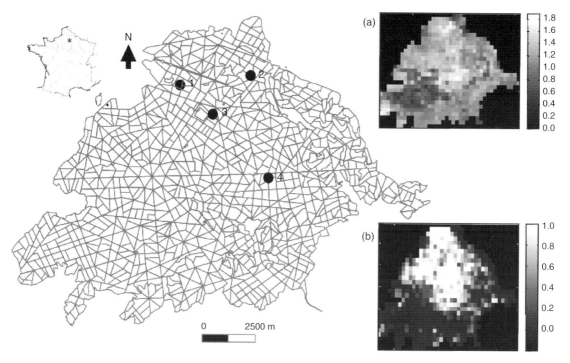

Fig. 12.4. The invasion of the Compiègne forest in France by American black cherry (*P. serotina*). The maps to the left indicate the location of the forest. To the right, (a) shows the conversion of the forest into an invasibility map, which is to be compared with (b) the current invasion state. Numbered locations on the large map indicate the sites of introduction used for simulations. The scale to the right of these two maps indicates (a) the invasibility index or (b) the proportion of a cell which is currently invaded; non-forest cells are in black (Sebert-Cuvillier *et al.*, 2008).

implications of connectivity, a researcher would have to test the ecological functions of the patches themselves, look at the importance of other factors such as edge effects or patch size, and consider environmental gradients (Bartuszevige *et al.*, 2006; Kupfer *et al.*, 2006; Lambrinos, 2006; Pauchard and Alaback, 2006; Gignac and Dale, 2007). This is reminiscent of the problems with the simpler models of source–sink dynamics, as once again there can be an implicit assumption that high quality (in this case, high connectivity) will prevent invasions by exotic species. If this is indeed not correct, the invasion risk will be underestimated. If connectivity, as measured here, is the sole arbiter of risk of invasion, and low connectivity prompts ecological restoration, a researcher would be advised to avoid becoming smug. This is because even if patches are structurally restored, this can facilitate the spread of already arrived invasives in the seed bank that are phenotypically plastic and adjust well to higher quality habitat. Longer-term landscape-scale success can be fleeting (Murphy, 2005).

Why not just measure movement of invasive species?

Why not just directly observe species' movements in sufficient detail, use a GPS (geographical positioning system) to tag movements, and then download data into the many GIS-based maps that are now available? Why go through the proxy data or the indirect approaches that have been described in detail? The real problem is getting sufficient data. Trying to tag the hundreds to millions of individuals that would be needed for a sufficiently large data set quickly becomes too expensive and requires too many people for most research grants to support. This should remind readers that, ironically, landscape connectivity is a prisoner of the species scale – the data used to measure connectivity depends on the movements within metapopulations of a species. This is one reason why researchers are considering more cross-scalar approaches, because landscape ecology is not just about how genes flow, populations interact, communities change or ecosystem processes move across space – it is about all of these scales.

12.5 Cross-scale Modelling of Landscape-scale Invasions

One of the struggles that ecologists have is the dichotomous need to focus and be an expert at one scale (or even one species in one region of the world) even though ecological processes, including those that facilitate invasive exotics, are multi-scale. It is difficult to be comfortable and effective with analyses at different scales, much more so in somehow combining them. However, the world is inherently cross-scalar and much literature has been written on how best to analyse this (Holling, 1992; Peterson and Parker, 1998). The ideal would be to have a model that could handle all scales (Boulant *et al.*, 2008) but there is rarely a simple linear relationship between similar processes across these scales (Holling, 1992; Peterson and Parker, 1998; Peters *et al.*, 2006). In retrospect, older examples of what appear to be analyses restricted to a single scale actually began to grapple with cross-scalar issues implicitly. The simulation by Lavorel *et al.* (1999) of the metapopulation dynamics of invasive mistletoe (*Amyema preissii*) could also examine the community scale, albeit in a simplified two-dimensional landscape space – they could predict where mistletoe might colonize. Cellular automata models also rest within the domain of populations as they predict how a species in one cell (representing a location in space) is likely to spread to adjacent cells. While these models ignore long-distance dispersal where an invasive might skip over many such cells, they can still be used to bridge the population–community–ecosystem gaps, as in Huang *et al.* (2008) who examined the historical spread of invasive smooth cordgrass (*Spartina alterniflora*).

More explicitly, Brown *et al.* (2008) used cross-scalar models to predict the population changes, the vulnerability of communities and the ecosystem processes that related to how and where Malabar plum (*Syzygium jambos*) was likely to invade. They started with population growth rate analyses, applied these to metapopulation models, and then used GIS to model habitat suitability and probabilities of how metapopulations would respond to different environmental variables. They found that at smaller spatial scales, past disturbances were most influential in determining invasion success. At larger scales, habitat similarity affected invasion success. The management prescription is to reduce disturbances and search for new invasions in habitats similar to the ones from which Malabar plum originated.

There also is an opportunity to use assembly as a cross-scale approach (Chapter 11). Filters are a broader theoretical framework, explicitly address population to ecosystem scales, and place emphasis on the order of invasion or colonization. Conceptually, one can think of assembly as relating how near-neighbour processes such as competition for resources are related to longer-distance processes such as dispersal. There is little literature on using assembly to address cross-scale issues related to landscape ecology of invasions. That laundry list of terms you just read should give you a good indication of why this is so. That is a complex order – made more difficult as analyses of studying assembly, metapopulations and explicit landscape metrics are developing rapidly and somewhat independently. None the less, some work has begun. One of the earliest studies was by Crawley *et al.* (1999) who determined that ecological traits rather than sheer numbers or biomass were key to understanding assembly by native species and disassembly by invasive species. Theoharides and Dukes (2007) used assembly to consider how invasions happen (Fig. 12.5). Given the interest in using assembly to repair and restore invasive-damaged communities (Temperton *et al.*, 2004), it seems that there is a useful avenue for research on the impacts of invasives in assembly.

12.6 Summary

Fragmentation is a natural process; however, human activities increase the extent of fragmentation. Landscape ecology is one way to study how invasive species are influenced by fragmentation and other changes. Over time there have been different ideas of how temporal and spatial patterns of invasion are influenced by landscape processes. There are many ways to explain colonization and extinction of habitat patches. Connectivity between habitats may provide dispersal routes for invasive species or increase ecological integrity, making invasions less likely. Community assembly is a cross-scale approach to understanding how

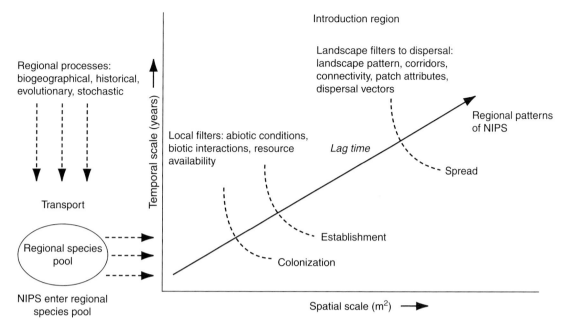

Fig. 12.5. The assembly of local communities is influenced by filters at local, landscape and regional scales. The regional species pool is assembled through speciation, migration, biotic exchange and geological events. Transport into this regional pool occurs on a much faster timescale than most natural movements of species. Here, transport is shown outside the axis of spatial and temporal scales. Following transport, colonizing non-native plant species (or non-indigenous plant species, NIPS) move through local abiotic filters to colonize, biotic filters to establish, and dispersal barriers to spread. As the non-native species moves from colonization to landscape spread, the temporal and spatial scales of processes underlying the invasion increase (Theoharides and Dukes, 2007).

ecosystem processes work at multiple scales. Landscape ecology is an emerging area of invasion ecology that provides a bridge between studies conducted at the micro-scale community level and at the global scale. In our next chapter we explore the emerging science of molecular ecology.

Box 12.1. Invasive species: landscape scales and invasive species.

- What traits might enable your invasive species to move across different landscapes?
- How might habitat fragmentation influence your species?

12.7 Questions

1. Why does fragmentation benefit invasive species?
2. How do humans fragment landscapes and what can we do to reduce our impact?
3. Explain why connectivity may or may not be beneficial to invasive species.
4. How do you measure connectivity between habitats?
5. What does it mean to examine invasive species at a landscape scale?
6. What does cross-scale modelling mean in practice?

Further Reading

Fortin, M.-J. and Dale, M.R.T. (2005) *Spatial Analysis: A Guide for Ecologists.* Cambridge University Press, Cambridge, UK.

Gergel, S.E. and Turner, M.G. (eds) (2001) *Learning Landscape Ecology: A Practical Guide to Concepts and Techniques.* Springer Verlag, New York.

Wiens, J.A., Moss, M.R., Turner, M.G. and Mladenoff, D. (eds) (2006) *Foundation Papers in Landscape Ecology.* Columbia University Press, New York.

Wu, J. and Hobbs, R.J. (eds) (2007) *Key Topics in Landscape Ecology.* Cambridge University Press, Cambridge, UK.

References

Andersen, M.D. and Baker, W.L. (2006) Reconstructing landscape-scale tree invasion using survey notes in the Medicine Bow Mountains, Wyoming, USA. *Landscape Ecology* 21, 243–258.

Andrew, M.E. and Ustin, S.L. (2008) The role of environmental context in mapping invasive plants with hyperspectral image data. *Remote Sensing of Environment* 112, 4301–4317.

Barnett, D.T., Stohlgren, T.J., Jarnevich, C.S., Chong, G.W., Ericson, J.A., Davern, T.R. and Simonson, S.E. (2007) The art and science of weed mapping. *Environmental Monitoring and Assessment* 132, 235–252.

Bartuszevige, A.M., Gorchov, D.L. and Raab, L. (2006) The relative importance of landscape and community features in the invasion of an exotic shrub in a fragmented landscape. *Ecography* 29, 213–222.

Boulant, N., Kunstler, G., Rambal, S. and Lepart, J. (2008) Seed supply, drought and grazing determine spatio-temporal patterns of recruitment for native and introduced invasive pines in grasslands. *Diversity and Distributions* 14, 862–874.

Brown, J.R. and Carter, J. (1998) Spatial and temporal patterns of exotic shrub invasion in an Australian tropical grassland. *Landscape Ecology* 13, 93–102.

Brown, K.A., Spector, S. and Wu, W. (2008) Multi-scale analysis of species introductions: combining landscape and demographic models to improve management decisions about non-native species. *Journal of Applied Ecology* 45, 1639–1648.

Burls, K. and McClaugherty, C. (2008) Landscape position influences the distribution of garlic mustard, an invasive species. *Northeastern Naturalist* 15, 541–556.

Calabrese, J.M. and Fagan, W.F. (2004) A comparison-shopper's guide to connectivity metrics. *Frontiers in Ecology and the Environment* 2, 529–536.

Collingham, Y.C., Wadsworth, R.A., Huntley, B. and Hulme, P.E. (2000) Predicting the spatial distribution of non-indigenous riparian weeds: issues of spatial scale and extent. *Journal of Applied Ecology* 37 (Suppl. 1),13–27.

Cook, A., Marion, G., Butler, A. and Gibson, G. (2007) Bayesian inference for the spatio-temporal invasion of alien species. *Bulletin of Mathematical Biology* 69, 2005–2025.

Crawley, M.J., Brown, S.L., Heard, M.S. and Edwards, G.R. (1999) Invasion-resistance in experimental grassland communities: species richness or species identity? *Ecology Letters* 2, 140–148.

Deckers, B., Verheyen, K., Vanhellemont, M., Maddens, E., Muys, B. and Hermy, M. (2008) Impact of avian frugivores on dispersal and recruitment of the invasive *Prunus serotina* in an agricultural landscape. *Biological Invasions* 10, 717–727.

Deines, A.M., Chen, V.C. and Landis, W.G. (2005) Modeling the risks of nonindigenous species introductions using a patch-dynamics approach incorporating contaminant effects as a disturbance. *Risk Analysis* 25, 1637–1651.

Diffendorfer, J.E. (1998) Testing models of source–sink dynamics and balanced dispersal. *Oikos* 81, 417–433.

Dillemuth, F.P., Rietschier, E. and Cronin, J.T. (2009) Patch dynamics of native grass in response to the spread of invasive smooth brome (*Bromus inermis*). *Biological Invasions* 11, 1381–1391.

Donald, P.F. and Evans, A.D. (2006) Habitat connectivity and matrix restoration, the wider implications of agri-environment schemes. *Journal of Applied Ecology* 43, 209–218.

Duguay, S., Eigenbrod, F. and Fahrig, L. (2007) Effects of surrounding urbanization on non-native flora in small forest patches. *Landscape Ecology* 22, 589–599.

Endress, B.A., Naylor, B.J., Parks, C.G. and Radosevich, S.R. (2007) Landscape factors influencing the abundance and dominance of the invasive plant *Potentilla recta*. *Rangeland Ecology and Management* 60, 218–224.

Fewster, R.M. (2003) A spatiotemporal stochastic process model for species spread. *Biometrics* 59, 640–649.

Gignac, L.D and Dale, M.R.T. (2007) Effects of size, shape, and edge on vegetation in remnants of the upland boreal mixed-wood forest in agro-environments of Alberta, Canada. *Canadian Journal of Botany* 85, 273–284.

Hanski, I. (1998) Metapopulation dynamics. *Nature* 396, 41–49.

Holling, C.S. (1992) Cross-scale morphology, geometry, and dynamics of ecosystems. *Ecological Monographs* 62, 447–502.

Huang, H.M., Zhang, L.Q., Guan, Y.J. and Wang, D.H. (2008) A cellular automata model for population expansion of *Spartina alterniflora* at Jiuduansha Shoals, Shanghai, China. *Estuarine Coastal and Shelf Science* 77, 47–55.

International Association of Landscape Ecology (2010). Landscape ecology: what is it? Available at http.//www.landscape-ecology.org/what_is.html, accessed 8 May 2010.

Jodoin, Y., Lavoie, C., Villeneuve, P., Thériault, M., Beaulieu, J. and Belzile, F. (2008) Highways as corridors and habitats for the invasive common reed *Phragmites australis* in Quebec, Canada. *Journal of Applied Ecology* 45, 459–466.

Jongejans, E., Shea, K., Skarpaas, O., Kelly, D., Sheppard, A.W. and Woodburn, T.L. (2008) Dispersal and demography contributions to population spread of *Carduus nutans* in its native and invaded ranges. *Journal of Ecology* 96, 687–697.

Kalwij, J.M., Milton, S.J. and McGeoch, M.A. (2008) Road verges as invasion corridors? A spatial hierarchical test in an arid ecosystem. *Landscape Ecology* 23, 439–451.

Kunin, W.E. (1998) Extrapolating species abundance across spatial scales. *Science* 281, 1513–1515.

Kupfer, J.A., Malanson, G.P. and Franklin, S.B. (2006) Not seeing the ocean for the islands: the mediating influence of matrix-based processes on forest fragmentation effects. *Global Ecology and Biogeography* 15, 8–20.

Lambrinos, J.G. (2006) Spatially variable propagule pressure and herbivory influence invasion of chaparral shrubland by an exotic grass. *Oecologia* 147, 327–334.

Lavorel, S., Smith, M.S. and Reid, N. (1999) Spread of mistletoes (*Amyema preissii*) in fragmented Australian woodlands, a simulation study. *Landscape Ecology* 14, 147–160.

Maheu-Giroux, M. and de Blois, S. (2007) Landscape ecology of *Phragmites australis* invasion in networks of linear wetlands. *Landscape Ecology* 22, 285–301.

Moilanen, A. and Hanski, I. (1998) Metapopulation dynamics: effects of habitat quality and landscape structure. *Ecology* 79, 2503–2515.

Müllerová, J., Pyšek, P., Jarošík, V. and Pergl, J. (2005) Aerial photographs as a tool for assessing the regional dynamics of the invasive plant species *Heracleum mantegazzianum*. *Journal of Applied Ecology* 42, 1042–1053.

Murphy, S.D. (2005) Concurrent management of an exotic species and initial restoration efforts in forests. *Restoration Ecology* 13, 584–593.

Pande, A., Williams, C.L., Lant, C.L. and Gibson, D.J. (2007) Using map algebra to determine the mesoscale distribution of invasive plants: the case of *Celastrus orbiculatus* in southern Illinois, USA. *Biological Invasions* 9, 419–431.

Parks, C.G., Radosevich, S.R., Endress, B.A., Naylor, B.J., Anzinger, D., Rew, L.J., Maxwell, B.D. and Dwire, K.A. (2005) Natural and land-use history of the northwest mountain ecoregions (USA) in relation to patterns of plant invasions. *Perspectives in Plant Ecology, Evolution and Systematics* 7, 137–158.

Pattison, R.R. and Mack, R.N. (2009) Environmental constraints on the invasion of *Triadica sebifera* in the eastern United States, an experimental field assessment. *Oecologia* 158, 591–602.

Pauchard, A. and Alaback, P.B. (2006) Edge type defines alien plant species invasions along *Pinus contorta* burned, highway and clearcut forest edges. *Forest Ecology and Management* 223, 327–335.

Peters, W.L., Meyer, M.H. and Anderson, N.O. (2006) Minnesota horticultural industry survey on invasive plants. *Euphytica* 148, 75–86.

Peterson D.L. and Parker, V.T. (1998) *Ecological Scale: Theory and Applications.* Columbia University Press, New York.

Predick, K.I. and Turner, M.G. (2008) Landscape configuration and flood frequency influence invasive shrubs in floodplain forests of the Wisconsin River (USA). *Journal of Ecology* 96, 91–102.

Pyšek, P. and Pyšek, A. (1995) Invasion by *Heracleum mantegazzianum* in different habitats in the Czech Republic. *Journal of Vegetation Science* 6, 711–718.

Pyšek, P., Kopecký, M., Jarošík, V. and Kotková P. (1998) The role of human density and climate in the spread of *Heracleum mantegazzianum* in the Central European landscape. *Diversity and Distributions* 4, 9–16.

Radosevich, S.R., Stubbs, M.M. and Ghersa, C.M. (2003) Plant invasions: process and patterns. *Weed Science* 51, 254–259.

Sebert-Cuvillier, E., Simon-Goyheneche, V., Paccaut, F., Chabrerie, O., Goubet, O. and Decocq, G. (2008) Spatial spread of an alien tree species in a heterogeneous forest landscape, a spatially realistic simulation model. *Landscape Ecology* 23, 787–801.

Stephenson, C.M., MacKenzie, M.L., Edwards, C. and Travis, J.M.J. (2006) Modelling establishment probabilities of an exotic plant, *Rhododendron ponticum*, invading a heterogeneous, woodland landscape using logistic regression with spatial autocorrelation. *Ecological Modelling* 193, 747–758.

Temperton, V.M., Hobbes, R.J., Nuttle, T. and Halle, S. (2004) *Assembly Rules and Restoration Ecology: Bridging the Gap Between Theory and Practice.* Island Press, Washington, D.C.

Theoharides, K.A. and Dukes, J.S. (2007) Plant invasion across space and time, factors affecting nonindigenous species success during four stages of invasion. *New Phytologist* 176, 256–273.

Thiele, J. and Otte, A. (2008) Invasion patterns of *Heracleum mantegazzianum* in Germany on the regional and landscape scales. *Journal for Nature Conservation* 16, 61–71.

Thiele, J., Schuckert, U. and Otte, A. (2008) Cultural landscapes of Germany are patch-corridor-matrix mosaics for an invasive megaforb. *Landscape Ecology* 23, 453–465.

Thomson, D.M. (2007) Do source–sink dynamics promote the spread of an invasive grass into a novel habitat? *Ecology* 88, 3126–3134.

Tiébré, M.S., Saad, L. and Mahy, G. (2008) Landscape dynamics and habitat selection by the alien invasive *Fallopia* (Polygonaceae) in Belgium. *Biodiversity and Conservation* 17, 2357–2370.

Valverde, T. and Silvertown, J. (1997) An integrated model of demography, patch dynamics and seed dispersal in a woodland herb, *Primula vulgaris. Oikos* 80, 67–77.

Verheyen, K., Vanhellemont, M., Stock, T. and Hermy, M. (2007) Predicting patterns of invasion by black cherry (*Prunus serotina* Ehrh.) in Flanders (Belgium) and its impact on the forest understorey community. *Diversity and Distributions* 13, 487–497.

Walsh, S.J., McCleary, A.L., Mena, C.F., Shao, Y., Tuttle, J.P., Gonzalez, A. and Atkinson, R. (2008) QuickBird and Hyperion data analysis of an invasive plant species in the Galapagos Islands of Ecuador: implications for control and land use management. *Remote Sensing of Environment* 112, 1927–1941.

Wangen, S.R., Webster, C.R. and Griggs, J.A. (2006) Spatial characteristics of the invasion of *Acer platanoides* on a temperate forested island. *Biological Invasions* 8, 1001–1012.

Zhu, L., Sun, O.J., Sang, W.G., Li, Z.Y. and Ma, K.P. (2007) Predicting the spatial distribution of an invasive plant species (*Eupatorium adenophorum*) in China. *Landscape Ecology* 22, 1143–1154.

13 Molecular Ecology: Applications for Invasive Plants

Concepts

- Molecular ecology can be used to study how genes respond to selection pressures during species invasions.
- New methods are arising to facilitate molecular research.
- Molecular ecology can track the origins, spread and genetic changes of an invading species.
- Knowledge of molecular structure may be used to manage invasive species.

13.1 Introduction

One of the fundamental goals in invasive plant ecology is to determine where invasions may arise and how populations are distributed in time and space. For example, we want to know what selection pressures will operate, especially where the pressures are human management techniques. These goals can be addressed at many scales but much work is being done at the molecular scale as new technologies allow us to do research and to link the results to population to landscape scales (Meekins *et al.*, 2001; Jackson *et al.*, 2002; Beebee and Rowe, 2008; Freeland, 2005; Broz *et al.*, 2007, 2008; Moody and Les, 2007; Slotta, 2008) – although there is some question as to whether there is much ecological linkage being tested (Johnson *et al.*, 2009). Molecular approaches are already being used to specify which genes are involved in more complex traits, such as colonization (Chapman *et al.*, 2004). Table 13.1 lists a number of species for which similar work has been done; there are few studies, but the trend is for more research and applications using molecular approaches. Indeed, there are questions that are either unique to that scale or informed better by work at the molecular level. Such questions include:

- Are certain genotypes of species likely to invade rapidly?
- How does one species replace another?

- If a native and a non-native species are closely related, are there genes present in the non-native that allow it to invade but are absent in the native species?
- Are new invasive species hybridizing with other invasive or native species?
- Once established, is the genetic diversity of a new invasive large or small?
- Where did the invasive species originate?
- If we attempt to manage invasives, do they have genes that allow rapid selection (counter-response)? Will this rapidly render management approaches ineffective or will they promote selection for traits that facilitate invasion?

13.2 Methods and Techniques in Molecular Research

One method of molecular research is to examine microsatellites. These are short sequence repeats of DNA that are tested. These repeats codes are neutral in the sense they do not usually code for any product that is beneficial or harmful to an organism. Hence, these microsatellites are useful in determining general genetic variation because there is little selection for changes in these sections of DNA between generations. Microsatellite research is useful because of another technical innovation known as polymerase chain reaction (PCR), which amplifies

Table 13.1. Examples of invasive exotic species used in molecular studies.

Common name	Latin name	Reference(s)
Velvetleaf	*Abutilon theophrasti*	Kurokawa *et al.*, 2004
Canada thistle	*Cirsium arvense*	Slotta *et al.*, 2006
Yellow nutnedge	*Cyperus esculentus*	Dodet *et al.*, 2008
Common St John's wort	*Hypericum perforatum*	Mayo and Langridge, 2003; Maron *et al.*, 2004
Pampas grass	*Cortaderia jubata*	Okada *et al.*, 2009
Japanese knotweed	*Fallopia japonica*	Gammon *et al.*, 2007; Grimsby *et al.*, 2007
Bladder campion	*Silene vulgaris*	McCauley *et al.*, 2003
Hogweeds	*Heracleum* spp.	Jahodová *et al.*, 2007
Hydrilla	*Hydrilla verticillata*	Hofstra *et al.*, 2000
Micronia	*Micronia calvescens*	Le Roux *et al.*, 2008
Passion flower	*Passiflora alata*	Koehler-Santos *et al.*, 2006
Barbados gooseberry	*Pereskia aculeata*	Paterson *et al.*, 2009

the numbers of copies of the microsatellites expressed. PCR not only can express large numbers of gene products to determine how they work, it can be part of the process that leads to inserting genes to produce genetically modified organisms. This means that traits such as cold tolerance or herbicide resistance found in one species (often bacteria) can be inserted and expressed in plants (e.g. in crops). There is concern, however, that genes for these traits may escape into related species, some of which may or may not become invasive.

A second molecular method is based on what is known as DNA barcoding. DNA barcoding is a method of species identification and recognition using sequence data for specific regions of DNA (Hebert *et al.*, 2003; Borisenko *et al.*, 2009). This technique examines chloroplast DNA, specifically the ribulose-bisphosphate carboxylase gene (*rbcL*) and the gene maturase K (*matK*) as a standard two-locus barcode to identify species at the molecular level (Newmaster *et al.*, 2006; CBOL Plant Working Group, 2009). As this technique develops further, we may reach a stage in the future where we are able to read these sequences in the field to determine the identity of a plant species. This would be useful in identifying invasive genotypes and invasive species in general.

Barcoding has been criticized in terms of both technical accuracy and possible misuse (Will *et al.*, 2005; Rubinoff, 2006; Larson, 2007). For example, to be accurate, barcoding requires enough genetic variability to distinguish among species. Identification of interspecific plant hybrids can be complicated because of this issue. This is why research should not abandon field identification via morphological traits (Gaskin, 2003; Bossdorf *et al.*, 2005; Ward, 2006; Ward *et al.*, 2008a,b). Nor should molecular approaches be viewed as simple novelty – an expensive and often hard-to-fund approach. Molecular approaches combined with field studies should be used to identify new invasives and test hypotheses regarding their mechanisms and likely success. Moody and Les (2007) did just this to identify hybridizing and adapting invasive species such as one of the hybrid (short-spike–Eurasian) water milfoils (*Myriophyllum spicatum* × *Myriophyllum sibiricum*).

A third molecular approach is to use microarrays. Microarray analysis is a method to interrogate gene activity (i.e. gene expression). The technique has been used effectively when comparing gene expression in plants responding to different selection pressures. More recently, microarrays have been used to isolate a conservative sequence of DNA (called a marker) that codes for proteins or other products that are then expressed in whole or in part as traits that confer some ability on a plant to become invasive. This means chemically testing tens of thousands of DNA sequences that are arranged on a chemical or physical surface – this is called the array. The goal of using the tests on an array is to find the conserved DNA sequences to be used as markers. This only works well if there already exists a library of DNA from a given species, i.e. have we sequenced much of the genome of that same species so we can use microarrays to search for those sequences. If so, we can determine how frequently a marker sequence is being expressed in a population and whether this has any implications for traits that confer a greater ability to be invasive. It is also possible to construct the

Biological processes

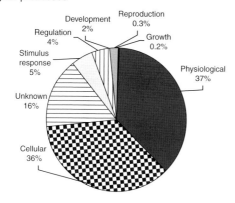

Molecular function

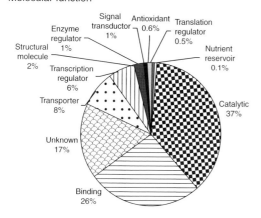

Cellular component

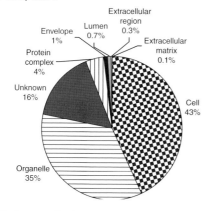

Fig. 13.1. An example of a categorized cDNA (complementary DNA) library for spotted knapweed (*C. maculosa*) from which unigenes (unique contiguous consensus sequences) were assembled and functionally categorized into biological processes, molecular functional and cellular components (redrawn from Broz *et al.*, 2007).

more modest and less labour-intensive expressed sequence tag (EST) libraries from invasive populations. EST allows researchers to identify and mark transcribed cDNA (complementary DNA) sequences with known expression, which can then be isolated, or at least used in experiments that combine molecular and field approaches to test for invasive success (Bossdorf *et al.*, 2005).

Despite their expense and complexity, molecular approaches have been used to study invasive plant ecology. One key is to relate the genes to the actual biochemical and physiological processes affecting plant response and fitness (Ross and Ague, 2008; Ross *et al.*, 2008). In that context, Broz *et al.* (2007) created an EST library (based on a cDNA library) for invasive spotted knapweed (*C. maculosa*) (Fig. 13.1). They focused on identifying genes that promote invasion, such as heat-shock proteins which help to withstand longer-term environmental stress, genes that produce allelochemicals for interspecific competition or genes that confer resistance to herbivores. Their library showed how DNA was categorized according to its functions, such as enzyme regulator or reproduction in spotted knapweed.

We caution that one cannot simply pick a DNA fragment, call it a marker, and blithely assume that it translates directly into fitness and a predictable response to management (Reed and Frankham, 2001; Müller-Schärer *et al.*, 2004; Ahmad *et al.*, 2008). The variability in gene expression that one would expect in a natural population of plants can limit the usefulness of this approach. In many cases, the expression of genes is affected by other genes and, generally, there will be interactions among genes and products such as enzymes or other proteins. Ultimately, these interactions produce a trait that may be visible on an everyday scale, such as the ability of an invasive species to grow tall. Rarely is that trait solely dependent on one gene. Thus, while it is crucial to study at molecular scales to isolate and characterize the sequence of DNA and genes, it is not enough to stop there.

13.3 Molecular Methods Can Help Determine Origins and Likely Spread of Invasive Species

Generally, molecular approaches can augment the broader knowledge about the geographical origin of an invasive species or a variety (Roche *et al.*,

2003; Schaal *et al.*, 2003), or how it may have been assisted by natural selection or slowed by genetic drift locally (Ryan *et al.*, 2007; Vilatersana *et al.*, 2007). Molecular studies can be used to compare the genetic diversity of populations to examine whether they have undergone adaptive differentiation once they establish away from their native range. From an initial population, mutations may increase under selection pressures and subpopulations may form, some of which may be differentially adapted to localized conditions.

Such population differentiation appears to have happened in the populations of some invasive species because genetic diversity has increased since the invasion occurred, e.g. white knapweed (*Centaurea diffusa*) (Marrs *et al.*, 2008). Conversely, in other invasive species, genetic diversity has decreased despite multiple origins and multiple introductions, e.g. purple loosestrife (*L. salicaria*). In North American populations of purple loosestrife, genetic drift appears to be the explanation for reduced genetic diversity (Chun *et al.*, 2009), and this means that the species may not be able to adapt to changing environmental conditions or management selection pressures. One reason for caution regarding this hypothesis is that diversity decline may be related to climate change or the introduction of biological control agents. Such factors make it difficult to untangle the web of multiple selection pressures. It is important to note that not everything is adaptive or related to natural selection; genetic drift is often an important cause of evolutionary responses. The explanations are further complicated because still other processes, such as phenotypic plasticity or asexual reproduction, may be more important in the success of invasive species, e.g. alligator weed (*Alternanthera philoxeroides*) (Xu *et al.*, 2003; Wang *et al.*, 2005; Geng *et al.*, 2007) and fountain grass (*Pennisetum setaceum*) (Poulin *et al.*, 2005).

None the less, molecular-based analyses have been used to show that building divided highways and arterial roads with wide medians, and shoulders with often wet shallow ditches, facilitated the spread of the exotic invasive genotype in common reed (*P. australis*) (Lelong *et al.*, 2007). This type of analysis has also been used to show where the original invasion points were for pontic rhododendron (*Rhododendron ponticum*) in the UK (Milne and Abbott, 2000) and cordgrass (*Spartina densiflora*) in California (Fortune *et al.*, 2008).

Molecular techniques allow for quite precise analysis, for example, they showed that the ornamental variety of Japanese barberry (*Berberis thunbergii* var. *atropurpurea*) was not invasive (Lubell *et al.*, 2008). This was important because it is easily confused, visually, with non-ornamental varieties that were, in fact, invasive. Knowing which varieties are invasive was only possible via use of molecular techniques. Identifying which varieties are a problem then leads to a shift in management efforts from the ornamental variety to varieties that actually are invasive. It is still possible that there is unexpressed genetic variation that may yet create future problems, even with the ornamental variety but, for now, efforts are best directed elsewhere.

13.4 Using Molecular Approaches to Study Invasive Plants

Case study I: invasive Canary Islands St John's wort

The relative importance of pre- and post-invasion selection, genetic drift and plasticity (as the main factors) may help us to understand and perhaps predict how invasive species may respond and succeed in a new habitat. Dlugosch and Parker (2007, 2008) examined aspects of this with the Canary Islands St John's wort (*H. canariense*). In its native habitat, this is an endemic – it is not common and is restricted to a few localized habitats. Yet when it was accidentally transported half a world away, Canary Islands St John's wort was aggressive enough to rapidly invade parts of Australia and New Zealand. How can a species with limited success in its native range be so successful in a new habitat?

In its original range, there are two genetic clusters of Canary Islands St John's wort. These clusters are expressed largely as different life histories, growth rates and flower size. They are restricted to different environments. The reason for an apparent lack of adaptation was that there was little environmental variation and therefore little selection for increased phenotypic expression from genetic diversity. There was no loss of genetic diversity nor was there a reduction in mutation rates. While there will be some genetic drift, the lack of strong selection in the native range has dampened the phenotypic expression of potentially large genetic variation. The capacity for adaptation via genetic mutation and increased diversity still existed. These conclusions were based on analysing the genetic diversity of the species in the Canary Islands: diversity was higher than expected based on the apparent lack of environmental variation and the inability of the St John's wort to expand its range locally.

Canary Island St John's wort was able to invade Australia and New Zealand because its relatively high (but unexpressed) genetic variation was selected upon after invasion. Generally, low genetic variation from founder effects and genetic bottlenecks might limit invasions in some species, such as wild fennel (*Verbascum thapsus*), but there can be unexpressed genetic variation (as in the Canary Island St John's wort) or sufficient gene flow between multiple introductions that often allows for rapid adaptation, such as that in barbed goatgrass (*A. triuncialis*) (Meimberg *et al.*, 2006; Dlugosch and Parker, 2008).

Case study II: invasive Brazilian pepper tree

To investigate how invasive species have or are likely to spread successfully, molecular techniques can be combined with other innovative approaches, such as landscape ecology. Williams *et al.* (2007) used chloroplast and nuclear satellite markers to show that the Brazilian pepper tree (*Schinus terebinthifolius*) invaded Florida from separate source populations when they were planted as ornamentals. In the area around the known historical introduction points, dispersal via birds and mammals facilitated spread of the species. There was intraspecific hybridization and spatial autocorrelation of genetic markers around the original sites. Spatial autocorrelation measures whether organisms are randomly distributed or whether they are clustered over a geographical region because of similar habitat or other preferences.

With some Brazilian pepper tree populations, the maps based on genetic markers show a clustering around a limited number of locations. Thus, autocorrelation of markers allowed the researchers to trace the invasion back to its initial point of introduction. Further, autocorrelation indicated that the spread of this particular Brazilian pepper tree was a gradual one facilitated by birds or mammals other than humans. However, there also were isolated, distant populations of Brazilian pepper tree that had the same markers found in the original source populations. These were nowhere near other Brazilian pepper tree infestations and there was little spatial autocorrelation. This means that it is likely that the isolated and distant infestations of Brazilian pepper tree still came from the same source as the clustered infestations because they all had the same markers. The difference was that the lack of autocorrelation means that there was a second mechanism of spread: the isolated, distant populations of Brazilian pepper tree originated via rapid transport by humans – not by other mammals or birds. Given that there were different avenues and rates of invasion, a state-wide management effort was recommended for Brazilian pepper tree (Williams *et al.*, 2007).

13.5   Using Molecular Methods to Manage Invasive Species

Molecular-scale approaches can help to identify the genetic variation of invasive species (Hollingsworth *et al.*, 1998; Saltonstall, 2003; Durka *et al.*, 2005). Greater genetic diversity may explain why some species are able to invade, persist and dominate habitats (Chapter 10). When a potentially invasive species is introduced multiple times across many locations, the total population will have more genetic diversity. There also may be more genetic differences between local populations, thereby increasing the likelihood that the newly invading species can adapt to environmental changes or attempts at management. This probably occurred in the cases of garlic mustard (*A. petiolata*), knotweed (*Fallopia* spp.), and common ragweed (*A. artemisiifolia*), but not in Asian giant reed (*A. donax*) (Meekins *et al.*, 2001; Genton *et al.*, 2005; Gammon *et al.*, 2007; Grimsby *et al.*, 2007; Tiébré *et al.*, 2008; Ahmad *et al.*, 2008).

This kind of knowledge assists us in forecasting which management methods may be effective. For example, common ragweed was introduced into Europe from multiple sources; as a result, it has high genetic diversity and therefore would be difficult to manage with biological control (Genton *et al.*, 2005). Similar issues arose with attempts to manage yellow toadflax (*Linaria vulgaris*) in North America once it became known that biological control was likely to be of no practical significance because of high genetic diversity (Ward *et al.*, 2008a). In contrast, Asian giant reed has invaded the southern USA but it has low genetic diversity. How low? This entire invasive population is non-sexual and, effectively, a widespread single genet (Chapter 5) so it is likely that biological control may work well, provided that the biocontrol species feeds exclusively on that genotype (Ahmad *et al.*, 2008). Biological control works best if the invasive species has lower genetic diversity because it is difficult for such populations to adapt to introduced natural enemies. The same logic could be used to justify other measures, such as the use of herbicides.

There are surprises revealed by molecular approaches. Tussock hawkeed (*Hieracium lepidulum*) and dandelion (*T. officinale*) are both agamospermous and the lack of genetic recombination means that genetic diversity was hypothesized to be low. However, the occasional recombination, the measured mutation rate and the molecular confirmation of multiple origins of invasive populations revealed that their diversity and ability to respond to environmental changes or management was higher than expected (Mes *et al.*, 2002; Chapman *et al.*, 2004).

Molecular-scale approaches should inspire researchers to think laterally about invasiveness. There is often too separate an approach to studying invasive ecology in the context of issues such as herbicide resistance. This is a false dichotomy because herbicide resistance is an expression of response to selection and will confer an important avenue of new invasion. More work is needed to weave herbicide resistance and invasion ecology together to determine how genetic variation may make populations more or less vulnerable to herbicides, and more or less able to be selected for resistance to herbicides. This has begun in some areas. For example, molecular approaches have revealed that multiple mutations in acetolactate synthase (ALS) within invasive species led to multiple paths of resistance to herbicides that target ALS (Michel *et al.*, 2004, Patzoldt and Tranel, 2007). Therefore, managers must anticipate that there will be selection for different forms of resistance, that alternatives will have to be planned in advance and that an integrated management programme is used to reduce continuous and directional selection pressures that would promote resistance. This idea is not new (Swanton and Murphy, 1996). What is new is that the molecular technology now exists to implement integrated management so long as we remember that the broad context of invasion ecology applies to specific issues such as herbicide resistance.

13.6   Summary

Molecular ecology is a new tool that will contribute to our understanding of invasive species. A number of methods are developing to study molecular ecology; each has its own uses and strengths. One use of molecular techniques is to track where an invasive species originated and how it spreads. Molecular studies, however, must be placed in the context of field observations. In the final chapter, we synthesize the many component processes to revisit the process of a plant invasion.

Box 13.1. Invasive species case study: molecular ecology.

- What molecular studies, if any, have been done on your species?
- What questions could molecular studies answer to explain the invasiveness of your species?

13.7   Questions

1. Explain how molecular techniques are used to determine the origin and spread of an invasive species.

2. How would an understanding of genetic diversity influence the management of an invasive species?

3. Why is it important to put molecular studies in the context of field studies?

Further Reading

Beebee, T. and Rowe, G. (2008) *An Introduction to Molecular Ecology*, 2nd edn. Oxford University Press, New York.

Bruford, M.W., Hewitt, G. and Hoelzel, A.R. (in press) *Molecular Ecology: Genes, Organisms and Processes.* Blackwell Publishing, New York.

Freeland, J. (2005) *Molecular Ecology*. Wiley, New York.

References

Ahmad, R., Liow, P.S., Spencer, D.F. and Jasieniuk, M. (2008) Molecular evidence for a single genetic clone of invasive *Arundo donax* in the United States. *Aquatic Botany* 88, 113–120.

Beebee, T. and Rowe, G. (2008) *An Introduction to Molecular Ecology*, 2nd edn. Oxford University Press, New York.

Borisenko, A.V., Sones, J.E. and Hebert, P.D.N. (2009) The front-end logistics of DNA barcoding, challenges and prospects. *Molecular Ecology Resources* 9, 27–34, Suppl. 1.

Bossdorf, O., Auge, H., Lafuma, L., Rogers, W.E., Siemann, E. and Prati, D. (2005) Phenotypic and genetic differentiation between native and introduced plant populations. *Oecologia* 144, 1–11.

Broz, A.K., Broeckling, C.D., He, J.B., Dai, X., Zhao, P.X. and Vivanco, J.M. (2007) A first step in understanding an invasive weed through its genes, an EST analysis of invasive *Centaurea maculosa*. *BMC Plant Biology* 7(25), doi:10.1186/1471-2229-725.

Broz, A.K., Manter, D.K., Callaway, R.M., Paschke, M.W. and Vivanco, J.M. (2008) A molecular approach to understanding plant-plant interactions in the context of invasion biology. *Functional Plant Biology* 35, 1123–1134.

CBOL (Consortium for the Barcode of Life) Plant Working Group (2009) A DNA barcode for land plants. *Proceedings of the National Academy of Sciences of the USA* 106, 12794–12797.

Chapman, H., Robson, B. and Pearson, M.L. (2004) Population genetic structure of a colonising, triploid weed, *Hieracium lepidulum*. *Heredity* 92, 182–188.

Chun, Y.J. Nason, J.D. and Moloney, K.A. (2009) Comparison of quantitative and molecular genetic variation of native vs invasive populations of purple loosestrife (*Lythrum salicaria* L., Lythraceae). *Molecular Ecology* 18, 3020–3035.

Dlugosch, K.M. and Parker, I.M. (2007) Molecular and quantitative trait variation across the native range of the invasive species *Hypericum canariense*, evidence for ancient patterns of colonization via pre-adaptation? *Molecular Ecology* 16, 4269–4283.

Dlugosch, K.M. and Parker, I.M. (2008) Founding events in species invasions, genetic variation, adaptive evolution, and the role of multiple introductions. *Molecular Ecology* 17, 431–449.

Dodet, M., Petit, R.J. and Gasquez, J. (2008) Local spread of the invasive *Cyperus esculentus* (Cyperaceae) inferred using molecular genetic markers. *Weed Research* 48, 19–27.

Durka, W., Bossdorf, O., Prati, D. and Auge, H. (2005) Molecular evidence for multiple introductions of garlic mustard (*Alliaria petiolata*, Brassicaceae) to North America. *Molecular Ecology* 14, 1697–1706.

Fortune, P.M., Schierenbeck, K., Ayres, D., Bortolus, A., Catrice, O., Brown, S. and Ainouche, M.L. (2008) The enigmatic invasive *Spartina densiflora*: a history of hybridizations in a polyploidy context. *Molecular Ecology* 17, 4304–4316.

Freeland, J. (2005) *Molecular Ecology*. John Wiley & Sons, New York.

Gammon, M.A., Grimsby, J.L., Tsfrelson, D. and Kesseli, R. (2007) Molecular and morphological evidence reveals introgression in swarms of the invasive taxa *Fallopia japonica, F. sachalinensis*, and *F. × bohemica* (Polygonaceae) in the United States. *American Journal of Botany* 94, 948–956.

Gaskin, J.F. (2003) Molecular systematics and the control of invasive plants: a case study of *Tamarix* (Tamaricaceae). *Annals of the Missouri Botanical Garden* 90, 109–118.

Geng, Y.P., Pan, X.Y., Xu, C.Y., Zhang, W.J., Li, B., Chen, J.K., Lu, B.R. and Song, Z.P. (2007) Phenotypic plasticity rather than locally adapted ecotypes allows the invasive alligator weed to colonize a wide range of habitats. *Biological Invasions* 9, 245–256.

Genton, B.J., Shykoff, J.A. and Giraud, T. (2005) High genetic diversity in French invasive populations of common ragweed, *Ambrosia artemisiifolia*, as a result of multiple sources of introduction. *Molecular Ecology* 14, 4275–4285.

Grimsby, J.L., Tsirelson, D., Gammon, M.A. and Kesseli, R. (2007) Genetic diversity and clonal vs. sexual reproduction in *Fallopia* spp. (Polygonaceae). *American Journal of Botany* 94, 957–964.

Hebert, P.D.N., Cywinska, A., Ball, S.L. and deWaard, J.R. (2003) Biological identifications through DNA barcodes. *Proceedings of the Royal Society of London B* 270, 313–321.

Hofstra, D.E., Clayton, J., Green, J.D. and Adam, K.D. (2000) RAPD profiling and isozyme analysis of New Zealand *Hydrilla verticillata*. *Aquatic Botany* 66, 153–166.

Hollingsworth, M.L., Hollingsworth, P.M., Jenkins, G.I., Bailey, J.P. and Ferris, C. (1998) The use of molecular markers to study patterns of genotypic diversity in some invasive alien *Fallopia* spp. (Polygonaceae). *Molecular Ecology* 7, 1681–1691.

Jackson, R.B., Linder, C.R., Lynch, M., Purugganan, M., Somerville, S. and Thayer, S.S. (2002) Linking molecular insight and ecological research. *Trends in Ecology and Evolution* 17, 409–414.

Jahodová, S., Trybush, S., Pyšek, P., Wade, M. and Karp, A. (2007) Invasive species of *Heracleum* in Europe, an insight into genetic relationships and invasion history. *Diversity and Distributions* 13, 99–114.

Johnson, J.B., Peat, S.M. and Adams, B.J. (2009) Where's the ecology in molecular ecology? *Oikos* 118, 1601–1609.

Koehler-Santos, P., Lorenz-Lemke, A.P., Muschner, V.C., Bonatto, S.L., Salzano, F.M. and Freitas, L.B. (2006)

Molecular genetic variation in *Passiflora alata* (Passifloraceae), an invasive species in southern Brazil. *Biological Journal of the Linnean Society* 88, 611–630.

Kurokawa, S., Shibaike, H., Akiyama, H. and Yoshimura, Y. (2004) Molecular and morphological differentiation between the crop and weedy types in velvetleaf (*Abutilon theophrasti* Medik.) using a chloroplast DNA marker: seed source of the present invasive velvetleaf in Japan. *Heredity* 93, 603–609.

Larson, B.M.H. (2007) DNA barcoding: the social frontier. *Frontiers in Ecology and the Environment* 5, 437–442.

Le Roux, J.J., Wieczorek, A.M. and Meyer, J.Y. (2008) Genetic diversity and structure of the invasive tree *Miconia calvescens* in Pacific islands. *Diversity and Distributions* 14, 935–948.

Lelong, B., Lavoie, C., Jodoin, Y. and Belzile, F. (2007) Expansion pathways of the exotic common reed (*Phragmites australis*): a historical and genetic analysis. *Diversity and Distributions* 13, 430–437.

Lubell, J.D., Brand, M.H., Lehrer, J.M. and Holsinger, K.E. (2008) Detecting the influence of ornamental *Berberis thunbergii* var. *atropurpurea* in invasive populations of *Berberis thunbergii* (Berberidaceae) using AFLP. *American Journal of Botany* 95, 700–705.

Maron, J.L., Vila, M., Bommarco, R., Elmendorf, S. and Beardsley, P. (2004) Rapid evolution of an invasive plant. *Ecological Monographs* 74, 261–280.

Marrs, R.A., Sforza, R. and Hufbauer, R.A. (2008) When invasion increases population genetic structure: a study with *Centaurea diffusa*. *Biological Invasions* 10, 561–572.

Mayo, G.M. and Langridge, P. (2003) Modes of reproduction in Australian populations of *Hypericum perforatum* L. (St. John's wort) revealed by DNA fingerprinting and cytological methods. *Genome* 46, 573–579.

McCauley, D.E., Smith, R.A., Lisenby, J.D. and Hsieh, C. (2003) The hierarchical spatial distribution of chloroplast DNA polymorphism across the introduced range of *Silene vulgaris*. *Molecular Ecology* 12, 3227–3235.

Meekins, J.F., Ballard, H.E. and McCarthy, B.C. (2001) Genetic variation and molecular biogeography of a North American invasive plant species (*Alliaria petiolata*, Brassicaceae). *International Journal of Plant Sciences* 162, 161–169.

Meimberg, H., Hammond, J.I., Jorgensen, C.M., Park, T.W., Gerlach, J.D., Rice, K.J. and Mckay, J.K. (2006) Molecular evidence for an extreme genetic bottleneck during introduction of an invading grass to California. *Biological Invasions* 8, 1355–1366.

Mes, T.H.M., Kuperus, P., Kirschner, J., Štepánek, J., Štorchová, H., Oosterveld, P. and den Nijs, J.C.M. (2002) Detection of genetically divergent clone mates in apomictic dandelions. *Molecular Ecology* 11, 253–265.

Michel, A., Arias, R.S., Scheffler, B.E., Duke, S.O., Netherland, M. and Dayan, F.E. (2004) Somatic mutation-mediated evolution of herbicide resistance in the nonindigenous invasive plant hydrilla (*Hydrilla verticillata*). *Molecular Ecology* 13, 3229–3237.

Milne, R.I. and Abbott, R.J. (2000) Origin and evolution of invasive naturalized material of *Rhododendron ponticum* L. in the British Isles. *Molecular Ecology* 9, 541–556.

Moody, M.L. and Les, D.H. (2007) Geographic distribution and genotypic composition of invasive hybrid watermilfoil (*Myriophyllum spicatum* × *M. sibiricum*) populations in North America. *Biological Invasions* 9, 559–570.

Müller-Schärer, H., Schaffner, U. and Steinger, T. (2004) Evolution in invasive plants: implications for biological control. *Trends in Ecology and Evolution* 19, 417–422.

Newmaster, S.G., Fazekas, A.J. and Ragupathy, S. (2006) DNA barcoding in land plants: evaluation of *rbcL* in a multigene tiered approach. *Canadian Journal of Botany* 84, 335–341.

Okada, M., Lyle, M. and Jasieniuk, M. (2009) Inferring the introduction history of the invasive apomictic grass *Cortaderia jubata* using microsatellite markers. *Diversity and Distributions* 15, 148–157.

Paterson, L.D., Downie, D.A. and Hill, M.P. (2009) Using molecular methods to determine the origin of weed populations of *Pereskia aculeata* in South Africa and its relevance to biological control. *Biological Control* 48, 84–91.

Patzoldt, W.L. and Tranel, P.J. (2007) Multiple ALS mutations confer herbicide resistance in waterhemp (*Amaranthus tuberculatus*). *Weed Science* 55, 421–428.

Poulin, J., Weller, S.G. and Sakai, A.K. (2005) Genetic diversity does not affect the invasiveness of fountain grass (*Pennisetum setaceum*) in Arizona, California and Hawaii. *Diversity and Distributions* 11, 241–247.

Reed, D.H. and Frankham, R. (2001) How closely correlated are molecular and quantitative measures of genetic variation? A meta-analysis. *Evolution* 55, 1095–1103.

Roche, C.T., Vilatersana, R., Garnatje, T., Gamarra, R., Garcia-Jacas, N., Susanna, A. and Thill, D.C. (2003) Tracking an invader to its origins: the invasion case history of *Crupina vulgaris*. *Weed Research* 43, 177–189.

Ross, C.A. and Auge, H. (2008) Invasive *Mahonia* plants outgrow their native relatives. *Plant Ecology* 199, 21–31.

Ross, C.A., Auge, H. and Durka, W. (2008) Genetic relationships among three native North-American *Mahonia* species, invasive *Mahonia* populations from Europe and commercial cultivars. *Plant Systematics and Evolution* 275, 219–229.

Rubinoff, D. (2006) Utility of mitochondrial DNA barcodes in species conservation. *Conservation Biology* 20, 1026–1033.

Ryan, F.J., Mosyakin, S.L. and Pitcairn, M.J. (2007) Molecular comparisons of *Salsola tragus* from

California and Ukraine. *Canadian Journal of Botany* 85, 224–229.

Saltonstall, K. (2003) Microsatellite variation within and among North American lineages of *Phragmites australis*. *Molecular Ecology* 12, 1689–1702.

Schaal, B.A., Gaskin, J.F. and Caicedo, A.L. (2003) Phylogeography, haplotype trees and invasive plant species. *Journal of Heredity* 94, 197–204.

Slotta, T.A.B. (2008) What we know about weeds: insights from genetic markers. *Weed Science* 56, 322–326.

Slotta, T.A.B., Rothhouse, J.M., Horvath, D.P. and Foley, M.E. (2006) Genetic diversity of Canada thistle (*Cirsium arvense*) in North Dakota. *Weed Science* 54, 1080–1085.

Swanton, C.J. and Murphy, S.D. (1996) Weed science beyond the weeds: the role of integrated weed management (IWM) in agroecosystem health. *Weed Science* 44, 437–445.

Tiébré, M.S., Saad, L. and Mahy, G. (2008) Landscape dynamics and habitat selection by the alien invasive *Fallopia* (Polygonaceae) in Belgium. *Biodiversity and Conservation* 17, 2357–2370.

Vilatersana, R., Brysting, A.K. and Brochmann, C. (2007) Molecular evidence for hybrid origins of the invasive polyploids *Carthamus creticus* and *C. turkestanicus* (Cardueae: Asteraceae). *Molecular Phylogenetics and Evolution* 44, 610–621.

Wang, B.R., Li, W.G. and Wang, J.B. (2005) Genetic diversity of *Alternanthera philoxeroides* in China. *Aquatic Botany* 81, 277–283.

Ward, S. (2006) Genetic analysis of invasive plant populations at different spatial scales. *Biological Invasions* 8, 541–552.

Ward, S.M., Reid, S.D., Harrington, J., Sutton, J. and Beck, K.G. (2008a) Genetic variation in invasive populations of yellow toadflax (*Linaria vulgaris*) in the western United States. *Weed Science* 56, 394–399.

Ward, S.M., Gaskin, J.F. and Wilson, L.M. (2008b) Ecological genetics of plant invasion: What do we know? *Invasive Plant Science and Management* 1, 98–109.

Will, K.W., Mishler, B.D. and Wheeler, Q.D. (2005) The perils of DNA barcoding and the need for integrative taxonomy. *Systematic Biology* 54, 844–851.

Williams, D.A., Muchugu, E., Overholt, W.A. and Cuda, J.P. (2007) Colonization patterns of the invasive Brazilian peppertree, *Schinus terebinthifolius*, in Florida. *Heredity* 98, 284–293.

Xu, C.Y., Zhang, W.J., Fu, C.Z. and Lu, B.R. (2003) Genetic diversity of alligator weed in China by RAPD analysis. *Biodiversity and Conservation* 12, 637–645.

14 Plant Invasions: a Synthesis

Concepts

- An invasion is the geographical expansion of a species into an area not previously occupied by it. Both native and non-native species can be invasive.
- Most invasions fail because the species do not possess the suite of traits necessary to transport, colonize, establish and spread into a habitat.
- Three habitat characteristics that are thought to encourage invasions are disturbances, low species richness and high resource availability.

14.1 Introduction

In Chapter 1, we briefly discussed the invasion process, noting that very few species that are introduced to a new habitat actually become invasive. We gave several possible definitions of the word invasive. In the remaining chapters we looked at how population-, community- and landscape-scale processes contribute to or hinder the invasions. In the first chapter, we introduced invasion, emphasizing that it is a process and that species must overcome specific filters for the invasion to proceed. In this chapter, we synthesize the information presented in the earlier chapters to highlight and understand invasions. We start by describing some conceptual models.

14.2 Using Conceptual Models to Understand Invasion

A successful invasion is a rare event. In Chapter 1 you learned that the 'tens rule' describes how approximately 10% of species pass through each transition from being dispersed, to colonizing, to becoming established and finally, becoming invasive (Williamson and Brown, 1986; Williamson, 1993, 1996) (Fig. 1.3). Conceptual models (e.g. Fig. 1.3), in general, are used to represent relationships between ideas. In invasion ecology, conceptual

models provide a way to visualize the invasion process, and help researchers to think about what factors are important to invasions and how they can be studied.

A successful invasion requires that a species arrives, colonizes, establishes, reproduces, spreads and integrates with other members of a community (Vermeij, 1996; Williamson, 1996; Richardson et al., 2000). A number of papers have used conceptual models to describe this process (Colautti and MacIsaac, 2004; Lockwood et al., 2005; Theoharides and Dukes, 2007; Hellmann et al., 2008; Prentis et al., 2008). In a simple form, Hellmann et al. (2008) showed four stages of invasions (transport, colonization, establishment and landscape spread) and the filters that a species faces at each transition (Fig. 14.1). Colautti and MacIsaac (2004) developed a new terminology based on the stages of invasion that emphasizes the ecological and biogeographical nature of invasions (Fig. 14.2). They categorized species based on the local population size and how widespread the species is. Other researchers have expanded basic models to include other factors such as spatial and temporal scales of the various stages and processes (Theoharides and Duke, 2007).

Invasion models have a number of things in common. First, they generally divide the invasion

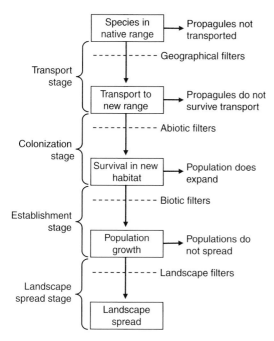

Fig. 14.1. Conceptual model of the four stages of invasions (transport, colonization, establishment and landscape spread) and the filters that a species faces at each transition (based on Hellmann *et al.*, 2008).

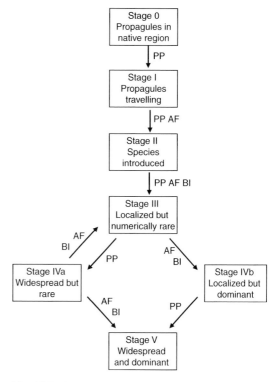

Fig. 14.2. Conceptual model of the invasion process which emphasizes the ecological and biogeographical nature of invasions. PP = propagule pressure; AF = abiotic filters; BI = biotic interactions (adapted from Colautti and MacIsaac, 2004).

process into four basic stages: transport, colonization, establishment and spread. Other researchers divide it into more stages (Catford *et al.*, 2008). A second common feature of invasion models is that a species faces a series of filters to pass through each of the invasion stages. As the invasion proceeds, different species will be lost at the transition of one stage to another. Over the course of an invasion, the filters change, thereby removing different species. We will look at each of these stages and the filters that species face.

14.3 Stages of Invasion

The invasion of a non-native species follows a characteristic curve (Chapter 1; Fig. 1.2). This curve is variable among species, but there are some generalities we can make about the processes responsible for the shape of the curve, as well as for the filters faced by invasive species and the evolutionary processes that act on them (Fig. 14.3) (Prentis *et al.*, 2008). We will examine these below in the context of the four general stages of invasion.

Transport

To invade, a species' propagules must first disperse to a novel recipient location (Chapter 6). Dispersal provides the opportunity for invasion. Long-distance dispersal mechanisms are important but are now augmented by human activities such as transportation. Humans are now the most important dispersers of introduced species (Williamson, 1999; Pauchard and Shea, 2006). In effect, human activity is expanding the total species pool (Hobbs, 2000). The major filter affecting dispersal to a new environment is the physical geographical distance. Species that become invasive have often been transported across oceans or continents.

Colonization

Once a propagule reaches a new habitat, there are many filters that it must withstand before it becomes a successful colonizer. This is primarily because the abiotic environment may not be suitable for it. As plants

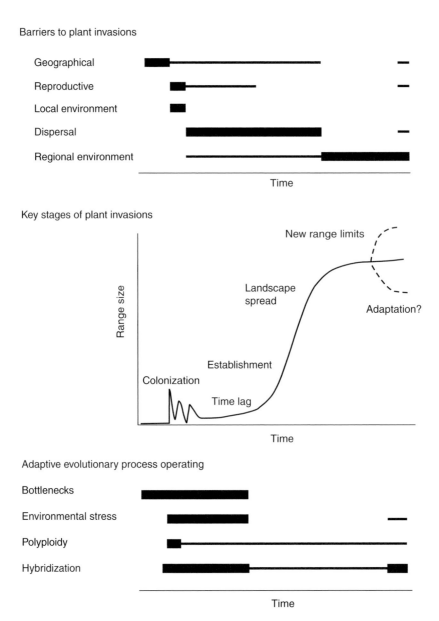

Fig. 14.3. Adaptive evolution during the stages of invasion. Key stages in plant invasions can be depicted with regard to the change in range size over time (centre), or with reference to sequential filters that the species must pass through (top). Invasion filters predispose plants to certain types of evolutionary change at different points in the invasion process (bottom). Elucidation of these dynamics is crucial for understanding adaptive evolution in plant invasions (Prentis *et al.*, 2008).

travel long distances, they may arrive in a habitat that is very different from their area of origin. Species that have broad ecological tolerances are therefore more likely to successfully tolerate the abiotic filters and successfully colonize a new habitat (Chapter 2).

Propagule pressure

An introduced species is more likely to colonize if it is introduced repeatedly. Propagule pressure is the number of new individuals that are transported to a new location (Lockwood *et al.*, 2009; Simberloff, 2009). Propagules may arrive all at once or in repeated introductions. Higher propagule pressure is likely to be the most important factor controlling whether a species will successfully colonize in a new habitat (Lockwood *et al.*, 2005; Von Holle and Simberloff, 2005; Colautti *et al.*, 2006). High propagule pressure increases colonization for several reasons (Lockwood *et al.*, 2005, 2009; Reaser *et al.*, 2008; Simberloff, 2009). First, it increases the likelihood that at least some individuals will land in a suitable biotic and abiotic environment. Second, Allee effects are reduced because more individuals are released into establishing populations (Chapter 4). Thus, it is easier for individual plants to find a compatible mate and less likely that they will succumb to demographic or environmental stochasticity. Third, repeated introductions will increase genetic variation, thus reducing inbreeding depression, founder effects and bottlenecks (Chapter 2). High genetic diversity increases invasiveness because it makes species more able to adapt to the local environment and gives them the potential to invade a wider variety of habits. Because propagule pressure is so important to the success of non-native species, some researchers have suggested that propagule-based policies would improve the management of invasive species (Reaser *et al.*, 2008).

Establishment

Once a species colonizes an area, it must establish self-sustaining, expanding populations if it is to be a successful invader (Theoharides and Duke, 2007). At this point, biotic filters become increasingly important (Chapters 8 and 9) because they influence survival, growth and reproduction (Theoharides and Duke, 2007). Therefore species with traits that confer competitiveness or resistance to herbivores will establish well. In addition, because populations are small, Allee effects may have a strong influence at this time (Chapter 4). Abiotic filters are still prevalent, but are less important during establishment.

Time lags

A striking characteristic of the establishment stage is that there is often a lag between the time when a species colonizes and when it spreads (Hobbs and Humphries, 1995; Kowarik, 1995; Crooks and Soulé, 1999; Crooks, 2005). Time lags can sometimes

be quite long; for example, white pine (*Pinus strobus*) was not considered invasive in Central Europe until more than 250 years after it was introduced for forestry (Rejmánek, 1996). In Britain, wild lettuce (*Lactuca virosa* and *Lactuca scariola*) were considered rare from the 1860s, when they were introduced, to the mid-1900s, when their abundance increased as they spread into gravel pits (Crooks and Soulé, 1999). The lag phase commonly lasts from 20 to 100 years (Hobbs and Humphries, 1995; Wade, 1997).

Time lags occur partly because population growth follows an exponential growth curve (Chapter 3). Small populations grow very slowly initially. The length of a time lag is dependent on the abiotic and biotic environment (Chapter 2), and on any genetic changes that occur (Chapter 4) (Crooks and Soulé, 1999). Also, small populations will have low seed set if pollinators are hard to attract or if little wind-dispersed pollen reaches plant stigmas (Chapter 4). Therefore, we can expect population expansion to be slow at first.

Time lags can also occur when species persist in small isolated pockets until a disturbance or a certain set of environmental conditions occurs that facilitates rapid expansion. This happened in the case of Oxford ragwort (*Senecio squalidus*). Its abundance increased after World War II because it invaded railway lines and disturbed areas created by bombings (Baker, 1965; Kowarik, 1995; Crooks and Soulé, 1999). Populations of agricultural weeds also change in response to farming practices; for example, the expansion of Canada horseweed (*Conyza canadensis*) was facilitated by the increase in reduced tillage.

A final reason why small populations may have a sudden rapid increase in population size is that they may undergo some genetic change that increases the population's fitness (Chapter 4). Hybridization and introgression (a hybrid backcrossing with one of the parent populations, which introduces new genes back into the species genome) are mechanisms by which genetic change can occur. For example, the introduced smooth cordgrass (*S. alterniflora*) hybridized with the native California cordgrass (*Spartina foliosa*), and the hybrid spreads more rapidly than either of the native species (Daehler and Strong, 1997).

Landscape spread

Spread refers to the dispersal of a species across a landscape (Chapter 12). Similar filters affect all species in the species spread stage because all species must disperse, survive and reproduce as they move between connected landscape fragments (Theoharides and Duke, 2007). The rate of spread is influenced by species' traits and by the pattern and heterogeneity of the landscape.

14.4 Why do Species Invade? Why are Habitats Invasible?

Regardless of the traits a species possesses, a community must still be able to be invaded. Species' traits, community characteristics, and the interactions between the community and the potential invader, as well as timing and chance will all determine whether an introduced species is successful (Lodge, 1993; Hobbs and Humphries, 1995).

Species' invasive ability

It would be useful if we could list the traits of a plant species and from that list determine when and where a species will invade. Unfortunately this is not possible, but there are some ways in which we can get a general idea of whether a species is likely to become invasive.

Life history traits

Life history traits, such as height and seed mass, are the characteristics that determine how a species will regenerate, persist and disperse (Chapter 6). Much of the early work on invasions tended to list traits likely to increase a species' invasive ability. Baker (1965) summarized life history traits of an ideal weed (Table 14.1); however, this overly simplifies the invasion process (Perrins *et al.*, 1992, 1993). Lodge (1993) summarized traits commonly associated with invasiveness but noted

that when tested, many of these were statistically rejected, or else there were too many exceptions to make them useful. A high rate of population increase (r) (Chapter 3), for example, does not necessarily increase invasiveness (Lawton and Brown, 1986). Trait lists only provide a few indicators that can help explain a weed's invasiveness, but they are not helpful in predicting which species will become invasive (Perrins *et al.*, 1992). Himalayan balsam (*I. glandulifera*), for example, is a problem weed in Britain and yet possesses only two of Baker's characteristics. Other weeds such as common field speedwell (*Veronica persica*) and common chickweed (*S. media*) possess many of Baker's traits but are not as problematic (Perrins *et al.*, 1993). Native and non-native species with expanding distribution in England, Scotland, the Republic of Ireland and the Netherlands were almost indistinguishable as far as trait lists were concerned (Thompson *et al.*, 1995).

Is it possible, then, to predict whether a species will invade based on its traits? Under specific circumstances, there has been limited success. For example, Rejmánek and Richardson (1996) were able to predict the invasive ability of pines (*Pinus* spp.) in the southern hemisphere. Using only three traits (seed mass, length of juvenile period and interval between seed mast years), they were able to create an invasiveness equation (Fig. 14.4). More recently, this equation has been found to apply to other woody plants (Table 14.2), and along with a decision tree (Fig. 14.5) can be used to decide whether to prohibit a woody species from being introduced into North America (Reichard and Hamilton, 1997).

Pyšek and Richardson (2007) compared the results of 18 comparative studies and found that it was

Table 14.1. Traits of an 'ideal' weed (based on Baker, 1965, 1974).

1. Germinates in a wide range of environmental conditions
2. Long-lived seeds that are internally controlled so that germination is discontinuous
3. Rapid growth from vegetative stage through to flowering stage
4. Self-compatible, but not completely autogamous or apomictic
5. Cross-pollination (when present) by wind or generalist insects
6. Seeds produced continuously throughout the growth period
7. Seed production occurs under a wide range of environmental conditions
8. High seed output when environmental conditions are favourable
9. Propagules (seeds) adapted to short- and long-distance dispersal
10. If perennial, has a high rate of vegetative reproduction or regeneration from fragments
11. If perennial, ramet attachments fragment easily, so the plant is difficult to pull from the ground
12. Strong potential to compete interspecifically via allelopathy, rosettes, rapid growth and other means

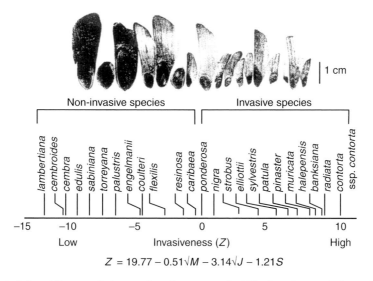

Fig. 14.4. Placement of pines (*Pinus* spp.) along an invasiveness gradient. The invasiveness (*Z*) equation was determined using a discriminant analysis based on mean seed mass in mg (*M*), mean interval between large seed crops (mast years) in years (*S*), and minimum juvenile period in years (*J*) (from Rejmánek, 1999).

difficult to summarize which traits confer invasiveness. Some invasive traits were evident; for example, specific leaf area and the timing of flowering were important traits, whereas the type of pollen vector and propagule size were not. Many traits gave conflicting results: for example, growth form (e.g. annual, biennial, perennial) was important, but only in relation to the growth form of the co-occurring native species; and being perennial was important only if coexisting native species were annuals.

Large geographical native range

Species with a large native range often have the potential to be successful invaders (Chapter 2). Goodwin

et al. (1999) tested this by comparing pairs of species from the same genera where one species had successfully invaded New Brunswick, Canada, and the other had not. They found that that the size of a species' native range was the single best predictor of invasiveness, and could predict whether a species would be invasive or non-invasive for about 70% of the species tested. Other traits, such as plant height and length of flowering period, were less useful.

Native range seems to be important for two reasons (Scott and Panetta, 1993; Rejmánek, 1995; Goodwin *et al.*, 1999). First, widespread species are more likely to be dispersed simply because they are in more locations and therefore are more likely to be transported by natural and human dispersal

Table 14.2. Rules for detecting invasive woody plants based on fruit and seed traits, the potential for vertebrate dispersal and a discriminant function '*Z*' (see Fig. 14.4) (Rejmánek, 1999).

Discriminant function, Z[a]	Fruit and seed traits	Opportunities of dispersal by vertebrates	
		Absent	Present
$Z > 0$	Dry fruit, large (>2 mg) seeds	Likely invasive	Very likely invasive
$Z > 0$	Dry fruit, small (<2 mg) seeds	Likely invasive in wet habitats	–
$Z > 0$	Fleshy fruit	Unlikely invasive	Very likely invasive
$Z < 0$	–	Non-invasive unless dispersed in water	Possibly invasive

[a] Z values >0 indicate invasiveness according to a discriminant function based on seed mass, intervals between large seed crops and minimum juvenile period (see Fig. 14.4).

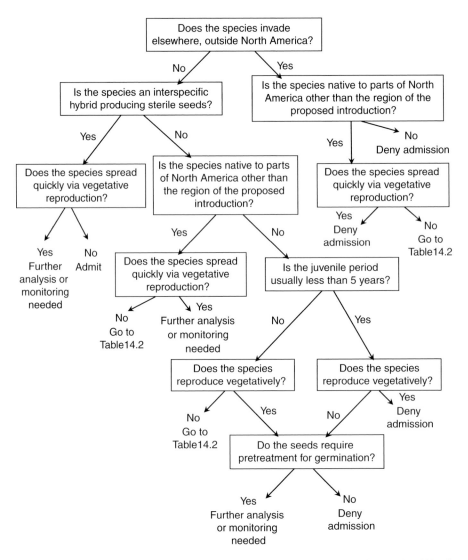

Fig. 14.5. Decision tree for admission of non-native trees into North America (redrawn from Reichard and Hamilton, 1997).

agents. Second, widespread species may be adapted to a wider range of environmental conditions and therefore are more likely to be able to survive in the abiotic environment of a new habitat following dispersal.

However, native range is not always a good indication of invasiveness. In cases where a species is controlled by biotic factors, such as herbivory or disease in its native range, it may spread rapidly when released from these in a new habitat (Scott and Panetta, 1993). Monterey pine (*P. radiata*) and Canary Islands St John's wort (*H. canariense*) are example of species with a small

native range, but which became weedy when dispersed to other habitats (Chapters 2 and 13) (Richardson and Bond, 1991; Dlugosch and Parker, 2007, 2008).

Habitat invasibility

Not all communities are equally invasible. Being able to characterize or predict which communities or habitat types are easier to invade would have practical benefits. We could work on protecting vulnerable habitats and not worry about invasion-resistant ones. Three habitat characteristics that are generally thought to encourage invasions are disturbances, low diversity and resource availability.

Role of disturbance

Any natural or managed ecosystem will experience periodic disturbances, and often specific types of disturbances may be required to maintain a community (Chapters 10 and 11). Fire, for example, is a natural part of the boreal forest ecosystem. Disturbance may even inhibit invasion; for example, fire is used to control the invasion of yellow star-thistle (*C. solstitialis*) in California grassland (Lodge, 1993; Hastings and DiTomaso, 1996). However, disturbance is often cited as a precursor to invasions because it provides plants with a window of opportunity (Baker, 1965; Fox and Fox, 1986; Orians, 1986; Crawley, 1987; Hobbs and Huenneke, 1992; Hobbs, 2000). In addition, human activities alter the natural disturbance regime and intensity (Hobbs, 2000).

Disturbance can change or remove the filters acting on a community. Therefore, disturbance alters habitat characteristics such that they become more conducive to the spread of plants (Vitousek *et al.*, 1997; Dukes and Mooney, 1999). A change in the type, frequency and extent of a disturbance in a community will create different invasion opportunities (Crawley *et al.*, 1999; Lonsdale, 1999; Stohlgren *et al.*, 1999) (Chapter 10). The expansion of foxtail barley (*Hordeum jubatum*) and dandelion (*T. officinale*), for example, were facilitated by an increase in reduced tillage (Derksen *et al.*, 1993; Shrestha *et al.*, 2002). Plants invading from the Mediterranean into similar climates in Chile and California had more impact in Chile than in California because the types of disturbances introduced were similar to those in California, but quite novel to Chile (Holmgren *et al.*, 2000).

Introducing a new species can alter the disturbance regime such that the extant species are no longer able to persist. Lehmann's lovegrass (*Eragrostis lehmanniana*) introduced into North America is more flammable than the native grasses and has changed the fire regime of the prairie community (Anable *et al.*, 1992). Disturbance, however, is not always necessary for an invasion to occur, and even intact natural ecosystems can be invaded. For example, King and Grace (2000) found that cogon grass (*Imperata cylindrica*) did not require disturbance gaps to invade wet pine savannas. Similarly, melastone (*Tibouchina herbacea*) is a small introduced perennial that invades the undisturbed wet native forests of Hawaii (Almasi, 2000).

Role of diversity

Elton (1958) was the first to hypothesize that there was a negative relationship between native species richness (diversity) and community invasibility (Chapter 10). Elton suggested that communities with many species would be invasion resistant, whereas species-poor communities will be highly invasible. There is evidence to both support and refute this hypothesis (Chapter 10). There may be no relationship between richness and invasibility, or it may be that we are asking too general a question. Species richness may simply be too broad a factor to explain the relative invasibility of communities (Levine and D'Antonio, 1999) because it is more likely to be an aggregation of other variables that are determinants of invasibility. Diverse communities may have more invasive species simply because the environment is favourable to many species (Levine and D'Antonio, 1999; Lonsdale, 1999; Levine, 2000). Factors that allow more species to coexist may also promote invasion, giving the appearance of cause and effect (Levine, 2000).

Resource availability

Davis *et al.* (2000) proposed that the availability of resources may be used to explain the invasibility of a community. They compared gross resource supply with resource uptake and proposed that communities are invasible when resource supply is greater than resource uptake (Fig. 14.6). Therefore, a nutrient-rich community will not be invasible as long as the current community is using most of the available nutrients. Conversely, a nutrient-poor community will be invasible if the community is not sequestering all of the nutrients. Davis and Pelsor (2001) found support for this hypothesis, showing that pulses of nutrients increased invasion success in controlled seed addition experiments over the short term (<1 year). Walker *et al.* (2005) tested this hypothesis by enriching resources (adding pulses of nutrients and water) in New Zealand tussock grassland. They found that over 6 years of nutrient pulses, habitat invasibility did not change.

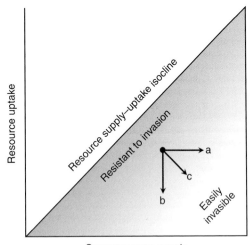

Fig. 14.6. Theory of fluctuating resource availability. The invasibility of a community is dependent on the ratio of gross resource supply to resource uptake. A community becomes more invasible if gross supply is increased (vector a), if resource uptake decreases (vector b) or if both occur (vector c) (redrawn from Davis *et al.*, 2000).

This hypothesis incorporates ideas of both species richness and disturbance in explaining invasibility. For example, disturbance has often been associated with invasion because it reduces or eliminates competitors or because it increases the availability of resources (Hobbs, 1989; D'Antonio, 1993). According to Davis *et al.* (2000), disturbance is only one mechanism that alters resource availability. When the community is disturbed, or when a species is removed, then the released resources become available to other invaders. Events other than disturbance can also release nutrients. For example, herbivory by a leaf beetle (*Trirhabda canadensis*) in experimental monocultures of prairie goldenrod (*Solidago missouriensis*) increased the availability of water, nitrate and light (Brown, 1994). Even though the relative growth rate of prairie goldenrod increased in response to herbivory, more species were able to invade compared with monocultures without the beetle.

Changing the abiotic environment can also alter the nutrient status of a community. Natural fluctuations in the abiotic environment will affect whether a species can invade and/or persist in a community. For example, a drought may create a window of opportunity for invasion when it kills species and more light becomes available. Alternatively, a flood may bring in an influx of nutrients.

The fact that resource availability fluctuates over time helps to explain why a community's invasibility changes over time. A community may have regular or intermittent periods of vulnerability to invasion dependent on resource availability and use. This is why the question 'Which communities are more invasible?' is misleading. Rather, we should be asking 'When is a community invasible?' (Davis *et al.*, 2000).

14.5 Summary

Is it better to have one highly abundant invasive species, or many invaders that are each less abundant, but have the same total biomass? There is no correct answer to this, but it does highlight some important issues in invasive ecology. Understanding how and why invasions occur is complicated by the fact that most invasions fail. Furthermore, it is difficult to identify successful invaders during the important early stages of an invasion. Once we recognise that both species' traits and habitat characteristics contribute to our understanding of why an invasion occurred, we can begin the next important step in weed ecology. This is to begin predicting when, how and why invasions of weeds will occur, what their impact may be, and what (if anything) can and should be done to manage invasions.

Box 14.1. Invasive species case study: invasions.

- What types of habitats does your selected invasive species invade?
- Why is it able to invade so many (or so few) types of habitats?
- Draw a general invasion curve for your selected species. Annotate it with specific filters that were important during its invasion. What traits does your species possess that allow it to pass through these filters?

14.6 Questions

1. Explain why conceptual models are used in ecology.

2. Why are some plant communities more vulnerable to invasion than others?

3. Look at the decision tree by Reichard and Hamilton (1997) (Fig. 14.5). Explain the importance of each decision.

4. How would climate change influence transport, colonization, establishment and spread of an invasive species?

5. Given the complexity of the interacting variables during invasion, do you think we will ever be able to predict which species will become invasive?

Further Reading

Mack, R.N., Simberloff, D., Lonsdale, W.M., Evans, H., Clout, M. and Bazzaz, F.A. (2000) Biotic invasions: causes, epidemiology, global consequences and control. *Ecological Applications* 10, 689–710.

Pyšek, P. and Richardson, D. (2007) Traits associated with invasiveness in alien plants: Where do we stand? In: Nentwig, W. (ed.) *Biological Invasions* (Ecological Studies 193). Springer-Verlag, Berlin and Heidelberg, pp. 87–125.

Richardson, D.M., Pyšek, P., Rejmánek, M., Barbour, M.G., Panetta, F.D. and West, C.J. (2000) Naturalization and invasion of alien plants: concepts and definitions. *Diversity and Distributions* 6, 93–107.

References

Almasi, K.N. (2000) A non-native perennial invades a native forest. *Biological Invasions* 2, 219–230.

Anable, M.E., McClaren, M.P. and Ruyle, G.B. (1992) Spread of introduced Lehmann lovegrass *Eragrostis lehmanniana* Nees. in Southern Arizona, USA. *Biological Conservation* 61, 181–188.

Baker, H.G. (1965) Characteristics and modes of origin of weeds. In: Baker, H.G. and Stebbins, G.L. (eds) *The Genetics of Colonizing Species*. Academic Press, New York, pp. 147–172.

Baker, H.G. (1974) The evolution of weeds. *Annual Review of Ecology and Systematics* 5, 1–24.

Brown, D.G. (1994) Beetle folivory increases resource availability and alters plant invasion in monocultures of goldenrod. *Ecology* 75, 1673–1683.

Catford, J.A., Jansson, R. and Nilsson, C. (2008) Reducing redundancy in invasion ecology by integrating hypotheses into a single theoretical framework. *Diversity and Distributions* 15, 22–40.

Colautti, R.I. and MacIsaac, H.J. (2004) A neutral terminology to define 'invasive' species. *Diversity and Distributions* 10, 135–141.

Colautti, R.I., Grigorovich, I.A. and MacIsaac, H.J. (2006) Propagule pressure: a null model for biological invasions. *Biological Invasions* 8, 1023–1037.

Crawley, M.J. (1987) What makes a community invasible? In: Gray, A.J., Crawley, M.J. and Edwards, P.J. (eds) *Colonization, Succession and Stability*. Blackwell Scientific, Oxford, pp. 429–453.

Crawley, M.J., Brown, S.L., Heard, M.S. and Edwards, G.R. (1999) Invasion resistance in experimental grassland communities: species richness or species diversity. *Ecology Letters* 2, 140–148.

Crooks, J.A. (2005) Lag times and exotic species: the ecology and management of biological invasions in slow motion. *Ecoscience* 12, 316–329.

Crooks, J.A. and Soulé, M.E. (1999) Lag times in population explosions of invasive species: causes and implications. In: Sandlund, O.T., Schei, P.J. and Vilken, A. (eds) *Invasive Species and Biodiversity Management*. Kluwer, Boston, Massachusetts, pp. 103–125.

D'Antonio, C.M. (1993) Mechanisms controlling invasion of coastal plant communities by the alien succulent *Carpobrotus edulis*. *Ecology* 74, 83–95.

Daehler, C.C. and Strong, D.R. (1997) Hybridization between introduced smooth cordgrass (*Spartina alterniflora*; Poaceae) and native California cordgrass (*S. foliosa*) in San Francisco Bay, California, USA. *American Journal of Botany* 84, 607–611.

Davis, M.A. and Pelsor, M. (2001) Experimental support for a resource-based mechanistic model of invasibility. *Ecology Letters* 4, 421–442.

Davis, M.A., Grime, J.P. and Thompson, K. (2000) Fluctuating resources in plant communities: a general theory of invasibility. *Journal of Ecology* 88, 528–534.

Derksen, D.A., Lafond, G.P., Thomas, A.G., Loeppky, H.A. and Swanton, C.J. (1993) Impact of agronomic practices on weed communities: tillage systems. *Weed Science* 41, 409–417.

Dlugosch, K.M. and Parker, I.M. (2007) Molecular and quantitative trait variation across the native range of the invasive species *Hypericum canariense*: evidence for ancient patterns of colonization via pre-adaptation? *Molecular Ecology* 16, 4269–4283.

Dlugosch, K.M. and Parker, I.M. (2008) Founding events in species invasions: genetic variation, adaptive evolution, and the role of multiple introductions. *Molecular Ecology* 17, 431–449.

Dukes, J.S. and Mooney, H.A. (1999) Does global change increase the success of biological invaders. *Trends in Ecology and Evolution* 14, 135–139.

Elton, C.S. (1958) *The Ecology of Invasions by Animals and Plants*. Methuen and Co., London.

Fox, M.D. and Fox, B.J. (1986) The susceptibility of natural communities to invasion. In: Groves, R.H. and Burdon, J.J. (eds) *Ecology of Biological Invasions*. Cambridge University Press, Cambridge, UK, pp. 57–66.

Goodwin, B.J., McAllister, A.J. and Fahrig, L. (1999) Predicting invasiveness of plant species based on biological information. *Conservation Biology* 13, 422–426.

Hastings, M.S. and DiTomaso, J.M. (1996) Fire controls yellow star thistle in California grasslands. *Restoration and Management Notes* 14, 124–128.

Hellmann, J.J., Byers, J.E., Bierwagen, B.G. and Dukes, J.S. (2008) Five potential consequences of climate change for invasive species. *Conservation Biology* 22, 534–543.

Hobbs, R.J. (1989) The nature and effects of disturbance relative to invasions. In: Drake, J.A., Mooney, H.A., di Castri, F., Groves, R.H., Kruger, F.J., Rejmánek, M. and Williamson, M. (eds) *Biological Invasions: A Global Perspective*. John Wiley, Chichester, UK, pp. 389–405.

Hobbs, R.J. (2000) Land-use changes and invasions. In: Mooney, H.A. and Hobbs, R.J. (eds) *Invasive Species in a Changing World*. Island Press, Washington, D.C., pp. 55–64.

Hobbs, R.J. and Huenneke, L.F. (1992) Disturbance, diversity, and invasion: implications for conservation. *Conservation Biology* 6, 324–337.

Hobbs, R.J. and Humphries, S.E. (1995) An integrated approach to the ecology and management of plant invasions. *Conservation Biology* 9, 761–770.

Holmgren, M., Avilés, R., Sierralta, L., Segura, A.M. and Fuentes, E.R. (2000) Why have European herbs so successfully invaded the Chilean matorral? Effects of herbivory, soil nutrients, and fire. *Journal of Arid Environments* 44, 197–211.

King, S.E. and Grace, J.B. (2000) The effects of gap size and disturbance type on invasion of wet pine savanna by cogongrass, *Imperata cylindrica* (Poaceae). *American Journal of Botany* 87, 1279–1286.

Kowarik, I. (1995) Time lags in biological invasions with regard to the success and failure of alien species. In: Pyšek, P., Prach, K., Rejmánek, M. and Wade, M. (eds) *Plant Invasions: General Aspects and Special Problems*. SPB Academic Publishing, Amsterdam, pp. 15–38.

Lawton, J.H. and Brown, K.C. (1986) The population and community ecology of invading insects. *Philosophical Transactions of the Royal Society of London B* 314, 607–617.

Levine, J.M. (2000) Species diversity and biological invasions: relating local process to community pattern. *Science* 288, 852–854.

Levine, J.M. and D'Antonio, C.M. (1999) Elton revisited: a review of evidence linking diversity and invasibility. *Oikos* 87, 15–26.

Lockwood, J.L., Cassey, P. and Blackburn, T. (2005) The role of propagule pressure in explaining species invasions. *Trends in Ecology and Evolution* 20, 223–228.

Lockwood, J.L., Cassey, P. and Blackburn, T.M. (2009) The more you introduce the more you get: the role of colonization pressure and propagule pressure in invasion ecology. *Diversity and Distributions* 15, 904–910.

Lodge, D.M. (1993) Biological invasions: lessons for ecology. *Trends in Ecology and Evolution* 8, 133–137.

Lonsdale, W.M. (1999) Global patterns of plant invasions and the concept of invasibility. *Ecology* 80, 1522–1536.

Orians, G.H. (1986) Site characteristics favoring invasions. In: Mooney, H.A. and Drake, J.A. (eds) *Ecology of Biological Invasions of North America and Hawaii*. Springer-Verlag, New York, pp. 133–148.

Pauchard, A. and Shea, K. (2006) Integrating the study of non-native plant invasions across spatial scales. *Biological Invasions* 8, 399–413.

Perrins, J., Williamson, M. and Fitter, A. (1992) Do annual weeds have predictable characters? *Acta Oecologica* 13, 517–533.

Perrins, J., Fitter, A. and Williamson, M. (1993) Population biology and rates of invasion of three introduced *Impatiens* species in the British Isles. *Journal of Biogeography* 20, 33–44.

Prentis, P.J., Wilson, J.R.U., Dormontt, E.E., Richardson, D.M. and Lowe, A.J. (2008) Adaptive evolution in invasive species. *Trends in Plant Science* 13, 288–294.

Pyšek, P. and Richardson, D. (2007) Traits associated with invasiveness in alien plants: Where do we stand? In: Nentwig, W. (ed.) *Biological Invasions* (Ecological Studies 193). Springer-Verlag, Berlin and Heidelberg, pp. 87–125.

Reaser, J.K., Meyerson, L.A. and Von Holle, B. (2008) Saving camels from straw: how propagule pressure-based prevention policies can reduce the risk of biological invasion. *Biological Invasions* 10, 1085–1098.

Reichard, S.H. and Hamilton, C.W. (1997) Predicting invasions of woody plants introduced into North America. *Conservation Biology* 11, 193–203.

Rejmánek, M. (1995) What makes a species invasive? In: Pyšek, P., Prach, K., Rejmánek, M. and Wade, M. (eds) *Plant Invasions: General Aspects and Special Problems*. SPB Academic Publishing, Amsterdam. pp. 3–13.

Rejmánek, M. (1996) A theory of seed plant invasiveness: the first sketch. *Biological Conservation* 78, 171–181.

Rejmánek, M. (1999) Invasive plant species and invasible ecosystems. In: Sandlund, O.T., Schei, P.J. and Vilken, A. (eds) *Invasive Species and Biodiversity Management*. Kluwer, Boston, Massachusetts, pp. 79–102.

Rejmánek, M. and Richardson, D.M. (1996) What attributes make some plant species more invasive? *Ecology* 77, 1655–1661.

Richardson, D.M. and Bond, W.J. (1991) Determinants of plant distribution: evidence from pine invasions. *American Naturalist* 137, 639–668.

Richardson, D.M., Pyšek, P., Rejmánek, M., Barbour, M.G., Panetta, F.D. and West, C.J. (2000) Naturalization and invasion of alien plants: concepts and definitions. *Diversity and Distributions* 6, 93–107.

Chapter 14

Scott, J.K. and Panetta, F.D. (1993) Predicting the Australian weed status of southern African plants. *Journal of Biogeography* 20, 87–93.

Shrestha, A., Knezevic, S.Z., Roy, R.C., Ball-Coelho, B.R. and Swanton, C.J. (2002) Effect of tillage, cover crop and crop rotation on the composition of weed flora in a sandy soil. *Weed Research* 42, 76–87.

Simberloff, D. (2009) The role of propagule pressure in biological invasions. *Annual Review of Ecology, Evolution, and Systematics* 40, 81–102.

Stohlgren, T.J., Binkley, D., Chong, G.W., Kalkham, M.A., Schell, L.D., Bull, K.A., Otsuki, Y., Newman, G., Bashkin, M. and Son, Y. (1999) Exotic plants invade hot spots of native plant diversity. *Ecological Monographs* 69, 25–46.

Theoharides, K.A. and Dukes, J.S. (2007) Plant invasion across space and time: factors affecting nonindigenous species success during four stages of invasion. *New Phytologist* 176, 256–273.

Thompson, K., Hodgson, J.G. and Rich, T.C.G. (1995) Native and alien invasive plants: more of the same? *Ecography* 18, 390–402.

Vermeij, G.J. (1996) An agenda for invasion biology. *Biological Conservation* 78, 3–9.

Vitousek, P.M., Aber, J.D., Howarth, R.W., Likens, G.E., Matson, P.A., Schindler, D.W., Schlesinger, W.H. and Tilman, D. (1997) Human alteration of the global nitrogen cycle: sources and consequences. *Ecological Applications* 7, 737–750.

Von Holle, B. and Simberloff, D. (2005) Ecological resistance to biological invasion overwhelmed by propagule pressure. *Ecology* 86, 3212–3218.

Wade, M. (1997) Predicting plant invasions: making a start. In: Brock, J.H., Wade, M., Pyšek, P. and Green, D. (eds) *Plant Invasions: Studies from North America and Europe.* Backhuys Publishers, Leiden, The Netherlands, pp. 1–18.

Walker, S., Wilson, J.B. and Lee, W.G. (2005) Does fluctuating resource availability increase invasibility? Evidence from field experiments in New Zealand short tussock grassland. *Biological Invasions* 7, 195–211.

Williamson, M.H. (1993) Invaders, weeds and the risk from genetically modified organisms. *Experientia* 49, 219–224.

Williamson, M. (1996) *Biological Invasions.* Chapman and Hall, New York and London.

Williamson, M. (1999) Invasions. *Ecography* 22, 5–12.

Williamson, M.H. and Brown, K.C. (1986) The analysis and modeling of British invasion. *Philosophical Transactions of the Royal Society of London B* 314, 505–522.

Glossary

Abiotic – non-living, physical or chemical environment

Abundance – a measure of a population's success in terms of numbers

Agamospermy – the production of seeds without fertilization

Allelopathy – an interaction where one individual has a direct effect on another through the release of chemical compounds from roots, shoots, leaves, or flowers

Amensalism – an interaction whereby only one individual is negatively affected and the other neither benefits nor is harmed

Apomixis – asexual reproduction

Apparent competition – an interaction that gives the appearance of being due to competition but is due to other factors

Asexual reproduction – the creation of offspring that are genetically identical to their parents, through a variety of mechanisms

Biological control – the management of weeds using introducing herbivores (often insects) as 'biological control agents'

Biomass – the weight of vegetation

Biotic – living

Clonal growth – the creation of new, potentially independent plants through vegetative growth (also vegetative reproduction)

Cohort – a group of individuals born within the same age class

Community – a group of populations that co-occur in the same space and at the same time

Competition – a negative interaction where individuals make simultaneous demands that exceed limited resources

Competitive ability – a combination of competitive effect and competitive response

Competitive effect – the ability of an individual to suppress the growth or survival of another

Competitive response – the ability of an individual to avoid being suppressed by another

Connectivity – the relative amount of linkages between fragments of habitat

Cover – the proportion of ground covered by a species when viewed from above

Demography – the study of a population's size and structure and how it changes over time

Density – the number of individuals in a given area

Dicotyledon – plants in the Class Magnoliopsida, meaning their seeds have two embryonic leaves known as cotyledons

Disturbance – any action that disrupts the structure or function of an ecosystem

Diversity – a measure of the number of taxa present (richness) and their relative abundances (evenness)

DNA barcoding – use of a short genetic (DNA) marker (from the total genome of a species) to help classify it taxonomically using molecular techniques

Dormant – in a resting state and unable to germinate

Ecological amplitude – the limits of environmental conditions that define where and how an organism may survive

Ecology – the study of organisms and their environment

Emergence – appearance of a shoot above the soil

Endosperm – the carbohydrate (with some proteins and oils) that 'feeds' the growing embryo in seeds

Epiphyte – a parasite that is dependent on its host for physical support

Establishment – occurs once a seedling no longer depends on seed reserves (endosperm and cotyledons), i.e. it is photosynthetically independent

Evenness – a measure of how similar the relative abundances of each species in a community are

Extirpation – local extinction of a genotype or species

Facilitation – a successional process whereby early invading species ameliorate the environment for later invaders

Facultative interaction – an interaction where both species can survive independently, but both benefit when they are found together

Fertility – the capacity of an organism to reproduce sexually

Fertilization – the fusion of a male and female gamete to form a zygote

Fitness – a relative measure of how well an individual succeeds at continuing its lineage

Fragmentation – the breaking up of a habitat from a contiguous landscape into smaller parts; usually related to human activities

Frequency – the proportion of sampling units (e.g. quadrats) that contain the target species

Fruit – structure formed from the flower ovary or receptacle and containing one or many seeds

Functional group – a groups of species with a similar set of traits

Gamete – the specialized haploid cells that fuse to become a zygote

Genet – an entire genetic individual, composed of ramets

Genetic drift – random changes in allele frequency, especially in small, isolated population

Genetic variation – the range of genes found in the genotypes or genomes of populations or species

Genome – the total genetic information available in a species

Genotype – the total genetic information available within an individual (genet)

Germination – the emergence of a root and/or shoot from a seed coat

Guerrilla-type growth – a type of clonal growth that results in loosely packed, often linear patches

Haploid – the condition, normally found in gametes, where meiosis has reduced the number of chromosomes to half the normal compliment found in somatic cells

Hemiparasite – parasite that relies on its host for only some resources

Herbaceous – plants that have no woody growth; in some usage, it also excludes any grass, sedge, or rush species that lack showy flowers or wide leaves

Herbivory – the consumption of plant tissue by animals

Holoparasite – parasite that is entirely dependent on its host

Inbreeding depression – reduced fitness that results from the accumulation of deleterious alleles caused by mating with a close relative

Indirect plant defences – plant uses another organism to defend itself against herbivory

Inhibition – a successional process whereby existing plants prevent or inhibit the establishment of subsequent species

Interspecific – between species

Intraspecific – with a species

Invasibility – the ease with which a habitat is invaded

Invasion – the expansion of a species into an area not previously occupied by it

Invasion meltdown – the acceleration of impacts on native ecosystems due to synergistic interactions of non-native species

Iteroparous – reproduction that occurs repeatedly through a plant's lifespan (often used as a synonym for polycarpic)

K – carrying capacity

Keystone species – a species that has a disproportionate effect on community function relative to its biomass

K-strategists – species that are large, have delayed reproduction, are long-lived, and are found in stable environments

Life history – the general description of a plant's life cycle and the more specific aspects of life cycles within population (age, stage, size)

Life table – a table summarizing the survival data of a population

Metapopulation – a group of spatially isolated populations that interact through migration or distant pollination

Microarray analysis – a method to determine gene expression; involves chemically testing thousands of DNA sequences that are arranged on a chemical or physical surface called an array

Monocarpic – sexual reproduction that occurs only once in a plant's lifespan (often used as a synonym for iteroparous)

Monocotyledon – organisms in the Class Liliopsida; in contrast to dicots, these only have one embryonic leaf (one cotyledon)

Morphology – the form or structure of an organism

Mutualism – an interaction that benefits both individuals

Non-dormant – able to germinate

Non-native plant – a plant whose presence is due to intentional or accidental introduction as a result of human activity, also exotic, alien, non-indigenous

Obligate interaction – an interaction where both partners of the association require each other in order to survive

Overcompensation – a type of tolerance to herbivory where the effect of herbivory benefits the plant

Parasitism – an interaction where an individual obtains nutrients, shelter, and/or support from its host

PCR – polymerase chain reaction; a technique used to increase the amount of DNA to help identify gene sequences

Persistence – a measure of how long a community remains the same

Phalanx-type growth – a type of clonal growth that results in slow-growing, branched clones that form dense patches

Phenology – the study of life cycle events and the environmental conditions that influence them

Phenotype – the expression of the genes in the genotype

Physiognomy – the general appearance of a community

Pollination – the transfer of pollen from an anther to a stigma

Polycarpic – sexual reproduction that occurs repeatedly throughout a plant's lifespan (often used a synonym for semelparous)

Polymorphism – when a structure produced has two or more morphology types; especially seeds

Polyploid – an organism with three or more sets of chromosomes

Population – a group of potentially inter-breeding individuals of the same species found in the same place at the same time

Post-dispersal seed predation – consumption of the seed after it has been dispersed

Potential distribution – the area in which a species is able to persist as determined by the abiotic environment (also physiological distribution or climatic range)

Pre-dispersal seed predation – consumption of the seed while it is still on the maternal parent plant

Primary dispersal – dispersal of seed from the parent plant to the ground

Primary dormancy – seeds that are unable to germinate when they first mature

Primary succession – succession that occurs on newly created land where no plants have grown previously or where there is no effective seed bank on site

Quiescent – seeds that are not dormant, but do not germinate because they have not encountered appropriate environmental conditions

r – the intrinsic rate of population growth

r-strategists – species that are small, annuals, have a rapid growth rate, reproduce early, and produce many small seeds and are therefore able to establish rapidly following a disturbance

Ramet – an individual which is genetically identical to the parent plant and capable of physiologically independent growth

Recruitment – the increase in population size through the introduction of new individuals

Resilience – a measure of a community's ability to return to its original state following a disturbance

Resistance – a measure of whether a community resists stress or disturbance

Richness – the number of taxa (i.e. species) present in an area or in a community

Ruderal – plant adapted to a frequently disturbed habitat

Safe site – a site that provides all the conditions necessary for the seed to germinate and emerge from the soil

Secondary dispersal – movement of seed subsequent to primary dispersal

Secondary dormancy – dormancy that is imposed after seeds have dispersed

Secondary succession – succession that occurs on land previously vegetated, but disturbed by natural or human-caused factors

Seed – the embryonic plant that develops from the fertilized ovule

Seed bank – seeds that become incorporated into the soil

Seed dispersal – movement of a seed or fruit away from the maternal parent plant

Seedling – a young plant

Self-compatible – individuals that can successfully mate with themselves if pollen is transferred from stigma to style

Self-incompatible – an individual that is not able to mate with itself

Semelparous – reproduction that occurs only once in a plant's lifespan (often used as a synonym for monocarpic)

Sexual reproduction – the creation of offspring via fusion of two gametes (a sperm and ovum) to form a zygote

Sporophyte – the diploid part of a plant's life cycle

Stability – a measure of how communities resist change in response to disturbance or stress, comprises persistence, resistance, and resilience

Strategy – a group of similar traits that causes species to exhibit ecological similarities

Stress – harsh environmental conditions

Succession – the directional change in community composition

Tens rule – describes how approximately 10% of species pass through each transition from being imported to becoming casual to becoming established, and finally becoming invasive

Time lag – the time between when a species is introduced and when its population growth explodes

Tolerance – a successional process whereby existing species have no effect on subsequent ones

Traits – the physical and physiological characteristics of a plant that determine its ecological function

Trajectory – a path through a series of community states

Transects – lines used to help determine where to locate quadrats to test for changes along environmental gradients

Water-use efficiency – the ability to minimize water use for a given amount of carbon assimilation

Weed – a native or introduced species that has a perceived negative ecological or economic effect, usually on agricultural or managed ecosystems

Woody – tissue containing secondary xylem

Zygote – diploid cell produced from the fusion of male and female gametes in sexual reproduction

Subject Index

Note: Page numbers in *italic* refer to tables or boxes; those in **bold** refer to figures

CLIMEX 13, 15
clonal plants
    classification of growth types **60**
    oldest living 94
    woody 59, 61, **62**, 64
    see also vegetative reproduction
cold 80, 86
collection bias 10
colonizers 144, 149–150, 187–188
    agamospermous species 66, 68
    availability 149–150
    defined 2–3
    non-native 150
commensalism *98*
community
    attributes 127
    composition changes 141, **142**
    definitions 124
    describing 125–127
    disturbance 141, 143
    equilibrium and non-equilibrium models 142–143
    invasibility 192–194
    stability 137, 143
community assembly 154–159, 172, **173**
compensation 111
competition
    above-ground 101–102, *104*
    apparent 100
    below ground 101–102, 104
    costs of 98
    definition 97
    evidence for 99–100
    interaction with herbivory 120
    for light 101
    for nutrients 100–101
    for pollen/pollinators 102
    and resource availability 98–99
    for space 101
    for water 101
competitive response 98
competitive species 38, 146
    ruderals (CR) 38
    traits 102–105
conceptual models 3, 186–187
connectivity, habitat 168–172
corms 59, *61*
costs of plant invasions 4
cover, estimation 17, *18–19*
Cowles, H.C. 141
cross-pollination 50
cross-scale modelling 172

dark period 89
data collection
    species distribution 10
    survivorship curves *33*

day length (photoperiod) 13, 88–91, *90*
'death hormone' hypothesis 94
Deevey curves 31, **32**
demographic stochasticity 43, 189
density, measures of 17, *18–19*
developmental stage 29, **30**
dicliny *47*, 48
dioecious plants *47*
disease transmission 65
dispersal 187
    see also seed dispersal
dissimilarity index 145, **146**
disturbance 134–135
    community responses 141, 143
    and developmental-stage distribution 29, **30**
    and diversity 135
    establishment following 36
    recolonization following 149–150
    and vegetative reproduction 59
diversity 127
    between community 132
    components 127–128
    and disturbance 135
    and ecosystem function 135–137
    measurement 128–132
        within community *129*
    patterns 133–135
diversity–stability hypothesis 136
DNA, telomere length 94
DNA bar coding 178
dormancy, *see* seed dormancy

earthworms 76
ecological corridors 165
ecological processes, interaction 120
ecology
    definition 1–2
    scales of 2
ecosystem function, and biodiversity 135–137
ecozones 90
ectomycorrhizal fungi 119
edge expansion 76
Egler, F.E. 147
elasticity 34
embryo dormancy 79
enemy release 114, 116
environmental cues 90–91
environmental factors
    effect on competition 105
    interaction with genetics 85
    and pollination 51
    seed germination 80–81
    and seed production 72
    and vegetative reproduction 63
epiphytes 117, *117*

monocliny *47*, 48
monoecy *47*
Mount St Helens eruption 144, 149, 152
movement of species, measurement
        171–172
mutualism *98*
    facultative 118–119
    obligate 118
mycorrhizal associations 119

native range 191–192
native species, hybridization with introduced
        species 53–54
natural ecosystems
    costs of plant invasions 4
    impact of plant invasions 4, 5
natural selection 19–20, 72–73, 98, 102
nearest-neighbour occupancy 168
nectar 48, 52
'neutral theory 20
neutralism *98*
nitrogen 100
nitrogen-fixation 119, 144, 149
no-till farming 37
nutrient-poor environments 98–99, 149
nutrient-rich environments 64, 65
nutrients, competition for 100–101

observer effect *112*
old-field succession 145, *153*,
    154–155
ornamental species/varieties 180, 181
overcompensation 113

palaeoecological records 10–11, **12**
parasitism *98*
    definitions 117, *117*
    examples of agricultural parasitic
        weeds 117–118, *117*
pasture management 27, 28–29
patch dynamics 165–166
phalanx strategy 61–62, 93
phenology
    definition 85
    effects of abiotic factors 85–91
    populations structure 29, **30**
phenotypic plasticity *68*, 93–94
photoperiod 13, 88–91, *90*
photosynthesis 90, 93, 104, *104*, *147*
phylogenetic constraint 68
physical space, competition for 101
physiological processes 86
physiological traits *147*
pigs 5

plant defences 20, 111, 113–114, *113*
plant size
    and competition 104–105
    measurement 29
    population structure 29
    and response to environmental cues 90
    and seed production 41
    and senescence 94
    vegetative reproduction 64
plant traits 156–159
    competitive species 102–105
    early/late succession species 146, *147*
    and invasive ability 190–191, **192**
    see also life history traits
'plantlets' 64
pollen
    allelopathic chemicals 102
    competition 102
    limitation *68*
    requirement in agamospermous
        species 67
pollen diagrams 11, **12**
pollination 46–52
    mechanisms 48–49, *49*
    problems 51–52
    self-compatibility/incompatibility 49–51
pollinators
    competition for 102
    insects 48, *49*
    specialized 51
    vertebrates 48, *49*
polycarpy 51, 52, 68, 93, 94
polymerase chain reaction (PCR) 177–178
polyploidy 67, *68*
population abundance 4, 8, 16–19, 127
    measures of 17, *18–19*
population differentiation 180
population distribution 8
    changes over time 8–10
    estimation and mapping 10–13
    models 13–16
    potential 13
    scales of 10, **11**, 15–16
    within-population patterns 20–21
population dynamics 41–43
population models 31, 34
    elasticity 34
    example of use 34–35
    matrix 31, 34
    sensitivity 34
population structure 25–35
    age/age-class 25–29
    developmental stage 29
    illustration of data 31, **32**, *33*
    and plant life cycles 35–41
    size 29
    value of 35

Species Index

strawberry guava (*Psidium cattleianum*) 119
sugar maple (*Acer saccharum*) 12, 29
sulfur cinquefoil (*Potentilla recta*) 28, 112
sunflowers (*Helianthus* spp.) 54, 79, 92, 117

tall cattail (*Typha* × *glaucaglauca*) 53, 100
tall goldenrod (*Solidago altissima*) 64
tansy ragwort (*Senecio jacobaea*) 80, 115, 116
Tasmanian blue gum (*Eucalyptus globulus*) 114
teasel (*Dipsacus fullonum*) 29
tick berry (*Chrysanthemoides monilifera*) 104
tobacco (*Nicotiana tabacum*) 111
tomatoes (*Solanum lycopersicum*) 71, 117
touch-me-not (*Impatiens glandulifera*) 75, 190
tree of heaven (*Ailanthus altissima*) 64, 120
trembling aspen (*Populus tremuloides*) 64
tropical soda apple (*Solanum viarum*) 13
trumpet honeysuckle (*Lonicera sempervirens*) 120
tussock hawkeed (*Hieracium lepidulum*) 182

velvetleaf (*Abutilon theophrasti*) 15, 79, 89, 93, 114,
    178
viper's bugloss (*Echium vulgare*) 47, 127, 128

water hyacinth (*Eichhornia crassipes*) 49
water milfoil (*Myriophyllum spicatum* × *Myriophyllum
    sibiricum*) 178
wheat (*Triticum aestivum*) 116
white clover (*Trifolium repens*) 62, 65, 66, 90

white knapweed (*Centaurea diffusa*) 180
white mulberry (*Morus alba*) 53, 54
white mustard (*Sinapis alba*) 87, 90
white oak (*Quercus alba*) 120
white pine (*Pinus strobus*) 189
whorled wood aster (*Oclemena acuminatus*) 64
wild buckwheat (*Polygonum convolvulus*) 41
wild burdock (*Arctium minus*) 75, 168
wild carrot (*Daucus carota*) 47, 48, 90, 128
wild fennel (*Verbascum thapsus*) 4, 20, 47, 85, 181
wild garlic (*Allium sativum*) 61, 65
wild ginger (*Asarum caudatum*) 113
wild gladiolus (*Gladiolus illyricus*) 169
wild lettuce (*Lactuca scariola*
    and *Lactuca virosa*) 189
wild mustard (*Sinapis arvensis*) 86–89
wild oat (*Avena fatua*) 79, 93
wild onion (*Allium vineale*) 59, 61
wingpetal (*Heterosperma pinnatum*) 73, 74
witchweed (*Alectra* spp.) 41, 117, 120
witchweed (*Striga lutea*) 41, 117, 120
witchweed (*Striga* spp.) 41, 117, 120

yarrow (*Achillea millefolium*) 115
yellow hawkweed (*Hieracium pratense*) 12
yellow nutnedge (*Cyperus esculentus*) 61, 178
yellow rattle (*Rhinanthus minor*) 117
yellow star-thistle (*Centaurea solstitialis*) 12, 51, 113,
    118, 119, 193
yellow toadflax (*Linaria vulgaris*) 181
yucca (*Yucca* spp.) 118